Radiogenic Isotopes in Geologic Processes

Radiogenic Isotopes in Geologic Processes

Sergei V. Rasskazov • Sergei B. Brandt
Ivan S. Brandt
Institute of the Earth's Crust, Siberian Branch of the Russian Academy of Sciences, Irkutsk, Russia

Springer

Prof. Sergei V. Rasskazov
Institute of the Earth's Crust
Russian Academy of Sciences
Lermontov Str. 128
Irkutsk, Russia 664033

Irkutsk State University
Lenin Str. 3
Irkutsk, Russia 664033
rassk@crust.irk.ru

Prof. Sergei B. Brandt
Institute of the Earth's Crust
Russian Academy of Sciences
Lermontov Str. 128
Irkutsk, Russia 664033

Dr. Ivan S. Brandt (deceased)

ISBN 978-90-481-2998-0 e-ISBN 978-90-481-2999-7
DOI 10.1007/978-90-481-2999-7
Springer Dordrecht Heidelberg London New York

Library of Congress Control Number: 2009938100

Cover design: Volcanoes in Middle Gobi (photograph taken by S. Rasskazov) and a fragment of Fig. 14.7 in this book.

Revised and updated edition of the book in Russian "Radioizotopnaya geologiya v zadachakh I primerakh, Novosibirsk, Academicheskoye izdatel'stvo Geo", S.V. Rasskazov, S.B. Brandt, I.S. Brandt et al./E.V. Sklyarov, S.V. Rasskazov (eds), 2005

Printed on acid-free paper

Springer is part of Springer Science+Business Media (www.springer.com)

Ivan S. Brandt 1943–2008

Ivan S. Brandt graduated from Moscow State University with a degree in physics and mathematics. At the Institute of the Earth's Crust SB RAS he studied kinetics of radiogenic isotopes in geologic processes, both theoretically and experimentally. He died of cancer when the manuscript of this book was under preparation.

Preface

Do we actually understand geologic processes? New technology brings new information and perceptions, which sometimes overturn imaginations based on simple observation and estimation, in conjunction with common sense inference. In 1902–1904, Pierre Curie and Ernest Rutherford first formulated the idea of using radioactive transformation of nuclides as a geologic chronometer. After a century of working with such tools, geology has advanced from a descriptive science to an analytic science that formulates conclusions based on exact values. The technology of radiogenic isotope geology has created a branch of science that considers the Earth as a planet generated within a Solar system and studies the subsequent evolution of geologic processes that has resulted in the present formation of our planet's continents and oceans.

The physicist Vitaly Ginsburg, Nobel Prize laureate, wrote recently: "If Kepler had been given information on orbital parameters of planets with modern precision, he would not have been able to formulate his laws". Indeed, after development of laws of celestial mechanics, methods of measurements became so advanced and such numerous secondary distortion effects were found that to describe an orbit of a cosmic body by a curve of the second order would appear impossible. But it does not mean that Kepler's laws are "cancelled"; they still occupy an honorable place in courses on celestial mechanics. A reasonable division into basic and secondary phenomena is accepted and the latter are entered as variations in the basic equations.

Application of radiogenic isotopes for study of geologic processes showed that accumulation of daughter substances, according to the Rutherford–Soddy Law of Radioactivity, in some cases was interfered with by other circumstances. Authors often gave only "verbal explanations" of these phenomena without appropriate differential equations. Present-day radiogenic isotope geology, however, expresses results in values that can claim a high degree of accuracy. Similar to the study of celestial mechanics, understanding of natural isotopic systems requires a deliberate selection in geologic processes of the main and secondary effects.

New data, technological breakthroughs, and changes of paradigms inevitably result in changes of approaches in geochronology and isotope geochemistry. Only the requirements for selection of material for an isotope analysis remain constant from the geologic point of view. A geologist aspires to obtain data most appropriate to satisfactory geochronological interpretations using isochron or Concordia–Discordia diagrams. Additional complexities arise in cases of deviations from the statistically substantiated models. Factually, every isotopic measurement is to some degree a compromise between a strict model and the probabilities of alternatives. In this book, we designate, therefore, results based on the principle approaches and on interpretations on a level of more or less approximate constructions.

Achievements of radiogenic isotope geology have been highlighted in several monographs (Faure 1989, 2001; Geyh and Schleicher 1990; Dickin 1997; McDougall and Harrison 1999; Ozima and Podosek 2002 etc.). Monographs that were published in Russian on this topic were listed by Shukolyukov in introductory comments to the Russian edition of the book on principles of isotope geology by Gunter Faure (1989). A few books have been published recently (Kotlyar et al. 2001; Titaeva 2005 etc.). Our Russian book "Radiogenic isotope geology in problems and examples" (Rasskazov et al. 2005a) was intended for students and geologists interested in modern approaches to applications of the principle geochronometric systems.

Respectively, comprehensive information on the theory, experiments, models and methods is presented in the first three chapters of this book, exhibiting modified compilations from the Russian monograph mentioned above. In the further chapters, some aspects of radiogenic isotope behavior are developed for understanding geologic processes. On the one hand, we substantiate theoretically the geochronological consequences of (1) possible fractionation of argon isotopes in the terrestrial atmosphere, (2) distribution of radiogenic argon in a mineral grain, (3) particularities of Ar–Ar spectra, (4) radiogenic isotope losses in an exocontact zone of an intrusive body and in a terrane affected by long-term elevated temperatures provided a Laplace regime, (5) decreasing isotope losses in a cooling dike with transition from an open to a closed isotope system, and (6) separation of leads in ore deposits according to the Concordia–Discordia model. On the other hand, we highlight geochronological data on processes of (1) the early Earth, (2) important Phanerozoic boundaries, and (3) the late Phanerozoic.

The book was prepared at the laboratory for isotopic and geochronological studies of the Institute of the Earth's crust, Siberian branch of the Russian Academy of Sciences, with the ultimate goal of establishing a special course on radiogenic isotopes in geologic processes for students of the Irkutsk State University, Irkutsk. The help of our colleagues occasionally contributed to this work is highly appreciated.

For promotion of this publication, we are especially grateful to Gillian R Foulger, Durham University and Petra D van Steenbergen, Springer Senior Publishing Editor. Drawings in the final manuscript version benefited from assistance of Irina Chuvashova and the text from corrections of Edwin Beschler, Hermine Vloemans, and the team at Crest Premedia Solutions (P) Ltd., Pune, India.

Irkutsk, Russia

Sergei Rasskazov
Sergei Brandt
Ivan Brandt

Contents

Chapter 1
Theory and Experience

We shall not discuss the philosophical contents of the concept "time". Geochronology deals with functional time, providing a measurement procedure not of the time itself, but of values that are included in a certain law that gives time through other parameters. For instance, the time of passage by an airplane at some distance is calculated by division of the distance of passage into speed. The time of a liquid transfusion from one vessel to another is calculated using the rate of this transfusion and ratios of liquid volumes in the vessels. The Roman water clock is based on this principle. In geochronometry, isotopes of natural radioactive elements (uranium, thorium, rubidium etc.) are used and the time is calculated by the formula:

$$t = \frac{1}{\lambda} \ln\left(\frac{D}{P} + 1\right)$$

where λ – constant of radioactive transformation, P – number of parent particles, left at present, D – number of daughter particles.

The formula is derived from the Law of Radioactivity of Rutherford–Soddy, which is applicable not only for today, but also for all geologic time, operating with constant rates. Such an approach is characteristic for application of the principle of uniformitarianism suggested by James Hutton: "the present is a key to the past". This principle has been criticized from various points of view. Nevertheless, in our case it is still valid. The rate of radioactive transformation of natural radioactive elements in terrestrial conditions has been found constant at the most extreme pressures and temperatures and also independent from chemical ties, in which the parent radioactive element plays a role. Numerous studies of rocks in various isotope systems have resulted in a mutually consistent system of ages defined for geologic events. The system itself is the proof of validity of uniformitarianism in application to the Law of Rutherford–Soddy.

A triumph of this principle in geochronometry has appeared nevertheless incomplete. Both parent (radioactive) and daughter (radiogenic) substances enter into minerals and rocks and, though the rate of radioactive transformation is constant, they, as the witnesses, participate in movements and changes of host compositions. Certainly, the isotope particles cannot simply disappear, but they can migrate from an analyzed structure to enter into an adjacent structure. Alternatively, particles of

S.V. Rasskazov et al., *Radiogenic Isotopes in Geologic Processes*,
DOI 10.1007/978-90-481-2999-7_1,

the adjacent structure can be introduced into the analyzed structure. Processes of migration depend on temperature, pressure, fluid regime, eutectics etc., i.e. on the geologic history of a sample. Thus, here arises a vicious circle: to make a geochronometric determination, one should know the history, and to recognize the latter, one should learn its duration. Models and approaches have been developed to get over this controversy and avoid interfering phenomena.

Certainly, each model is based on abstraction and takes into account only essential features, neglecting minor ones. A geochronologist is responsible for proving conformity of a selected series of probes to hypothetical predictions of a model. Geochronometry deals with a typical inverse problem of natural sciences: a consequence observed today serves to restore a cause and effect relation that operated in the geologic past. Solution of such problems is not unequivocal (Tikhonov and Goncharskii 1987). Geochronometric models are also not unequivocal. Nevertheless, geologic, geochemical, and petrologic arguments usually provide sufficient proofs for validity of the chosen model.

1.1 The Law of Rutherford–Soddy

Differentiating types of radioactivity is not a subject of our consideration. Their individual characteristics are available in appropriate textbooks and manuals. We note only that in some cases a nucleus loses elementary particles (n, α, β), i.e. radioactive decay occurs; in other cases, on the contrary, a nucleus captures a particle (K-capture). The latter type of radioactive process cannot be called "decay", but is more appropriately referred to as "radioactive transformation".

Radioactivity is responsible for transition of a certain number of particles of a radioactive isotope P (Parent) into particles of a radiogenic isotope D (Daughter). Assume that the small quantity of decreasing particles $-dP$ for a small interval of time dt is proportional to the total quantity of particles P:

$$-dP = \lambda P dt$$

Coefficient of proportionality λ is called a constant of radioactive transformation (decay). Change the expression to:

$$\frac{dP}{P} = -\lambda dt$$

and integrate the latter. The left-hand part is a differential of the logarithm and the right-hand one is a differential of t:

$$\ln P = -\lambda t + c$$

or

$$P = c \times e^{-\lambda \cdot t}$$

where c – constant. What does the c mean? Suppose that at the beginning of a process, at $t = 0$, the number of parent particles was equal to P_0. Then $c = P_0$ and finally:

$$P = P_0 e^{-\lambda \cdot t} \tag{1.1}$$

This is the law of Radioactivity of Rutherford–Soddy showing that the number of particles of a parent radioactive isotope decreases in time exponentially.

Note that, besides a constant of radioactive transformation (decay), one more parameter—period of half-decay—is used. This parameter represents a time interval τ, during which a parent substance decreases by half. From formula (1.1), we get:

$$\frac{P}{P_0} = e^{-\lambda \cdot \tau} = \frac{1}{2}; \quad \lambda\tau = \ln 2 \quad \text{or} \quad \tau = \frac{\ln 2}{\lambda} = \frac{0.6931}{\lambda}$$

Hence, a period of half-decay is inversely proportional to a constant of radioactive transformation.

Look now, how the daughter isotope accumulates. It is clear that the number of particles of a daughter isotope D plus the number of particles of the left-hand parent isotope in expression (1.1) always should be equal to the initial number of particles (particles cannot simply disappear). The formula:

$$D + P_0 e^{-\lambda \cdot t} = P_0 \quad \text{or} \quad D = P_0(1 - e^{-\lambda \cdot t}) \tag{1.2}$$

describes accumulation of the daughter substance. Practically, the initial number of particles of the parent substance P_0 is unknown, but its present-day quantity P can be measured. Obtain from (1.1) $P_0 = Pe^{\lambda \cdot t}$ and from substituting this expression in (1.2):

$$D = P(e^{\lambda \cdot t} - 1) \tag{1.3}$$

These relations assume important space-(geo)-chronological consequences. Natural radioactive substances cannot exist eternally. There should exist a moment of the nucleogenesis, i.e., formation of radioactive nuclei by absorption of neutrons, protons, and electrons. Indeed, substitution into formula (1.1) $t = \infty$ yields $P = 0$. If the parent substance had existed eternally, the radioactive elements would have completely decayed. And on the contrary, assuming final P means indefinitely large P_0. In this case, the radiogenic substance would have filled up the Universe leaving no place for other substances.

Hence, there must have been an instant of nucleogenesis that initiated the "radioactive clock". Some researchers attribute it to a huge explosion (the big bang), which gave birth to the Universe, others to the appearance of a Super New Star. Geochronology starts from a certain instant of time, when the natural radioactive elements appeared. This appearance preceded the formation of planets, meteorites, and the Solar system as a whole. Radioactive elements were spatially distributed on various objects (Earth, Moon, asteroids, meteorites etc.). Constant rate of radioactive transformation leads to a remarkable fact: any objects with radioactive elements show at present identical isotopic compositions. For instance, in all accessible terrestrial and space objects, $^{238}U/^{235}U = 137.88$. The only natural exclusion—the effect of the Oklo natural nuclear reactor (Shukolyukov 1982)—has no essential meaning for geochronometry. Even if isotope compositions of radioactive elements in objects had not been constant, it would not have destroyed, but only slightly complicated, geochronometry.

1.2 Determination of Decay Constants

Constants of radioactive transformation are defined by (1) counting, (2) radiochemical, and (3) geochemical methods (Begemann et al. 2001). In a counting experiment, the direct measurement of activity of a sample enriched with a radioactive element is carried out. The constant of decay (λ, year^{-1}) is calculated according to the equation:

$$\lambda = AWY/fN_0$$

where A – activity of a sample in pulses per one second on one gram of substance, W – nuclear weight of a radioactive element, f – fraction of a radioactive isotope in a sample, Y – amount of seconds in a tropical year, and N_0 – Avogadro's number. Among radioactive isotopes, used in geochronology, the counting experiment has yielded the most accurate constants of decay ^{235}U and ^{238}U (Jaffey et al. 1971).

If activity of a sample is low, the radiochemical experiment yields more accurate results than the counting. In this case, the amount of a daughter element is measured in a probe after its enrichment by a radioactive isotope and several years of endurance. If activity of a parent nuclide is low and it has a long period of half-decay, the constant of decay is defined geochemically, i.e. by measurements of the daughter isotope concentration in a sample reliably dated by any radioisotope method (for example, U–Pb). The radiochemical and geochemical approaches have been successfully used for determination of the decay constant for ^{187}Re characterized by low activity (2.47 KeV) and long periods of half-decay ($\sim$40 Ga) (Lindner et al. 1989; Smoliar et al. 1996).

1.3 Mass Spectrometry

Timing of a geologic process is determined from relations between parent and daughter substances, expressed by isotope quantities as extensive values and isotope ratios as intensive parameters (i.e. not connected with isotope concentrations).

Isotopes are measured by mass spectrometers, the working principle of which is the following. A mixture of isotopic particles of an element studied is placed in an ion source, where the neutral particles are ionized, concentrated into a narrow ion beam, accelerated to some speeds (different isotopes show equal kinetic energies), and "shot" in space (Fig. 1.1).

A free run of a particle at normal pressure has an order of 10^{-5} cm. Therefore, the measuring system is placed in an ultrahigh vacuum. After flying some distance with

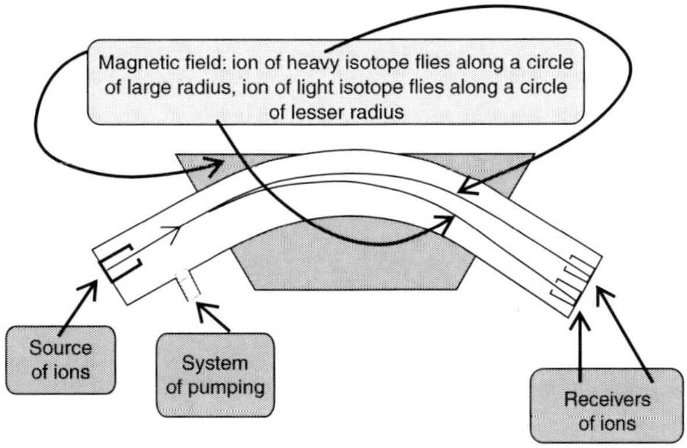

Fig. 1.1 Schematic illustration of a mass spectrometer

constant speed, the ion enters a homogeneous magnetic field with lines perpendicular to the direction of the ion beam. A Lorenz force begins to act:

$$F = e \cdot v \cdot H$$

where e – charge of the ion, v – its speed, H – intensity of a magnetic field. This force is directed to the third perpendicular. It is known that the force, directed perpendicular to the speed of a particle, does not change its kinetic energy, but creates centripetal acceleration. The particle begins to move on a circle of a radius:

$$R = \frac{m \cdot v}{e \cdot H}$$

The ion returns the charge to an electrode–receiver. There are two methods of further measurements. In one method, the electrode–receiver, shaped like a Faraday cup, is placed on the trajectory of each beam (multicup system). In another method, an immobile electrode–receiver is reached by ion beams through their adjustment in a varying magnetic field. So, the beam of each mass is connected to this electrode (scanning system).

Measurement of ion beam intensities, data processing, and ultrahigh vacuum control are fully automated. Detailed descriptions of mass spectrometers are presented in specialized publications (Sysoev et al. 1993; White and Wood 1986).

1.4 Isotope Dilution

A mass spectrometer is ideally suited for measurement of isotope ratios (i.e. masses of the same elements): $^{40}Ar/^{36}Ar$, $^{87}Sr/^{86}Sr$, $^{207}Pb/^{204}Pb$ etc. However, geochronometric practice requires also measurements of mass ratios for nuclides of different elements, for example, daughter–parent ratios: $^{87}Sr/^{87}Rb$, $^{207}Pb/^{235}U$, $^{143}Nd/^{147}Sm$ etc. Such ratios are calculated using the nuclide quantities measured by the method of isotope dilution using an internal standard.

Suppose we wish to determine the amount of He released by heating of a certain sample in a vacuum vessel (reactor). Open the input valve and tune the mass spectrometer precisely to mass 4. Note that the pointer of the amplifier shows an angle α. Further, without closing the input valve, add into the reactor a known reference amount of helium ($^4He_{ref}$). Respectively, the pointer of the device will increase value to α_1. The required quantity of ^{4}He will be defined by proportion:

$$\frac{^4He}{^4He + {}^4He_{ref}} = \frac{\alpha}{\alpha_1} \quad \text{or} \quad {}^4He = \frac{\frac{\alpha}{\alpha_1}{}^4He_{ref}}{1 - \frac{\alpha}{\alpha_1}}$$

It is noteworthy that dilution is made for a particular element by the same element with a known concentration. It is impossible to dilute helium, for example, by argon or nitrogen. The conditions of ionization in an ion source and introduction through the valve of these elements are different and the proportional ratios are not valid.

Consider now the isotope dilution in application to ^{40}Ar, the quantity of which should be known for calculating an age of a sample. Due to melting, a probe releases all volatile components, including radiogenic ^{40}Ar. In this case, measurement is not as simple as that of ^{4}He. Along with the radiogenic component, the probe contains air argon (Ar_{air}) that is a mixture of isotope masses 36, 38, and 40: $(^{40}Ar/^{36}Ar) = 295.5$; $(^{36}Ar/^{38}Ar) = 5.32$ (Ozima and Podosek 2002). Frequently, probes contain considerably more air argon than radiogenic. The latter "masks" radiogenic argon and sometimes makes measurement impossible.

We prefer to work with a two-beam mass spectrometer; one beam is adjusted on the 40th mass, and another one on the 36th mass. First of all, we measure the isotope ratio of reference argon (Ar_{ref}), used for isotope dilution:

$$k_1 = \frac{^{36}Ar_{ref}}{^{40}Ar_{ref}} \tag{1.4}$$

and also air argon:

$$k_{air} = \frac{^{36}Ar_{air}}{^{40}Ar_{air}} \tag{1.5}$$

Then, after melting of a probe in a reactor, we measure the isotopic composition of the air and radiogenic argon mixture:

$$k_2 = \frac{^{36}Ar_{air}}{^{40}Ar_{air} + {}^{40}Ar_{rad}} \tag{1.6}$$

At last, we do isotope dilution by putting into the reactor a known (for example, measured by means of a Mac-Leod manometer) amount of reference gas (Ar_{ref}):

$$k_3 = \frac{^{36}Ar_{air} + {}^{36}Ar_{ref}}{^{40}Ar_{air} + {}^{40}Ar_{ref} + {}^{40}Ar_{rad}} \tag{1.7}$$

The measured isotope ratios (1.4)–(1.7) enable us to establish a system of equations and to solve this system relative to $^{40}Ar_{rad}$ (Rasskazov et al. 2000b):

$$^{40}Ar_{rad} = Ar_{ref}\frac{(k_1 - k_3)}{(k_3 - k_2)}\left(1 - \frac{k_2}{k_{air}}\right) \tag{1.8}$$

Notice that the formula (1.8) is homogeneous; we can multiply numerators and denominators by any number, not affecting the result (certainly, except Ar_{ref}). It is important, because in a two-beam mass spectrometer, which has for each channel its own coefficient of amplification, we have no necessity to calibrate channels. The isotopic ratios can be expressed in any units.

Alternatively, measurement by one-beam mass spectrometers, which is carried out by switching of the receiver of the sole amplifier through a magnetic field from a mass 36 onto a mass 40, requires absolute calibration and a special procedure to eliminate the memory effect. During scanning, masses 36 and 40 yield ~1 and ~300 v alternating with partial distortion of results, similar to an eye looking at solar light, which loses the ability for a while to see items at room illumination.

As an example, we present results of radiogenic argon measurement in the Paleoproterozoic phlogopite from the Fedorov deposit (Aldan shield). For initial mass of a probe (0.35 g), we have obtained values: $k_1 = 347.8$; $k_2 = 16$; $k_3 = 97$; $k_{air} = 347.8$ with the reference argon 0.1087 nmm^3 (normal cubic mm). Each gram of the probe corresponds to the added reference argon 0.1087/0.35 = 0.312 13 nmm^3/g. Using the formula (1.8), we obtain:

$$^{40}Ar_{rad} = 0.3113\frac{347.8 - 97}{97 - 16}\left(1 - \frac{16}{347.8}\right) = 1.07653\,\text{nmm}^3/g$$

From calculating the quantity of the air argon admixture by the ratio:

$$\frac{k_2}{k_{air}} = \frac{16}{347.8} = 0.046$$

we infer that this admixture is small (4.6%).

Different parameters have been measured for determination of radiogenic argon in the Quaternary basalt TV–98–102 from the East Tuva volcanic field (Rasskazov et al. 2000a). The values, obtained for a probe of 12.1 g: $k_1 = 357.4$; $k_2 = 340$; $k_3 = 351.4$; $k_{air} = 348$; $Ar_{ref} = 4.872 \cdot 10^{-2}$ nmm^3. One gram of the probe yields $4.872 \cdot 10^{-2}/12.1 = 4.026 \cdot 10^{-3}$ nmm^3/g.

From the formula (1.8), we calculate:

$$^{40}Ar_{rad} = 4.062 \cdot 10^{-3}\frac{357.4 - 351.4}{351.4 - 340.0}\left(1 - \frac{340}{348}\right) = 4.87 \cdot 10^{-5}\,\text{nmm}^3/\text{g}$$

and

$$\frac{k_2}{k_{air}} = \frac{340}{348} = 0.977$$

In this case, the measured argon is strongly dominated by admixture of air argon (97.7%).

The measured quantities of argon isotopes in the two given examples differ by 5 logarithmic orders. The latter measurement is made at a limit of device sensitivity expressed in small differences of isotope coefficients.

Isotope dilution of Ar is often done also using the reference of mono-isotope ^{38}Ar. The isotope coefficients l are measured:

$$l_3 = \frac{^{36}Ar_{air}}{^{38}Ar_{air} + {}^{38}Ar_{ref}}; \quad l_4 = \frac{^{38}Ar_{ref} + {}^{38}Ar_{air}}{^{40}Ar_{air} + {}^{40}Ar_{rad}}; \quad l_1 = \frac{^{36}Ar_{air}}{^{40}Ar_{air}}; \quad l_2 = \frac{^{38}Ar_{air}}{^{40}Ar_{air}}$$

The amount of radiogenic argon is calculated using a formula:

$$^{40}Ar_{rad} = {}^{38}Ar_{ref}\frac{l_1 - l_3 l_4}{l_4(l_1 - l_2 l_3)}$$

Similarly, the method of isotope dilution is applied for determination of isotope concentrations in Rb–Sr, Sm–Nd, Lu–Hf, and other isotope systems (Zhuravlev et al. 1987; Kostitsyn and Zhuravlev 1987; Mel'nikov 2003; Galer 1999; Faure 2001).

1.5 Statistical Estimation of Results

Frequently researchers have to compare ages, subdivide them into groups, calculate mean values etc. Each measured age is characterized by an analytical error. For processing a set of such values, methods of mathematical statistics are described in many textbooks and manuals (Müller et al. 1982; Rumshinskii 1976 etc.).

We consider here only calculation of a mean value from results of several measurements. Let samples from different parts of a lava flow yield dates $t_1, t_2, \ldots, t_n$ with deviations $\sigma_1, \sigma_2, \ldots, \sigma_n$. To calculate the mean age, use the equation:

$$\bar{t} \pm \sigma_{\bar{t}} = \frac{\sum_{i=1}^{n} t_i w_i}{\sum_{i=1}^{n} w_i} \pm \sqrt{1 \Big/ \sum_{i=1}^{n} w_i} \tag{1.9}$$

where w – weighing factor $1/\sigma^2$.

The probability for a set of values is calculated through a test:

$$\chi^2 = \sum_{i=1}^{n} \frac{(\bar{t} - t_i)^2}{\sigma_{t_i}^2} \tag{1.10}$$

If the calculated χ^2 is less than (or equal to) the value specified in Table 1.1, the complete set of ages belongs to the uniform distribution. If χ^2 exceeds the tabulated value, calculation of the mean age using Eq. (1.9) is not valid statistically.

Table 1.1 Values of χ^2 for given probability and numbers of measurements

Probability (%)	Number of measurements								
	2	3	4	5	6	7	8	9	10
99.0	6.63	9.21	11.34	13.28	15.08	16.81	18.48	20.09	21.67
99.5	7.88	10.60	12.84	14.86	16.75	18.55	20.28	21.95	23.59

For instance, impact glasses from the Chicxulub Crater and tektites from Mexico and Haiti yielded four $^{40}Ar/^{39}Ar$ ages: 64.98 ± 0.05 Ma, 65.2 ± 0.4 Ma, 65.07 ± 0.10 Ma, and 65.01 ± 0.08 Ma, which mark the Cretaceous–Paleogene boundary (Chap. 11). From formulas (1.9) and (1.10), the mean weighted age of the boundary corresponds to 65.00 ± 0.04 Ma with $\chi^2 = 0.9$. The latter value is less than 11.34 (Table 1.1). We infer that all the ages belong to a common set with probability of 99.9%.

1.6 Summary

The basis of isotope geology is the Law of Radioactivity of Rutherford–Soddy. The methods developed for geochronological studies—recording of the isotope ratios and measurements of isotope concentrations by isotope dilution using sophisticated mass spectrometers—define reliable ages of geologic events with statistically reasonable errors.

Chapter 2
Geochronometric Models

2.1 Isochron Model

At closure of a parent–daughter system some (initial) quantity of daughter substance is frequently observed. The age definition becomes impossible. To make a detour around this difficulty, an isochron model is applied.

2.1.1 Isochron Plane

Consider the isochron model for the Rb–Sr method that implies:

$$^{87}Sr = {}^{87}Sr_0 + {}^{87}Rb(e^{\lambda \cdot t} - 1)$$

The expression represents extensive values (i.e. concentrations). To get intensive values (i.e. ratios), divide the equation term by term into the quantity of the stable isotope ^{86}Sr:

$$\frac{^{87}Sr}{^{86}Sr} = \left(\frac{^{87}Sr}{^{86}Sr}\right)_0 + \frac{^{87}Rb}{^{86}Sr}(e^{\lambda \cdot t} - 1)$$

Obtain:

$$\frac{^{87}Sr}{^{86}Sr} - \left(\frac{^{87}Sr}{^{86}Sr}\right)_0 - \frac{^{87}Rb}{^{86}Sr}(e^{\lambda \cdot t} - 1) = 0 \tag{2.1}$$

The modern $^{87}Sr/^{86}Sr$ and $^{87}Rb/^{86}Sr$ ratios are accessible to measurements. The other two members of Eq. (2.1) remain unknown. Therefore, as a function of two variables, the age of a probe may be presented in a three-dimensional space:

$$\mathrm{z = F(x, y) \quad or \quad \Phi(x, y, z) = 0}$$

Assume that measured ratios in three probes differ on initial strontium isotope ratios, but have an identical age, i.e. that they belong to a single isochron. Three data points P_1, P_2, and P_3 (Fig. 2.1) will define an isochron plane in space:

$$Cz - By - Ax = 0 \tag{2.2}$$

S.V. Rasskazov et al., *Radiogenic Isotopes in Geologic Processes*,
DOI 10.1007/978-90-481-2999-7_2,

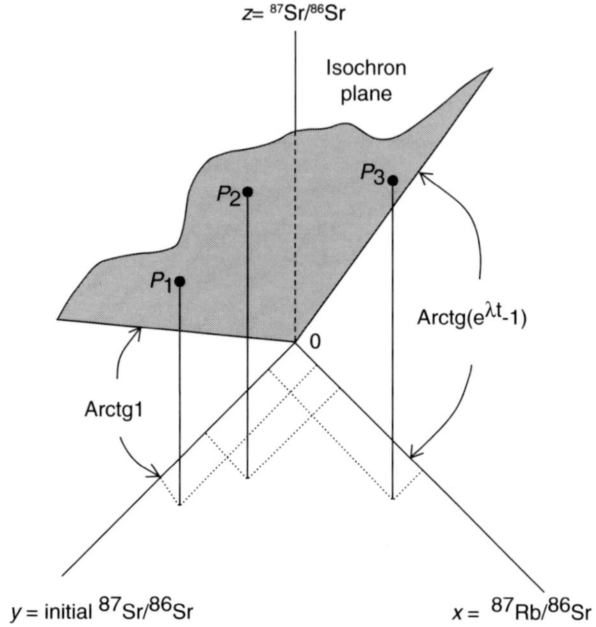

Fig. 2.1 All data points of coeval probes lie on the isochron plane. Initial ^{87}Sr/^{86}Sr of individual probes may differ

where

$$C = 1; \quad z = \frac{^{87}Sr}{^{86}Sr}; \quad B = 1; \quad y = \left(\frac{^{87}Sr}{^{86}Sr}\right)_0; \quad A = (e^{\lambda \cdot t} - 1); \quad x = \frac{^{87}Rb}{^{86}Sr}$$

The track of the isochron plane on a coordinate plane $y0z$ corresponds to $^{87}Sr/^{86}Sr =$ initial $^{87}Sr/^{86}Sr$, i.e. a straight line, which goes through the beginning of the coordinates with a slope equal to 1. Respectively, the track of the isochron plane on a coordinate plane $x0z$ corresponds to $^{87}Sr/^{86}Sr = (^{87}Rb/^{86}Sr)(e^{\lambda \cdot t} - 1)$, i.e. a straight line, which goes through the beginning of the coordinates with a slope $arctg(e^{\lambda \cdot t} - 1)$. Hence, *any data points of coeval probes always belong to a single isochron plane*.

Is it possible to determine an isochron plane by measurements of $z = {^{87}Sr}/{^{86}Sr}$ and $x = {^{87}Rb}/{^{86}Sr}$ for points $x_1z_1; x_2z_2; x_3z_3$?

Substitute the measured coordinates in Eq. (2.2):

$$\begin{aligned} z_1 - y_1 - Ax_1 &= 0 \\ z_2 - y_2 - Ax_2 &= 0 \\ z_3 - y_3 - Ax_3 &= 0 \end{aligned} \tag{2.3}$$

The point (0, 0, 0) (i.e. the origin of coordinates) lies in the isochron plane. The system of three equations (2.3) contains four unknowns: y_1, y_2, y_3, and A. While adding equations, we always get one more unknown. This approach to an isochron plane measurement cannot yield positive results. For solution, an additional ratio is required.

Take into account that the value of y_i cannot be less than the value of any other primary object in the Solar system. Such a value of 0.698 990 ± 0.000 047 was referred to as BABI (Basaltic Achondrite Best Initial) by Papanastassiou and Wasserburg (1969). The initial $A_{max} = (z_0 - 0.699)/x_0$, therefore, the least value among the obtained initial A_{max} will define the largest possible age.

Further, note that the differences y_i are maintained at any y:

$$\frac{z_0 - 0.699}{x_0} = \frac{z_k - y_k}{x_k}$$

or

$$\frac{z_0}{x_0} - \frac{z_k}{x_k} = \frac{0.699}{x_0} - \frac{y_k}{x_k}$$

Thus, at any chosen A, one can estimate whether a difference in initial strontium ratio is available, i.e. whether the assumption about sin-genetic relations of samples is valid.

Further, we explain the specified features of the isochron model by a numerical example. Take a set of probes having the common age of 1 Ga $A = e^{1.42 \cdot 10^{-10} \times 1 \cdot 10^9} - 1 = 1.4301 \cdot 10^{-2}$ and the following precisely calculated data:

	x	y	z	A
1	0.1	0.7	$7.0143 \cdot 10^{-1}$	$1.4301 \cdot 10^{-2}$
2	0.5	0.73	$7.3715 \cdot 10^{-1}$	$1.4301 \cdot 10^{-2}$
3	0.8	0.78	$7.8144 \cdot 10^{-1}$	$1.4301 \cdot 10^{-2}$

Make a projection of the data points of this system on a plane $z0x$. It will yield the usual rubidium–strontium isochron (Fig. 2.2). The data points are scattered and the regression line $y = 0.1129x + 0.6873$ is defined as a fictitious value of the age:

$$t = \frac{1}{1.42 \cdot 10^{-11}} \ln(0.1142 + 1) = 7\,\text{Ga with initial } {}^{87}Sr/{}^{86}Sr = 0.6873$$

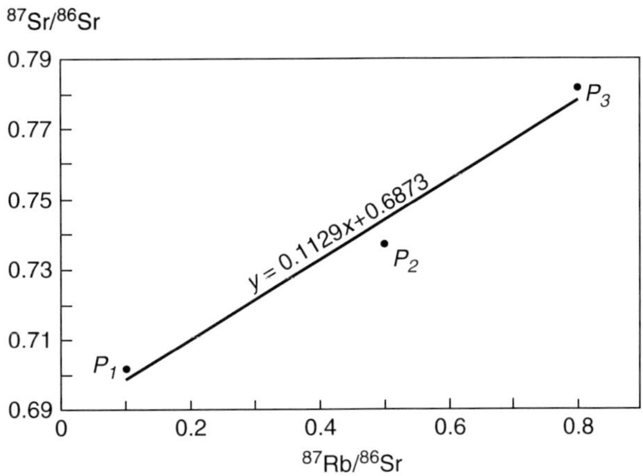

Fig. 2.2 Projection of points of the isochron plane on the $x0z$ plane. This is an ordinary Rb–Sr isochron. In our case, due to extremely contrasting data, it determines a meaningless age of ca. 7 Ga and an underestimated initial strontium isotope ratio. These features may more or less exhibit themselves in cases of ordinary distorted Rb–Sr isochrons. The points do not lie on a straight line

Solve a reverse task:

$$A_{\max 1} = \frac{z_1 - 0.7}{x_1} = \frac{7.0143 \cdot 10^{-1} - 0.7}{0.1} = 1.43 \cdot 10^{-2}$$

$$A_{\max 2} = \frac{z_2 - 0.7}{x_2} = \frac{7.3715 \cdot 10^{-1} - 0.7}{0.5} = 7.43 \cdot 10^{-2}$$

$$A_{\max 3} = \frac{z_3 - 0.7}{x_3} = \frac{7.8144 \cdot 10^{-1} - 0.7}{0.8} = 1.018 \cdot 10^{-1}$$

$A_{\max 1}$—the least of three values—lies close to the true value (in our case, it coincides with the true one, but this does not influence our conclusions).

The age $t = (1/1.42 \cdot 10^{-11}) \ln(1.43 \cdot 10^{-2} + 1) = (1/1.42 \cdot 10^{-11}) \ln(1.43 \cdot 10^{-2} + 1) \approx 1$ Ga

The closeness to the true value is caused by specific initial data. Find the initial isotope ratios:

$$y_0 = z_0 - Ax_0$$
$$y_1 = z_1 - Ax_1 = 7.0143 \cdot 10^{-1} - 1.4301 \cdot 10^{-2} = 0.7$$
$$y_2 = 7.371 \cdot 10^{-1} - 7.1506 \cdot 10^{-3} = 0.73$$
$$y_3 = 7.8144 \cdot 10^{-1} - 1.144 \cdot 10^{-2} = 0.78$$

Find a track of the isochron plane, defined by new points, on a plane $y0z$. The regression line $y = 0.0143x + 6 \cdot 10^{-5}$ (Fig. 2.3) with a value of multiplier $x = 0.0143$, designating a slope of a line relative to the axis of abscissa, determines an age $t = (1/1.42 \cdot 10^{-11}) \ln(0.0143 + 1) \approx 1$ Ga. The small difference of the free member from zero is caused, obviously, by low precision of calculation. The initial isotope ratios with precisely fixed differences coincide with the given values.

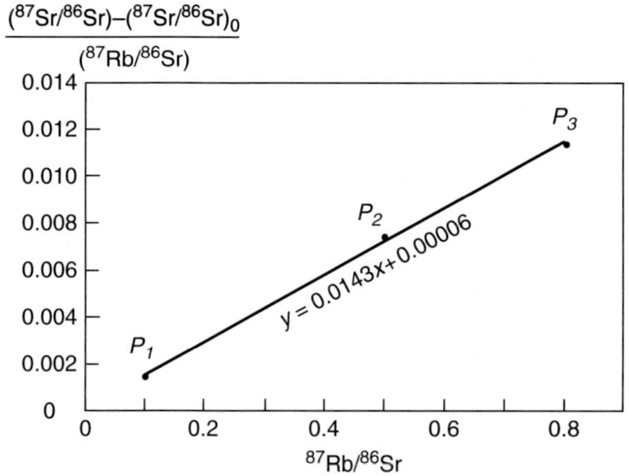

Fig. 2.3 A track of an isochron plane on the $y0z$ plane. Three data points lie on a straight line, which is directed to the origin of the coordinates. The slope of the line determines an age of 1 Ga. A satisfactory approximation is obtained

At any quantity of coeval data points, the number of unknowns will exceed by one the number of equations. It is necessary to involve an age obtained by some other method or at least one initial ratio. Even the approximate solution is valuable in reaching a conclusion on differences of the initial isotope ratios that remain constant at any fictitious age. Therefore, the method of the isochron plane allows checking sin-genetic series of data points.

2.1.2 Common Rb–Sr Isochron

We have seen that after introduction of the requirement for synchronism to a set of data points, the system of the equations remains insoluble all the same. Enter one more condition: all points of the system should be sin-genetic, i.e. should have the same initial strontium isotope ratio:

$$z_1 - y - Ax_1 = 0$$

$$z_2 - y - Ax_2 = 0$$

$$z_n - y - Ax_n = 0$$

All points of this system should lie on one straight line:

$$z = Ax + y$$

The line has a slope $A = e^{\lambda \cdot t} - 1$ and cuts a segment y on the axis of ordinate. The system becomes soluble. For its definition, one should know two points: $x_1 z_1$ and $x_2 z_2$. Then:

$$A = \frac{z_1 - z_2}{x_1 - x_2} \quad \text{and} \quad y = \frac{z_2 x_1 - z_1 x_2}{x_1 - x_2}$$

The isochron represents a projection of the data points on a plane $x0z$. This kind of two-dimensional isochron has been widely used for age determinations in all isotope systems.

Examine the geochronometric sense of isochrones, derived from whole-rock and mineral fractions (for example, granites and their micas, amphiboles etc.).

The locus of data points $D_1, D_2, \ldots, D_n$ of magmatic differentiates in the isochron coordinates of Fig. 2.4 is a straight line subjected to the equation ${}^{87}Sr/{}^{86}Sr = ({}^{87}Sr/{}^{86}Sr)_0 + ({}^{87}Rb/{}^{86}Sr)(e^{\lambda \cdot t} - 1)$. The intersection of the line with the axis of ordinates determines the initial strontium isotope ratio D_0 and the slope of the line yields an age of the series. Whole-rock compositions define an external isochron that is not consistent with distribution of mineral points. The latter, designated as phases N_1, N_i, N_n, and M_1, M_i, M_n for respective samples, show isochrons with a gentle slope and elevated initial values. The slope of the mineral isochrons determines the moment of metamorphism, which has resulted in homogenization of the strontium isotope system in all mineral fractions of each probe D_i. Respectively, intersection of each mineral isochron with an axis of ordinates reflects the strontium isotope composition of the probe D_i at the moment of metamorphism.

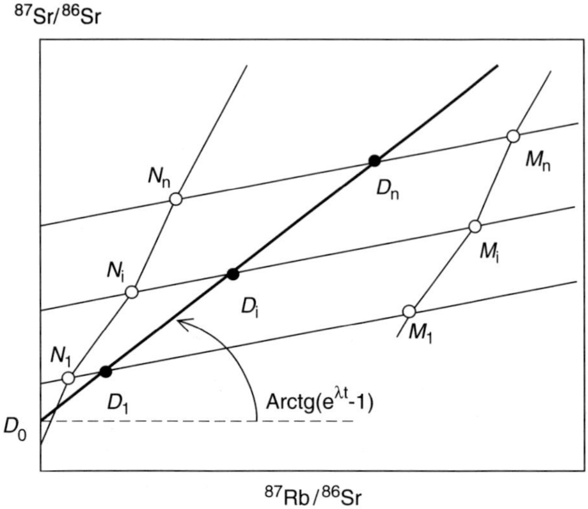

Fig. 2.4 Relation between whole-rock and intrinsic (mineral) isochrons in coordinates of the Rb–Sr isotope system

Hence, the isochron construction of minerals (for example, micas) from various differentiates is unacceptable. Crystallization of different fractions might differ and connecting data points of their minerals (N_1–N_i–N_n; M_1–M_i–M_n) by common lines can show misleading results.

We have considered ideal cases of points with fixed values. Actually, a researcher receives a series of points, each of which is characterized by an analytical error. In a regression line, the statistical model (most often the model of York (1969)) is used:

$$A = \frac{\sum_i Z_i^2 (y_i - \overline{y}) \left[\frac{x_i - \overline{x}}{\omega(x_i)} + \frac{A(y_i - \overline{y})}{\omega(x_i)} - \frac{r_i(y_i - \overline{y})}{\alpha_i} \right]}{\sum_i Z_i^2 (x_i - \overline{x}) \left[\frac{X_i - \overline{x}}{\omega(y_i)} + \frac{A(y_i - \overline{y})}{\omega(x_i)} - \frac{Ar_i(y_i - \overline{y})}{\alpha_i} \right]}$$

where $\omega_i(x_i) = 1/\sigma_{x_i}^2$, $\omega_i(y_i) = 1/\sigma_{y_i}^2$, r_i- parameter of correlation between errors of the i-th measurements (calculated similar to the coefficient of linear correlation (Müller et al. 1982; Size 1987)):

$$\alpha_i = \sqrt{\omega(x_i)\omega(y_i)},\ \overline{x} = \frac{\sum Z_i x_i}{\sum Z_i},\ \overline{y} = \frac{\sum Z_i y_i}{\sum Z_i},\ Z = \frac{\alpha_i^2}{\omega(x_i) + \omega(y_i) - 2Ar_i}$$

This equation is solved by iteration. Programs for use of the York algorithm are available in easily approached form (Ludwig 1990, 1999).

The isochron quality is estimated from values of Mean Squared Weighted Deviates (MSWD), representing the relation given from a line regression to the expected disorder calculated from correlation of errors. The data are interpreted in the framework of the isochron model, if MSWD = 1. Frequently, MSWD value exceeds 1, which means distortion of the measured values in comparison with their analytical errors, probably because of unknown analytical problems or natural geologic scattering of data points. For rocks and minerals with distorted Rb–Sr isotope systems, more than four data points cannot yield an isochron with an acceptable MSWD value and, on the contrary, if MSWD $\sim$1, the isotope system of the analyzed series of samples is not distorted (Kostitsyn 2002).

Brooks et al. (1972) showed that a regression line cannot be accepted as an isochron with a confidence interval of 95% at MSWD > 2.5. Taking into account this rule, researchers accept frequently as the true isochron a regression line with MSWD = 2.5. Actually, it is not correct, because this regression line corresponds to only 5% confidence (e.g. Dickin 1997). At a quantity of data points less than four, the MSWD value reveals a tendency to artificial understating depending on the particular algorithm used.

Besides the York algorithm, other ways of age calculations have been proposed (e.g. Faure 1989; Dickin 1997). For instance, Lepin and Brandt have suggested calculating ages for each separate pair of data points, taking into account their errors and displaying the received age values as probability functions (Makagon et al. 2000). Satisfactory results are obtained by simultaneous application of the York and Lepin–Brandt methods (Vladimirov et al. 2003).

2.2 Pb–Pb Isochron

In the Rb–Sr isotope system, we have seen that the isochron defines the initial $^{87}Sr/^{86}Sr$ with a low limit determined by BABI. For the Th, U–Pb system, the least isotope ratios were measured in a troilite of the meteorite Canyon Diablo (Tatsumoto et al. 1973):

$$\left(\frac{^{206}Pb}{^{204}Pb}\right)_0 = 9.307; \quad \left(\frac{^{207}Pb}{^{204}Pb}\right)_0 = 10.294; \quad \left(\frac{^{208}Pb}{^{204}Pb}\right)_0 = 29.476$$

This lead, referred to as primeval, has never been in contact with uranium or thorium.

Assume that radiogenic leads began accumulating t years ago and before this moment had only primeval composition. Then the equations are valid:

$$\frac{^{207}Pb}{^{204}Pb} - 10.294 = \frac{^{235}U}{^{204}Pb}(e^{\lambda_5 t} - 1)$$

$$\frac{^{206}Pb}{^{204}Pb} - 9.307 = \frac{^{238}U}{^{204}Pb}(e^{\lambda_8 t} - 1)$$

Dividing the former ratio by the latter and taking into account $^{238}U/^{235}U = 137.88$, we obtain a transcendent equation:

$$\frac{^{207}Pb - 10.294}{^{206}Pb - 9.307} = \frac{e^{\lambda_5 t} - 1}{137.88(e^{\lambda_8 t} - 1)}$$

For age definition, this method requires measurements of only one pair of isotope ratios for one probe: $^{206}Pb/^{204}Pb$ and $^{207}Pb/^{204}Pb$.

In this respect, the model is elementary. Assumption of equality of the initial lead to primeval values is not substantiated, however. Therefore, the method has not received a wide distribution. The only exception is dating of Ca–Al inclusions in the meteorite Allende on fractions enriched by radiogenic isotopes (Allègre et al. 1995).

Assume now that a captured lead had some composition at the moment of a system closure. Enter designations:

$$\frac{^{207}Pb}{^{204}Pb} = a; \quad \frac{^{206}Pb}{^{204}Pb} = b; \quad \frac{^{238}U}{^{204}Pb} = c$$

Designate the appropriate initial ratios by an index "0". Then:

$$a - a_0 = \frac{c}{137.88}(e^{\lambda_5 t} - 1); \quad b - b_0 = c(e^{\lambda_8 t} - 1)$$

Express c and combine the equations:

$$c = \frac{137.88\,(a - a_0)}{e^{\lambda_5 t} - 1} = \frac{b - b_0}{e^{\lambda_8 t} - 1}$$

After transformations, we obtain:

$$a = kb + a_0 - kb_0 = kb + B$$

The equation corresponds to a straight line that has a slope $k = (e^{\lambda_5 t} - 1)/137.88(e^{\lambda_8 t} - 1)$ and cuts a segment B on an axis of ordinates (Fig. 2.5).

The locus of data points $a = f(b)$ in the isochron and sin-genetic uranium–lead system is a straight line that can be defined by two pairs of data points: $^{207}Pb/^{204}Pb = a$; $^{206}Pb/^{204}Pb = b$. Such an isochron does not give, however, an opportunity to define values, which are included in the expression for initial values: a_0 and b_0. The latter has no practical application.

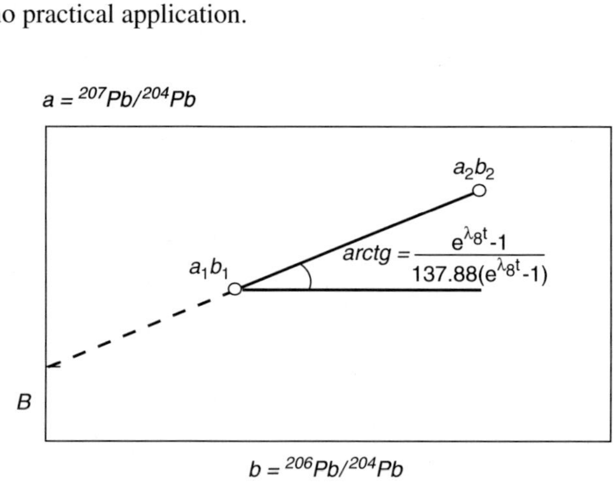

Fig. 2.5 The slope of the straight line a_1b_1–a_2b_2 is a function of age and its intercept with the ordinate axis equals B (see text)

2.3 Concordia–Discordia Model

This model has been developed for isotopes of U and Pb. Uranium consists of two radioactive isotopes: ^{235}U and ^{238}U. The former turns to ^{207}Pb through a series of radioactive nuclides, releasing $(235-207)/4 = 7\alpha$-particles with the constant $\lambda_5 = 9.8 \cdot 10^{-10}$. The latter turns to ^{206}Pb with similar releasing $(238-206)/4 = 8\alpha$-particles with the constant $\lambda_8 = 1.55 \cdot 10^{-10}$ (Jaffey et al. 1971). The present-day uranium has $^{238}U/^{235}U = 137.88$. Temporally, this ratio is growing. Both isotopes of uranium are closely connected to each other in kinetic processes due to the value $\sqrt{238/235} = 1.0063$, which is close to 1.

Similarity of the isotope properties assumes coincidence of age values calculated from $^{207}Pb/^{235}U$ and $^{206}Pb/^{238}U$ ratios. Discordant values, obtained in practice, have resulted in development of the Concordia–Discordia model.

This model describes an open system with losses of radiogenic lead in accessory minerals (zircons, monazites, apatites etc.) assuming their crystallization during any geologic process (magmatic, metamorphic etc.).

We deduce the basic relations. As a parametric curve of the consent, the Concordia means a locus of equal ages in coordinates:

$$x = \frac{^{206}Pb}{^{238}U} = e^{\lambda_8 t} - 1; \quad y = \frac{^{207}Pb}{^{235}U} = e^{\lambda_5 t} - 1 \tag{2.4}$$

The parameter is the age (Fig. 2.6). This line has essential curvature.

Practical measurements have revealed the tendency to a discordant arrangement of data points (points showing various ages on axes x and y) along straight lines intersecting the Concordia that is unexpected for the system subjected to complex transcendent dependences. The suggested explanations are known as the Wetherill and Wasserburg models.

The Wetherill model assumes that zircons, accumulated radiogenic lead since t_1 (Fig. 2.6), were subjected to a metamorphic impact t_2, at which they lost radiogenic lead partially or completely, whereas some zircons kept the radiogenic lead of the point t_1.

Where will be the point of the zircon in case of complete loss of radiogenic lead? To find out, consider the expression (2.4) as:

$$^{206}Pb = {}^{238}U_0(1 - e^{-\lambda_8 t}) \text{ and } {}^{207}Pb = {}^{235}U_0(1 - e^{-\lambda_5 t}) \tag{2.5}$$

The loss of lead t_2 years ago resulted in:

$$^{238}U_0(1 - e^{-\lambda_8(t_1 - t_2)}); \quad {}^{235}U_0(1 - e^{-\lambda_5(t_1 - t_2)})$$

Subtract the loss from the expression (2.5):

$$^{206}Pb = {}^{238}U_0(1 - e^{-\lambda_8 t_1} - 1 + e^{-\lambda_8(t_1 - t_2)}) = {}^{238}U_0(e^{-\lambda_8(t_1 - t_2)} - e^{-\lambda_8 t_1})$$

and

$$^{207}Pb = {}^{235}U_0(1 - e^{-\lambda_5 t_1} - 1 + e^{-\lambda_5(t_1 - t_2)}) = {}^{235}U_0(e^{-(t_1 - t_2)} - e^{-\lambda_5 t_1})$$

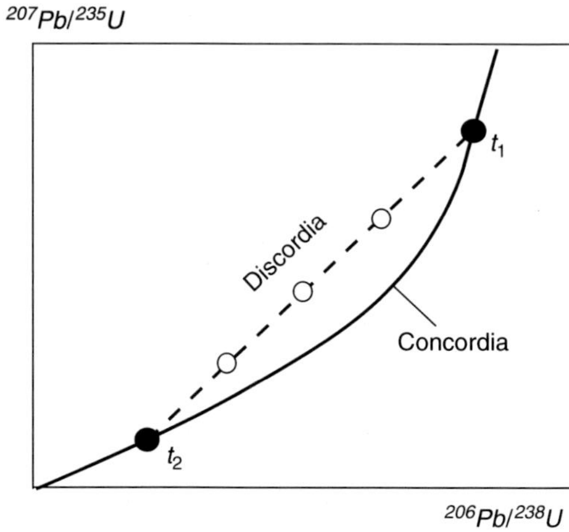

Fig. 2.6 Concordia diagram. The Concordia is a parametric locus of equal $^{207}Pb/^{235}U$ and $^{206}Pb/^{238}U$ ages. Filled circles on the Concordia show an age t_1 for crystallization of a system that had experienced a metamorphic impact at an instant t_2 with total loss of radiogenic lead. Open circles on a straight line correspond to portions of 25, 50, and 75% of radiogenic lead losses

Proceed now to modern uranium: ${}^{i}Ue^{\lambda_i t} = {}^{i}U_0$ and we get:

$$\frac{^{206}Pb}{^{238}U} = e^{\lambda_8 t_2} - 1; \frac{^{207}Pb}{^{235}U} = \mathrm{e}^{\lambda_5 \mathrm{t}_2} - 1$$

We see that the point of the probes, which have lost all radiogenic lead at the time of the metamorphic impact, lies on the Concordia and marks the moment of loss t_2.

Where will be the intermediate points that have lost 25, 50, and 75% of radiogenic lead? According to the accepted hypothesis, the loss of the isotopes ^{206}Pb and ^{207}Pb took place instantly and congruently. Therefore, allocation will occur along a straight line connecting the points t_1 and t_2 (Fig. 2.6). In general, lead, possessing a small ion radius, is extremely mobile in host Si-, OH-structures, entering as a ligand into an alloy of intrusion, not an alloy of replacement. This very property is used in the model. The manuals on geochronometry show many examples of complete conformity of the model to the requirements. However, experience has shown even more often deviations from the model.

First of all, it concerns the lower intersection of the Concordia and Discordia (point t_2 in Fig. 2.6). It is not observed frequently and is not substantiated by any geologic evidence. No traces of the late metamorphic influence are detected. Respectively, the alternative (Wasserburg) model that was based on diffusion of lead from zircons without the lower intersection has been proposed (Chap. 14).

Data points could reveal also significant dispersion without reference to any line. This requires explanations by involvement of multiple metamorphic impacts, which can complicate searching of possible concordant ages.

Now consider a series of probes corresponding to an isochron in coordinates: $^{206}Pb/^{204}Pb$; $^{207}Pb/^{204}Pb$. What geometrical image will the series represent in coordinates: $^{206}Pb/^{238}U$; $^{207}Pb/^{235}U$? And on the contrary, consider a series of probes that precisely meets the Discordia by two intersections with the Concordia. What

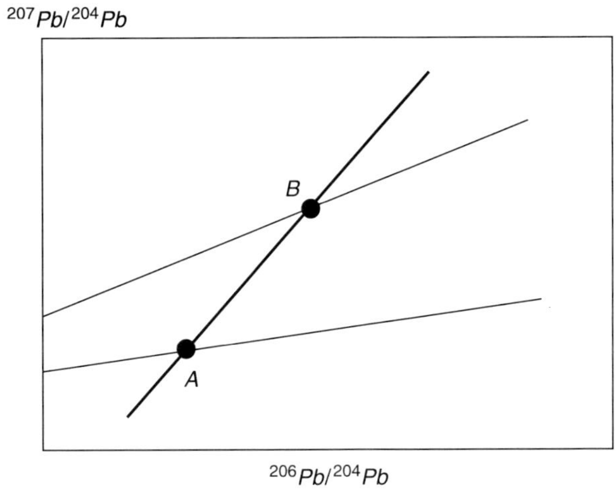

Fig. 2.7 Interpretation of data points of the U–Pb coordinates on the Pb–Pb diagram shows a relation between isochrones corresponding to the older and younger Concordia ages (respectively, thin lines B and A) and fictitious overestimated age derived from the Discordia (heavy line *AB*)

geometrical image will it represent in coordinates: $^{206}Pb/^{204}Pb$; $^{207}Pb/^{204}Pb$? Can we tell whether these images make sense?

The name "isochron" itself presumes a locus of data points displaying the same age. From definition of the Discordia by a parameter "age", a set of isochron points will represent a unique point of repeatability *i*.

Consider a case when the points belong to the two-stage Discordia t_1t_2 (Fig. 2.6). The point t_2 corresponds to the younger concordant age and will belong to an isochron with a smaller slope *A* in Pb–Pb coordinates (Fig. 2.7). The point t_1 corresponds to the older concordant age and will lie, therefore, on the steeper isochron B. The line AB has a steeper slope than either isochron. If the experimenter derives an age from the line AB in Fig. 2.7, he will receive some fictitious overestimated age value, which has no geochronological sense.

Radiogenic Pb is formed in monazite, zircon, apatite, and other accessory minerals due to radioactive decay of Th and U. The lead quantity depends on time. Is it possible to calculate age precisely by measurements in such minerals with only U, Th, and Pb concentrations?

An isotopic age was first calculated this way by Boltwood in 1907. Later on, ages were calculated from Th, U, and Pb concentrations using the procedure of Th translation to an equivalent quantity of U by multiplying constants λ_U and λ_{Th}. Up to 200 million years, approximate ages were obtained with diversity about 2%. The older ages were calculated with corrections for: (1) daughter nuclides formed due to decay in the actinium series, which ends with stable ^{207}Pb, (2) decay of parent substance for the passed time, and (3) various rates of decay for U and Th, resulting in wrong values of their equivalent expression. From experience, it was inferred that age determinations by the U–Pb method without isotope analysis of Pb did not yield reliable results (Starik 1961).

Recently, age determinations were reported again using high quality microprobe analyses. The developed Chemical Isochron Method (CHIME) was based on assumed relative variations of Th, U, and Pb on the Concordia. The complete deduction of equations for age calculations and results of dating were reported by Suzuki et al. (1991, 1994, 1996) and Suzuki and Adachi (1994):

$$^{208}Pb_{\mathrm{rad}} = {}^{208}Pb_{\mathrm{init}} + {}^{232}Th(\exp(\lambda_{232}t) - 1) \tag{2.6}$$

$$^{207}Pb_{\mathrm{rad}} = {}^{207}Pb_{\mathrm{init}} + {}^{235}U(\exp(\lambda_{235}t) - 1) \tag{2.7}$$

$$^{206}Pb_{\mathrm{rad}} = {}^{206}Pb_{\mathrm{init}} + {}^{238}U(\exp(\lambda_{238}t) - 1) \tag{2.8}$$

where λ_{232}, λ_{235}, and λ_{238} are decay constants of ^{232}Th, ^{235}U, and ^{238}U, respectively. The total concentration of lead in a mineral is the sum of all lead isotopes:

$$Pb = {}^{208}Pb_{\mathrm{init}} + {}^{208}Pb_{\mathrm{rad}} + {}^{207}Pb_{\mathrm{init}} + {}^{207}Pb_{\mathrm{rad}} + {}^{206}Pb_{\mathrm{init}} + {}^{206}Pb_{\mathrm{rad}} \tag{2.9}$$

Taking into account the ratio ^{235}U/^{238}U = 137.88 (except for natural nuclear reactor Oklo), the Eqs. (2.6–2.9) are transformed to equation:

$$Pb = Pb_{ini} + Th \times \left[\frac{\exp(\lambda_{235}t) + 137.88\exp(\lambda_{238}t)}{138.88} - 1\right] \tag{2.10}$$

For the measured concentrations of UO_2, ThO_2, and PbO (wt%) in the mineral (neglecting initial concentration of lead (Pb_{init}), at first approximation), the Eq. (2.10) is transformed to the equation:

$$\frac{PbO}{W_{Pb}} = \frac{ThO_2}{W_{Th}}(\exp(\lambda_{232}t_{app}) + 1) + \frac{UO_2}{W_U} \times \left(\frac{\exp(\lambda_{235}t_{app}) + 137.88\exp(\lambda_{238}t_{app})}{138.88} - 1\right) \tag{2.11}$$

where W – molecular mass of each oxide ($W_{Pb} = 224$, $W_{Th} = 264$, and $W_U = 270$), and t_{app} – apparent age estimate without taking into account initial lead concentrations. The designed apparent age is used for estimation of initial concentration of ThO_2 or UO_2, respectively, for monazite or zircon:

$$^{*}ThO_2 = ThO_2 + \frac{UO_2 W_{Th}}{W_U(\exp(\lambda_{232}t_{app}) - 1)} \times \left(\frac{\exp(\lambda_{235}t_{app}) + 137.88\exp(\lambda_{238}t_{app})}{138.88} - 1\right) \tag{2.12}$$

or

$$^{*}UO_2 = UO_2 + \frac{138.88 \times ThO_2 W_U(\exp(\lambda_{232}t_{app}) - 1)}{W_{Th}(\exp(\lambda_{235}t_{app}) + 137.88\exp(\lambda_{238}t_{app}) - 138.88)} \tag{2.13}$$

If coeval monazites or zircons have identical concentrations of Pb_{init}, in coordinates PbO (wt%) – $^{*}ThO_2$ (wt%) or PbO (wt%) – $^{*}UO_2$ (wt%), data points fall on a straight line, where $^{*}ThO_2$ – sum of ThO_2 that was measured and calculated as equivalent to the quantity of formed radiogenic particles of lead to the measured UO_2 or, respectively, $^*UO_2$ – sum of UO_2 that was measured and calculated as equivalent to the quantity of formed radiogenic particles of lead to the measured ThO_2.

For the first approximation, the slope of a line (m) corresponds to the age of a mineral crystallization or time of complete homogenization of the isotope system:

$$t = \frac{1}{\lambda_{232}} \ln\left(1 + m\frac{W_{Th}}{W_{Pb}}\right) \tag{2.14}$$

or

$$m\frac{W_U}{W_{Pb}} = \frac{\exp(\lambda_{235}t) + 137.88\exp(\lambda_{238}t_{app})}{138.88} \tag{2.15}$$

For the second approximation, the obtained age t is substituted in the appropriate equation (2.12) or (2.13). Iteration is done up to coincidence of t_{app} and t. Concentration of initial lead in a mineral is obtained from intersection of the regression line with an axis PbO (wt%). Any distortion of Th, U–Pb isotope system results in loss of data point linearity.

Monazites of a gneiss, sampled from the Ryoke metamorphic belt (southwest Japan), showed intervals of ThO_2 2.7–9.8 wt%, UO_2 0.4–3.3 wt%, and PbO 0.020–0.056 wt%. Concentrations of thorium and uranium oxides were recalculated to conventional units $^*ThO_2$. Linearity of points in PbO – $^*ThO_2$ coordinates was obtained with a low MSWD value (Fig. 2.8). The slope of the line with the age of 98.9 ± 2.1 Ma was interpreted as timing of metamorphism. Similar data were obtained for zircons in coordinates of lead and conventional uranium (Suzuki et al. 1996).

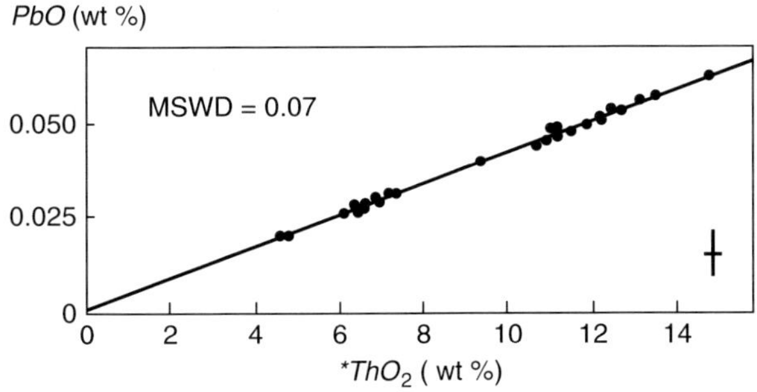

Fig. 2.8 Isochron plot for monazites of a gneiss from the Ryoke metamorphic belt (SW Japan) (Suzuki et al. 1996). $^*ThO_2$ is the sum of ThO_2 measured and of ThO_2 equivalent to UO_2. 36 analyses on 20 monazite grains were performed. In the lower right corner of the diagram, the interval of analytic uncertainty (2σ) of measurements is shown. For age calculations without determination of Pb isotope ratios on presented relations between parent and daughter substances, however, additional corrections are required (Starik 1961). This isochron age is overestimated as compared with the Concordia age

A prominent feature of the Th, U–Pb system is its open evolution. The CHIME measurements do not provide any criterion for concordant age relations. Therefore, linearity might have a discordant nature. As shown on Fig. 2.7, U and Pb isotopes, controlled in accessory minerals by the Concordia–Discordia model, yield discordant linearity of data points with the fictitious overestimated age.

Direct comparisons between results obtained by the Concordia–Discordia and Chemical Isochron methods have never been done. For such purpose, we used concentrations of U, Th, and Pb measured in five monazites of the Proterozoic migmatites from the Grand Canyon, Arizona, USA (UG–78.0–6) (Hawkins and Bowring 1999). The data points of the fractions, obtained by air abrasion of grains, have given an interval of the oldest $^{207}Pb/^{206}Pb$ ages from 1706 to 1698 Ma. Using the Eqs. (2.12–2.15) on a linear series of points, we have calculated an unreal age of 17 Ga.

On the one hand, the CHIME dating does not assume corrections required for age calculations from element concentrations. On the other hand, the results of age calculations from microprobe data are useful for age estimates of samples younger than 200 Ma (Starik 1961). The results of measurements, performed by this method by Suzuki et al, largely for the late Mesozoic magmatic and metamorphic rocks of southwest Japan, have been interpreted in general agreement with results obtained by other isotope systems.

2.4 Summary: Models for Open and Closed Parent–Daughter Systems

Geochronometry is based on the preserved conformity between parent and daughter nuclides, satisfying the Law of Radioactivity of Rutherford–Soddy. The conformity must be substantiated by models applied both to open and closed isotope systems.

For the open system, reliable dating is provided by the U–Pb method with interpretation of data points for accessory minerals in the Concordia–Discordia model. This model is applied to an isotope system, in which not less than two nuclides of one element transform to nuclides of another element. Ages are determined through comparisons of results on kinetically equivalent pairs of nuclides. In interpretation of the Discordia as an isochron, the age is overestimated.

For the closed system, data points are interpreted in terms of an isochron model that assumes two unknown variables: the age and the initial isotope ratio. In three-dimensional coordinates, all data points of coeval probes belong to a single isochron plane, which, however, cannot be constructed a priori. We have not found a way of direct calculations of the plane from the measured data, because the number of unknowns exceeds by one the number of equations at any quantity of probes. To present interpretation at the isochron plane, one should know the age or the initial ratio for at least one probe. Conventionally, data on the closed system are interpreted in isochron coordinates representing a projection of data points on the plane $x0z$. For practical purposes, reliability of the isochron model is estimated statistically in terms of data point deviations using the algorithm of York and other approaches.

Chapter 3
Principle Geochronometric Isotope Systems

3.1 Th, U–Pb Isotope System

The main advantage of the system—two series of radioactive decay: $^{238}U \rightarrow {}^{206}Pb$ and $^{235}U \rightarrow {}^{207}Pb$—provides reliable results of dating (Faure 1989; Dickin 1997). The ages are calculated in the U–Pb, Pb–Pb, and mixed coordinates with simultaneous measurements of $^{206}Pb/^{238}U$, $^{207}Pb/^{235}U$, and $^{207}Pb/^{206}Pb$ ratios (Ludwig 2000).

Unlike uranium, thorium has a single radioactive isotope ^{232}Th, α-converted to ^{208}Pb with the decay constant $\lambda = 4.9 \cdot 10^{-11}$. Both U and Th have a valence of +4 and geochemically are close to each other, but can be separated, however, in oxidizing conditions due to occurrence of the uranyl ion (UO_2^{2+}) with a valence of +6 (Faure 1989). U can be separated from Th due to different distribution coefficients of these elements in zircon or garnet relative to magmatic liquid. The radioactive decay $^{232}Th \rightarrow {}^{208}Pb$ is used in studies of time-integrated geochemical effects.

3.1.1 *Decay Constants of ^{235}U and ^{238}U*

The decay constants $\lambda_{235} = 1.551\ 25(\pm 0.000\ 83) \cdot 10^{-10}$ years^{-1} and $\lambda_{238} = 9.8485\,(\pm 0.0067) \cdot 10^{-10}$ years^{-1}, accepted in geochronology (Steiger and Jäger 1977), have been determined experimentally by Jaffey et al. (1971). The errors are specified with a 68.3% confidence interval (i.e. 1σ). The systematic errors, if present, do not exceed an analytical deviation 1σ multiplied by four (Jaffey et al. 1971). Therefore, while compared to other isotope systems, it is necessary to use, at least 2σ or, to be more correct, 4σ. The currently accepted decay constants of ^{235}U and ^{238}U are confirmed by recent experimental studies (Schön et al. 2004).

3.1.2 *U–Pb Dating*

Accessory minerals such as zircon, baddeleyite, monazite, titanite, perovskite, and apatite are dated using the Concordia–Discordia model. Zircon is the most popular, especially in dating igneous rocks of silicic to intermediate compositions.

S.V. Rasskazov et al., *Radiogenic Isotopes in Geologic Processes*,
DOI 10.1007/978-90-481-2999-7_3, © Springer Science+Business Media B.V. 2010

A significant role is also played by dating of monazite and baddeleyite (Heaman and LeCheminant 1993; Bayanova 2003 etc.).

The method is labor-intensive. Its technical problems have been discussed in special publications (Krough 1973, 1982; Kotov et al. 1995; Habfast 1998; Galer 1999). The procedure consists of (1) extraction of accessory zircons from a sample, (2) choice of suitable grains for the analysis and their preliminary processing, (3) introduction of a spike and decomposition of grains by strong acids, (4) extraction of U and Pb on chromatographic columns, and (5) mass spectrometric measurements of the isotope ratios. To get several grams of zircon grains with magmatic habit, one has to process tens of kilograms of a granite. To get concordant age of a crystal nucleus, outer parts of zonal grains are removed by acid or abrasion treatment. To avoid contamination, chemical preparation of probes for isotope analysis is performed in ultra clean laboratories.

Stages 1, 3, and 4 are avoided in case of direct U and Pb release by electro-thermal evaporation, ion-probe or laser ablation (Compston et al. 1985; Horn et al. 2000; Kober 1987). In the two latter techniques, grains are prepared for measurements similar to microprobe analysis of minerals. The isotope ratios are measured in the most appropriate part of a crystal. Accuracy of these techniques is appreciably lower than the accuracy of dating on mineral fractions. Therefore, these techniques are used mainly in dating of Precambrian rocks with a complex geologic history (Wilde et al. 2001; Bowring and Williams 1999; Bibikova et al. 2003). Electro-thermal evaporation of a thin grain of zircon is provided directly on the cathode of a mass spectrometer with measurements of only the $^{207}Pb/^{206}Pb$ ratio (Kober 1987). This is effective only for concordant zircons or for those with recent losses of lead. In other cases, the measured age can strongly differ from the true value. The method is effective in combination with the classical method of U–Pb dating (Khain et al. 2002).

As an example, Fig. 3.1 presents the Concordia diagram for baddeleyite and zircon crystals of leucogabbro from the intrusion Norilsk–1, Siberian trap province. These minerals show both measured $^{206}Pb/^{238}U$ and $^{207}Pb/^{235}U$ ratios close to the Concordia. The former yields an age of 251.2 ± 0.3 Ma accepted as the crystallization age (Kamo et al. 1996). With uncertainties in constants of uranium decay at a level of 1σ, the measured values become concordant at 251.5 ± 0.4 Ma (Ludwig 2000).

3.1.3 Pb–Pb Dating

The method of Pb–Pb isochrons is applied for dating of U-bearing rock series: meteorites, metamorphic complexes, deep-seated parts of terranes exhibited by inclusions in basalts etc. (Patterson 1956; Tera et al. 1997). Reliable results have been obtained by stepwise dissolution of titanite and apatite grains (Frei et al. 1997; Berger and Braun 1997).

The Pb–Pb isochron method provides dating of marine carbonaceous rocks that cannot be dated by other isotope methods (Ovchinnikova et al. 2000a; Semikhatov et al. 2000). Selection of material for this purpose requires preliminary petrographic and geochemical analyses to avoid samples with admixtures of detrital

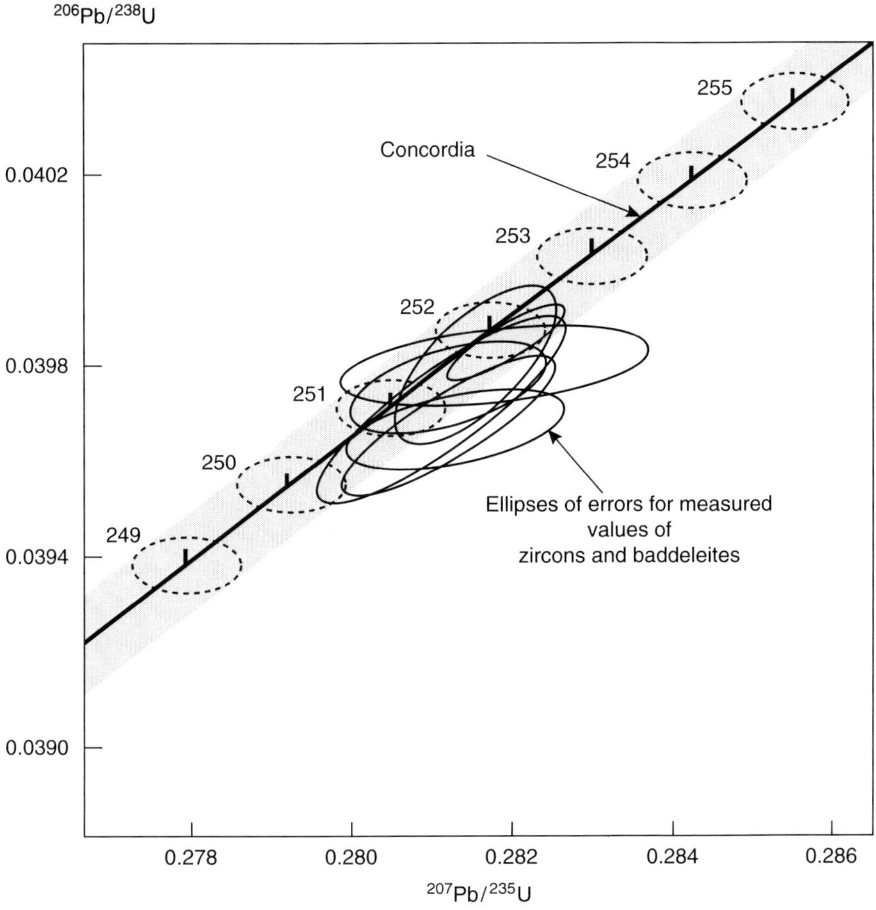

Fig. 3.1 The Concordia diagram for baddeleyites and zircons of a leucogabbro from the intrusion Norilsk–1, Siberian trap province (Kamo et al. 1996). The *dotted ellipses* show concordant age values with uncertainties (1σ) of decay constants for uranium isotopes (Ludwig 2000). The $^{206}Pb/^{238}U$ age of 251.2 ± 0.3 Ma corresponds to the concordant age of 251.5 ± 0.4 Ma

substance or secondary alteration. The selected carbonate samples yield an isochron for samples subjected to differential dissolution Ovchinnikova et al. 2000a, b). The obtained regressions yield, however, large scattering of data points with MSWD > 2.5. The isochron for metamorphosed apatite-bearing carbonates corresponds to the metamorphic impact (Fefelov et al. 2000).

3.2 K–Ar Isotope System

The isotope system is characterized by transformation of radioactive ^{40}K to radiogenic ^{40}Ar. Argon is a noble gas that has no chemical ties with other elements and can easily leave a crystal lattice of a mineral at elevated temperature. For age determination of a mineral, the isotope system must be closed, i.e. $^{40}Ar/^{40}K$ ratio in the mineral must depend only on time. Due to the long period for half-decay of ^{40}K and wide distributed of potassium-bearing rocks, K–Ar dating is applied in a range from the early Precambrian through the Holocene. Age determinations are performed on K-bearing minerals of magmatic, metamorphic, metasomatic, volcanic rocks, authigenic minerals from sediments, minerals from weathering crust etc.

3.2.1 Constants of Radioactive Transformation

Constants of radioactive transformation of radioactive ^{40}K, proposed by the Subcommittee on Geochronology of the IUGS (Steiger and Jäger 1977), were based on compilation of the direct measurements of potassium activity (Beckinsale and Gale 1969). Statistical estimates of these experiments were presented by Endt and Van der Leun (1973). Other constants, accepted in nuclear physics, were derived from more representative compilation of experimental data (Audi et al. 1997).

Discrepancy of the constants requires updating values, accepted in geochronology, in connection with specifications for the nuclear mass of potassium, $^{40}K/K_{total}$ ratio, and Avogadro's number (Min et al. 2000). The constants, specified in Table 3.1, differ from those accepted in geochronology. Because of inexact values of the constants, $^{40}Ar/^{39}Ar$ dates are systematically underestimated relative to the U–Pb ages for ca. 1% (Chap. 11).

Despite the inconsistency, geochronologists continue using the accepted constants (Steiger and Jäger 1977), favored for comparisons of data with those obtained earlier by conventional K–Ar and $^{40}Ar/^{39}Ar$ techniques. The uncertainty in the constants complicates, however, age interpretations in integrated studies performed in K–Ar, U–Pb, and other isotope systems (Renne et al. 1998).

3.2.2 Conventional K–Ar Method

An age is calculated from relative quantities of radioactive ^{40}K and ^{40}Ar in a probe. Concentration of the former nuclide is defined by any suitable method (isotope dilution, flame photometry etc.) with accuracy 1.0–3.0%. Concentration of ^{40}K

Table 3.1 Constants for radioactive transformation of ^{40}K proposed by different authors

Compilation	Beckinsale and Gale (1969)		Endt and Van der Leun (1973)	Audi et al. (1997)
Decay constant, year^{-1}	Recalculated (Steiger and Jäger 1977)	Recalculated (Min et al. 2000)		
λ_β	$4.962 \cdot 10^{-10}$	$4.950 \cdot 10^{-10}$	$4.884 \cdot 10^{-10}$	$4.845 \cdot 10^{-10}$
λ_e	$0.581 \cdot 10^{-10}$	$0.580 \cdot 10^{-10}$	$0.580 \cdot 10^{-10}$	$0.5814 \cdot 10^{-10}$
$\lambda = \lambda_e + \lambda_\beta$	$5.543 \cdot 10^{-10}$	$5.530 \cdot 10^{-10}$	$5.463 \cdot 10^{-10}$	$5.428 \cdot 10^{-10}$

λ_e – the constant for electron capture by ^{40}K, λ_β – the constant of β-decay

is calculated from total concentration of potassium using the ratio $^{40}K/K_{total}$ = 0.000 116 7, which is accepted in geochronology (Steiger and Jäger 1977). The latter is comparable within an error with the value accepted in physics (0.000 117 ± 0.000 002) (Min et al. 2000).

For argon analysis, a probe is melted in a high vacuum furnace, the released gas is extracted by means of chemical ties, and the separated argon fraction is introduced for measurements by (1) isotope dilution, (2) Cassignol–Gillot unspiked technique, or (3) absolute calibration of a scale in a mass spectrometer. The former technique is widely used (Chap. 1). The second is based on measurement of isotope composition of argon in a sample and its comparison with composition of air argon. The measured volume is calibrated by standard samples with known concentrations of radiogenic argon (Gillot and Cornette 1986). In the third technique, the scale of the mass spectrometer is calibrated systematically between measurements of probes (Pasteels et al. 1989; Boven 1992).

An analytical error depends on argon concentration in a sample; the lower the concentration, the higher the error. The highest accuracy of argon measurement has been achieved by the unspiked technique. For Pleistocene basalts from the Canary and Hawaiian islands, the analytical error was in a range of 0.7–2.7% (Guillou et al. 1996, 1997).

Age measurements by the conventional K–Ar method have been successfully applied for dating of volcanic rocks crystallized in the past 15 Ma. The older rocks are often subjected to argon losses and are required substantiation of undistorted relations between radioactive ^{40}K and radiogenic ^{40}Ar.

A K–Ar age can be calculated in three ways. To demonstrate general geochronometric principles, consider a numerical example for a probe, which yields K = 1.5% and $^{40}Ar_{rad} = 7.98 \cdot 10^{-4}$ nmm^3/g radiogenic argon. For calculation, use equation:

$$t = \frac{1}{\lambda} \ln \left(\frac{\lambda_e + \lambda_\beta}{\lambda_e} \times \frac{^{40}Ar}{^{40}K} + 1 \right) \tag{3.1}$$

where λ_e, λ_β, and λ – the IUGS constants (Table 3.1).

Firstly, compare numbers of particles. A probe of 1 g contains $1.5 \cdot 10^{-2}$ g of potassium (nuclear mass 39.098). Respectively, 1 g of potassium contains $6.0221367 \cdot 10^{23}$ of particles (Avogadro's number). The probe contains:

$$K = 6.0221367 \cdot 10^{23} \frac{1.5 \cdot 10^{-2}}{39.098} = 2.310 \cdot 10^{20} \text{ particles}$$

Only 0.011 67% ($1.167 \cdot 10^{-4}$) of this potassium is radioactive ^{40}K. Hence, in our case $^{40}K = 2.310 \cdot 10^{20} \times 1.167 \cdot 10^{-4} = 2.6958 \cdot 10^{16}$ particles.

Proceed now to argon. A volume of 22.4 l ($22.4 \cdot 10^6$ nmm^3) of argon contains $6.0221367 \cdot 10^{23}$ particles. From the proportion, it follows that:

$$^{40}Ar = 6.0221367 \cdot 10^{23} \frac{7.98 \cdot 10^{-4}}{22.4 \cdot 10^6} = 2.145 \cdot 10^{13} \text{ particles}$$

Put these values in formula (3.1) and obtain the age:

$$t = \frac{1}{5.543 \cdot 10^{-10}} \ln\left(9.54 \frac{2.145 \cdot 10^{13}}{2.6958 \cdot 10^{16}} + 1\right) = 13.64\,\mathrm{Ma}$$

Calculate an age by the second way from the concentrations of argon and potassium given in mol/g. The ratio of concentrations will be equal to the ratio of particles.

The concentration of ^{40}K (mol/g):

$$^{40}K = (1.5 \cdot 10^{-2}/39.098) \times 0.01167 \cdot 10^{-2} = 4.4772 \cdot 10^{-8}$$

The concentration of ^{40}Ar (mol/g):

$$^{40}Ar = 7.98 \cdot 10^{-4}/22.4 \cdot 10^{6} = 3.562 \cdot 10^{-11}$$

Put these values in the formula (3.1) and obtain the age:

$$t = \frac{1}{5.543 \cdot 10^{-10}} \ln\left(9.54 \frac{3.562 \cdot 10^{-11}}{4.4772 \cdot 10^{-8}} + 1\right) = 13.64\,\mathrm{Ma}$$

Calculate now and age by the third way using all coefficients for transfer K to wt% and for $^{40}Ar_{rad}$ to nmm^3/g. In this case, the formula (3.1) looks like:

$$t = 1.80408 \times 10^{9} \ln\left(14.269 \frac{^{40}Ar_{rad}}{K} + 1\right) = 13.64\,\mathrm{Ma} \tag{3.2}$$

This simple method yields the same result as the two previous ways.

3.2.3 ^{40}Ar/^{39}Ar Method

Unlike the conventional K–Ar method, the ^{40}Ar/^{39}Ar method requires irradiation of samples by a flow of fast neutrons (>1.5 MeV) in a nuclear reactor before measurements with transition (Merrihue and Turner 1966):

$$^{39}\mathrm{K}(n, p)\ ^{39}\mathrm{Ar} \tag{3.3}$$

The quantity of created $^{39}Ar_K$ is proportional to ^{39}K and depends on time of irradiation (Δ), neutron flow ($\phi_{(E)}$), energy E, and section of this energy capture (σ_E) (Mitchel 1968):

$$^{39}Ar_k = {}^{39}K \Delta \int \phi_{(E)} \sigma_{(E)} dE \tag{3.4}$$

Estimation of a total neutron flow in a nuclear reactor is complicated by neutrons with different energies. The problem of a neutron flow measurement is solved by using a monitor of a geologic sample with the known K–Ar age t. A coefficient of proportionality between quantity of created ^{39}Ar and quantity of ^{40}K, referred to as the J-factor, is calculated by the following expression:

$$J = \frac{^{40}K}{^{39}K}\frac{\lambda_{Ar}}{\lambda}\frac{1}{\Delta\int\phi_{(E)}\sigma_{(E)}dE} \tag{3.5}$$

Respectively,

$$^{40}\mathrm{K} = {}^{39}\mathrm{Ar_K}/\mathrm{J} \times \lambda/\lambda_{\mathrm{Ar}} \tag{3.6}$$

and the age equation (3.1) takes the form:

$$t = (1/\lambda) \times \ln(1 + J^{40}Ar_{rad}/^{39}Ar_K) \tag{3.7}$$

The J-factor is derived from the isotope ratios measured in the monitor:

$$J = (\exp(\lambda t) - 1)/(^{40}Ar_{rad}/^{39}Ar_K) \tag{3.8}$$

Besides the nuclear reaction (3.3), the irradiated sample is affected by other transformations, among which five reactions with Ca, K, and Cl are used for corrections (McDougall and Harrison 1988). The significant quantity of ^{36}Ar, ^{37}Ar, and ^{39}Ar is formed in three reactions of calcium isotopes:

$$^{40}\mathrm{Ca(n, n, \alpha)}, {}^{36}\mathrm{Ar} \tag{3.9a}$$

$$^{40}\mathrm{Ca(n, \alpha)}, {}^{37}\mathrm{Ar} \tag{3.9b}$$

$$^{42}\mathrm{Ca(n, \alpha)}, {}^{39}\mathrm{Ar} \tag{3.9c}$$

Natural samples do not contain ^{37}Ar and, therefore, allow corrections on calcium-produced argon isotopes ^{36}Ar and ^{39}Ar by means of measured $(^{36}\mathrm{Ar}/^{37}\mathrm{Ar})_{Ca}$ and $(^{39}\mathrm{Ar}/^{37}\mathrm{Ar})_{Ca}$ in a calcium glass or CaF_2, irradiated together with samples. Isotope ^{37}Ar is radioactive and decays to ^{37}Cl with the period of half-decay 35.1 days. Therefore, a number of days (D) between irradiation and measurement of a sample should be known:

$$\left(\frac{^{36}Ar}{^{37}Ar}\right)_{Ca} = \frac{^{36}Ar_{measured} - {}^{40}Ar_{measured}/295.5}{^{37}Ar_{measured} \times 2D/35.1},$$

$$\left(\frac{^{39}Ar}{^{37}Ar}\right)_{Ca} = \frac{^{39}Ar_{measured}}{^{37}Ar_{measured} \times 2D/35.1} \tag{3.10}$$

Similar to the basic reaction (3.3), irradiation results in creation of a significant amount of ^{40}Ar from ^{40}K:

$$^{40}\mathrm{K(n, p)}, {}^{40}\mathrm{Ar} \tag{3.11}$$

The factor of correction for transformation $^{40}\mathrm{K} \rightarrow {}^{40}\mathrm{Ar}$ is received by measurement of argon isotope ratios in K_2SO_4 irradiated together with samples:

$$\left(\frac{^{40}Ar}{^{39}Ar}\right)_K = \frac{^{40}Ar_{measured} - {}^{36}Ar_{measured} \times 295.5}{^{39}Ar_{measured}} \tag{3.12}$$

The factor of correction $(^{40}Ar/^{39}Ar)_K$ is much lower in a case of sample irradiation in a cadmium capsule (Tetley et al. 1980). Cadmium prevents penetration of slow neutrons, favored to the reactions (3.9). Using cadmium capsules is especially effective in irradiation of young samples with low concentrations of radiogenic ^{40}Ar.

Two nuclear reactions with isotopes of chlorine are known (Mitchel 1968):

$$^{35}Cl(n,\gamma), {}^{36}Cl \tag{3.13a}$$

$$^{37}Cl(n,\gamma), {}^{38}Cl \tag{3.13b}$$

Both are caused only by thermal neutrons. Therefore, the cadmium capsule protects a sample from creation of chlorine-derived argon isotopes.

The isotopes ^{36}Cl and ^{38}Cl are radioactive and decay, respectively, to ^{36}Ar and ^{38}Ar with release of β-particles. The period of half-decay for ^{36}Cl is $3{\cdot}10^5$ years. Hence, a significant quantity of ^{36}Ar will be collected in a sample with high $Cl/^{40}Ar_{rad}$ ratio in a case of a long (year and more) interval of time between irradiation and measurement of a sample. The period of half-decay for ^{38}Cl is only 37.3 min. By the moment of measurement of a sample, all ^{38}Cl turns to ^{38}Ar that allows one to make a correction on chlorine-derived ^{36}Ar:

$$\left(\frac{^{36}Ar}{^{38}Ar}\right)_{Cl} = \frac{^{36}Ar_{measured} \times (2 \times D/365)/(3 \times 10^5) - {}^{40}Ar_{measured}/295.5}{^{38}Ar_{measured} - {}^{40}Ar_{measured}/(295.5 \times 5.35)} \tag{3.14}$$

The measured peak of ^{40}Ar in a sample after irradiation, in addition to radiogenic, excessive, and atmospheric (air) argon, contains nucleogenic argon:

$$^{40}Ar_{measured} = {}^{40}Ar_{rad} + {}^{40}Ar_{excess} + {}^{40}Ar_{air} + {}^{40}Ar_{nucl} \tag{3.15}$$

Consider a case, when $(^{40}Ar/^{36}Ar)_{excess} = 0$. Then:

$$^{36}Ar_{measured} = {}^{36}Ar_{air} + {}^{36}Ar_{Ca} + {}^{36}Ar_{Cl} \tag{3.16a}$$

$$^{38}Ar_{measured} = {}^{38}Ar_{air} + {}^{38}Ar_{Cl} \tag{3.16b}$$

$$^{39}Ar_{measured} = {}^{39}Ar_{K} + {}^{39}Ar_{Ca} \tag{3.16c}$$

The Eq. (3.15) is changed to equation:

$$^{40}Ar_{rad} = {}^{40}Ar_{measured} - 295.5(^{36}Ar_{measured} - {}^{36}Ar_{Ca} - {}^{36}Ar_{Cl}) - {}^{40}Ar_{K} \tag{3.17}$$

By dividing the Eq. (3.17) on the Eq. (3.16c), we get:

$$\frac{^{40}Ar_{rad}}{^{39}Ar_K} = \frac{^{40}Ar_{measured} - 295.5 \times (^{36}Ar_{measured} - {}^{36}Ar_{Ca} - {}^{36}Ar_{Cl}) - {}^{40}Ar_K}{^{39}Ar_{measured} - {}^{39}Ar_{Ca}} \tag{3.18}$$

Including all corrections in Eq. (3.18), we have:

$$\frac{^{40}Ar_{rad}}{^{39}Ar_K} = \frac{^{40}Ar - 295.5 \times \left(^{36}Ar - {^{37}Ar} \times \left(\frac{^{36}Ar}{^{37}Ar}\right)_{Ca} - {^{38}Ar} \times \left(\frac{^{36}Ar}{^{38}Ar}\right)_{Cl}\right) - {^{39}Ar} \times \left(\frac{^{40}Ar}{^{39}Ar}\right)_K}{^{39}Ar - {^{37}Ar} \times \left(\frac{^{39}Ar}{^{37}Ar}\right)_{Ca}} \tag{3.19}$$

where argon isotopes without symbols in the right-hand part designate the measured values corrected on the appropriate background values. It is convenient to express the right-hand part of the equation through the isotope ratios by dividing into ^{39}Ar:

$$\frac{^{40}Ar_{rad}}{^{39}Ar_K} = \frac{\frac{^{40}Ar}{^{39}Ar} - 295.5 \times \left(\frac{^{36}Ar}{^{39}Ar} - \frac{^{37}Ar}{^{39}Ar} \times \left(\frac{^{36}Ar}{^{37}Ar}\right)_{Ca} - \frac{^{38}Ar}{^{39}Ar} \times \left(\frac{^{36}Ar}{^{38}Ar}\right)_{Cl}\right) - \left(\frac{^{40}Ar}{^{39}Ar}\right)_K}{1 - \frac{^{37}Ar}{^{39}Ar} \times \left(\frac{^{39}Ar}{^{37}Ar}\right)_{Ca}} \tag{3.20}$$

Equation (3.20)—basic for age calculation in the ^{40}Ar/^{39}Ar method—involves only isotope ratios. Ages are measured in gas fractions released from a sample at stepwise heating within a high vacuum furnace. A sample can be thermally treated also in a vacuum chamber by a laser of constant action.

The stepwise heating of a sample is performed with isotope analysis at each temperature step up to complete melting. In a system of released gas, the laser technique provides background values 1–2 orders lower than in the furnace technique. The laser is used, however, for analysis of a small sample (individual grain or some grains) within limits of a laser beam (up to hundreds of microns). Such a sample contains a small quantity of radiogenic argon and can produce few steps. Annealing within a high vacuum furnace allows processing through more heating steps. For example, 10.39 mg of a nepheline separate from a nepheline syenite (Zadoi massif) was heated through 58 steps in a temperature range of 670–1330°C (Chap. 11).

The results obtained by stepwise heating experiments are displayed in coordinates of the apparent (measured) age versus released ^{39}Ar (Fig. 3.2). Age values for each temperature step are calculated using Eqs. (3.7) and (3.20).

Reliability of age determination depends on isotope composition of initial argon. The age will be overestimated, if $(^{40}Ar/^{36}Ar)_{init} > 295.5$ and underestimated, if $(^{40}Ar/^{36}Ar)_{init} < 295.5$ (Chap. 6). The results of argon measurements in a technique of stepwise heating can be displayed in so-called inverse isochron coordinates ^{36}Ar/^{40}Ar–^{39}Ar/^{40}Ar. If data points have uniform linear dependence, the interceptions between a regression line and the coordinate axis's yield an initial ^{36}Ar/^{40}Ar and $^{40}Ar_{rad}/^{39}Ar_K$. The latter is used for age calculation. The regression line is treated using a technique of York (1969). The isochron age of 3.97 ± 0.11 Ma for the anorthoclase U–96–75 is comparable within error with the plateau age of 3.94 ± 0.04 Ma. The initial ^{36}Ar/^{40}Ar ratio of 284 ± 35 of the sample corresponds to an air value within error (Fig. 3.3).

A measured ^{40}Ar/^{39}Ar age is calculated relative to the monitor (standard sample) with a known age. If a reference age for a standard sample is changed, results of

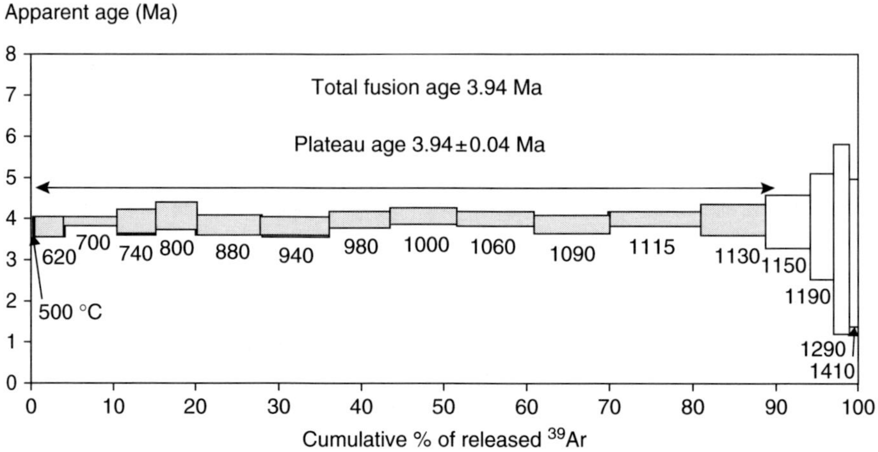

Fig. 3.2 Results of $^{40}Ar/^{39}Ar$ dating in an argon-release diagram for anorthoclase megacryst U–96–75 of a basaltic lava from the Udokan range, Siberia (Rasskazov et al. 2000b). A plateau age is consistent with a total fusion age. χ^2 value of 1.6 calculated for 12 steps of the plateau is low (see Table 1.1) demonstrating high degree of statistical confidence of the results

$^{40}Ar/^{39}Ar$ measurements can be recalculated:

$$T_{\text{new}} = 1/\lambda \ln[(\exp\{\lambda t_{\text{former}}\} - 1)\text{R} + 1] \tag{3.21}$$

$$\text{R} = (\exp\{\lambda\tau_{\text{former}}\} - 1)/(\exp\{\lambda\tau_{\text{new}}\} - 1) \tag{3.22}$$

where t and τ – ages of a sample, subjected to recalculation, and a monitor, respectively.

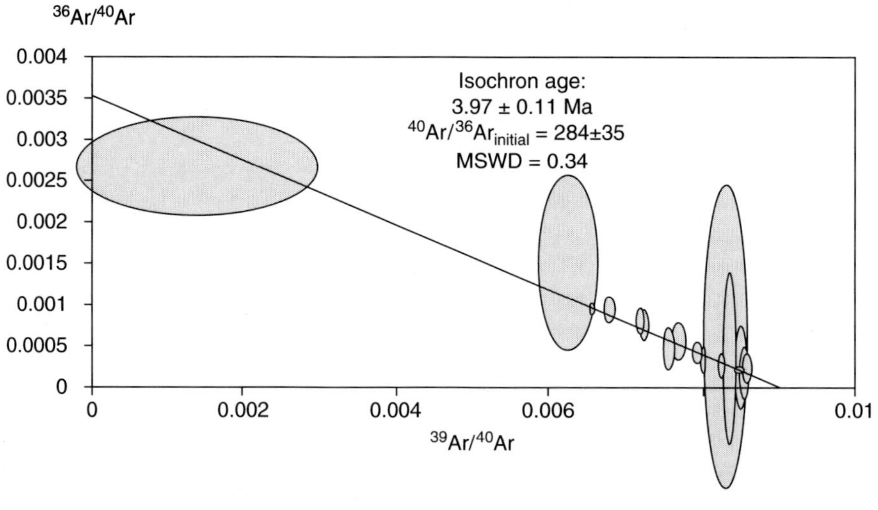

Fig. 3.3 Plot of $^{40}Ar/^{39}Ar$ dating by stepwise heating on an inverse isochron diagram. The isochron age of the anorthoclase megacryst U–96–75 is in agreement with the plateau age (see Fig. 3.2). The initial isotope ratio of argon corresponds to the air value within error

For example, Baksi et al. (1996) reported the age of 128.1 Ma for the standard LP–6, consistent with the age of 97.8 Ma for the standard GA–biotite 1550. The new age of the standard GA–1550 was defined as 98.79 Ma (Renne et al. 1998). Using Eqs. (3.21) and (3.22), the new age for LP–6 is recalculated to the value 129.39 Ma.

3.3 Rb–Sr Isotope System

The β-decay constant of ^{87}Rb $1.42 \cdot 10^{-11}$ year^{-1} (Steiger and Jäger 1977) is based on results of counting and radiochemical experiments, respectively: $1.42(+0.03/-0.02) \cdot 10^{-11}$ year^{-1} (Neumann and Huster 1976) and $1.42 \pm 0.01 \cdot 10^{-11}$ year^{-1} (Davis et al. 1977). From results of Rb–Sr and U–Pb dating of chondritic and granite fragments in lunar rocks, other values were proposed: $1.402\,(\pm 0.008) \cdot 10^{-11}$ year^{-1} (Minster et al. 1982) and $1.402\,(\pm 0.011) \cdot 10^{-11}$ year^{-1} (Shih et al. 1985). Experimental data of Davis et al. (1977) and Neumann and Huster (1976) were used by Begemann et al. (2001) for calculating values $\lambda = 1.403\,(\pm 0.009) \cdot 10^{-11}$ year^{-1} for a counting experiment and $\lambda = 1.406(\pm 0.007) \cdot 10^{-11}$ year^{-1} for a radiochemical one. A mean value of the four ^{87}Rb decay constants $1.404\,(\pm 0.004) \cdot 10^{-11}$ year^{-1} can be used at present as the best estimate.

Application of the Rb–Sr isotope system has been explored in several monographs (Faure and Powel 1972; Gorokhov 1985 etc.) and numerous papers. The full procedure of Rb–Sr dating consists of (1) selections of suitable rocks and minerals, (2) chemical processing, and (3) mass spectrometric measurements. The altered samples are excluded at a stage of preliminary petrographic analysis. The Rb–Sr method yields precise timing of crystallization for magmatic rocks of silicic and intermediate compositions from the late Precambrian through the late Tertiary. Dating of clay slates, glauconites, and illites assumes dating processes of early digenesis, which might differ sufficiently from timing of sedimentation (Rublev 1998).

The high precision of isotope analysis is achieved by qualitative separating of strontium from Rb, K, Na, Ca, Ba (Brick 1986). The most effective separation is performed by using an anion exchange resin Sr. spec, EIChroM Industries (Horwitz et al. 1992; Pin and Bassin 1992). The procedure is quick and requires a small quantity of HNO_3 and H_2O. A drawback is its single expendable use.

The strontium isotope composition is measured on a multi-collector mass spectrometer with thermal ionization (TIMS). Mass fractionating is corrected during the analysis from natural value $^{86}Sr/^{88}Sr = 0.1194$. The unspiked technique for definition of Rb/Sr ratio by inductively coupled mass spectrometry (ICP MS) was developed by Montero and Bea (1998).

3.4 Sm–Nd Isotope System

The isotope system is based on α-decay $^{147}Sm \rightarrow {}^{143}Nd$ with $\lambda = 6.54(\pm 0.05) \cdot 10^{-12}$ years^{-1} ($\pm 1\sigma$) which resulted from compilation of four counting experiments performed in 1961–1970 (Lugmair and Marti 1978). The subsequent experiments by Martins et al. (1992) gave three values, two of which were not consistent with the accepted decay constant. Begemann et al. (2001) have revised these experimental data and proposed values, consistent with those accepted in geochronology (Table 3.2). In an experimental study of Kinoshita et al. (2003), the obtained period of half-decay exceeded the accepted value by approximately 10%. The currently accepted constant provides age calculations in the Sm–Nd system consistent with those in the U-Pb system (Lugmair and Marti 1978). Table 3.2 presents the most probable parameters of half-decay and λ calculated as weighed mean values used by Lugmair and Marti (1978) and consistent results obtained by Al-Bataina and Jänecke (1987) and Begemann et al. (2001).

The basic procedures of the method—selections of suitable rocks and minerals, chemical processing, and mass spectrometric measurements—are stated in relevant publications (Wasserburg et al. 1981; Zhuravlev et al. 1987a; Faure 1989; Dickin 1997 etc.). The method yields satisfactory results for Precambrian magmatic rocks of mafic and ultramafic compositions. The age is defined by an isochron model. The required condition for dating is closure of the Sm–Nd isotope system from the moment of crystallization of magmatic rocks or from the moment of complete isotope homogenization of metamorphic rocks. In conditions of high temperatures and pressures, rare earth elements (including Sm and Nd) concentrate in garnet paragenetically connected with omphacite.

Neodymium consists of isotopes ^{142}Nd, ^{143}Nd, ^{144}Nd, ^{145}Nd, ^{146}Nd, ^{148}Nd, and ^{150}Nd. At mass spectrometric measurements, the masses are imposed by isobar masses ^{142}Ce, ^{144}Sm, ^{148}Sm, and ^{150}Sm. Neodymium is separated from samarium and cerium on chromatographic columns (Zhuravlev et al. 1987). The neodymium fraction contains an admixture of ^{142}Ce, complicated definition of ^{142}Nd with necessary accuracy (Dickin 1997). The measurement of this isotope improves reproducibility of results due to secondary normalization on $^{142}Nd/^{144}Nd$ (Thirwall 1991). The effective separation of neodymium from cerium is provided by resin Ln. Spec (EIChroM Industries) (Pin et al. 1994; Pin and Zalduegui 1997).

The isotope ratios in the Sm–Nd system are measured on a multi-collector mass spectrometer. Mass fractionating is corrected during the analysis from natural values $^{146}Nd/^{144}Nd = 0.7219$. It is accompanied by a distinct correlation between $^{143}Nd/^{144}Nd$ and $^{142}Nd/^{144}Nd$ (Thirwall 1991). The reason for this correlation is not

Table 3.2 Compilation of decay constants for ^{147}Sm

$T_{1/2}$, 10^9 years	λ, 10^{-12} years^{-1}	Data source	Note
1.060 ± 0.008	6.54 ± 0.05	Lugmair and Marti (1978)	Values accepted in geochronology
1.05 ± 0.04	6.6 ± 0.3	Al-Bataina and Jänecke (1987)	
1.23 ± 0.04	5.6 ± 0.2	Martins et al. (1992)	
1.06 ± 0.04	6.5 ± 0.3	Begemann et al. (2001)	Recalculated from (Martins et al. 1992)
1.17 ± 0.02	5.9 ± 0.1	Kinoshita et al. (2003)	
1.06 ± 0.01	6.54 ± 0.06		Probable values

clear, however. Reproducibility of results is improved by additional correction of natural value $^{142}Nd/^{144}Nd = 1.4117$.

Unlike TIMS, multi-collector ICP–MS measurements are performed with more complex mass fractionation resulting in non-comparable $^{143}Nd/^{144}Nd$ ratio at normalization on natural $^{146}Nd/^{144}Nd$ ratio. The problem is solved by using for normalization $^{145}Nd/^{142}Nd$ ratio (Vance and Thirwall 2002). The Sm/Nd ratio is determined also by unspiked ICP–MS measurements with sufficient accuracy (Montero and Bea 1998).

3.5 Re–Os Isotope System

The radioactive β-decay of ^{187}Re to ^{187}Os is accompanied by release of a relatively small amount of energy, 2.47 KeV, which is not enough for definition of the decay constant by direct counting. The rate of decay has been determined by radiochemical and geochemical methods. The most reliable values are listed in Table 3.3.

Molybdenite—the basic mineral for direct Re–Os dating of ore mineralization—is enriched by ^{187}Re and does not contain primary ^{187}Os (Stein et al. 2001). The Re–Os method was applied successfully for dating of ultramafic and mafic rocks as well as iron meteorites (Shen et al. 1996; Smoliar et al. 1996).

The value $\lambda = 1.666 \cdot 10^{-11}$ year^{-1}, used in geochronology (Smoliar et al. 1996), was obtained by using the angrite age for the Re–Os isochron of the iron meteorites. The accuracy of the constant is 0.6% (Horan et al. 1998). Galeazzi et al. (2001) and Arnaboldi et al. (2003) used the method of cryogenic bolometry that fixed increasing temperature of a sample caused by release of energy due to the decay of ^{187}Re. The most exact $\lambda = 1.605\ (\pm\ 0.012) \cdot 10^{-11}$ year^{-1} (Arnaboldi et al. 2003) differed notably from the value accepted in geochronology. The Re–Os isochron for iron meteorites yielded an erroneous date more than 100 Ma older than the earliest (dated by U–Pb isotope system) objects of the Solar system. On the contrary, data of Galeazzi et al. (2001) were consistent with the data on Re–Os dating of iron meteorites at a level of analytical and systematic uncertainness. The obtained $\lambda = 1.682\ (\pm\ 0.008_{stat} \pm 0.04_{system}) \cdot 10^{-11}$ year^{-1} might be the closest to the true value. One more value ($\lambda = 1.64\ (\pm 0.05) \cdot 10^{-11}$ year^{-1}), consistent with the Re–Os dating of iron meteorites, was obtained by Lindner et al. (1989) by radiochemical experiments.

Table 3.3 Selective data on decay constants of ^{187}Re

$T_{1/2}$, 10^9 years	λ, 10^{-11} years^{-1}	Data source
42.3 ± 1.3	1.64 ± 0.05	Lindner et al. (1989)
41.6 ± 0.25*	1.67 ± 0.01*	Smoliar et al. (1996)
$41.2 \pm 0.2_{stat.} \pm 1.1_{system}$	$1.682 \pm 0.008_{stat} \pm 0.04_{system}$	Galeazzi et al. (2001)
$43.2 \pm 0.2_{stat} \pm 0.1_{system}$	$1.605 \pm 0.008_{stat} \pm 0.004_{system}$	Arnaboldi et al. (2003)

*Error according to (Horan et al. 1998)

A chemical procedure of the method consists of (1) decomposition of a sample, (2) equilibration between a spike and a probe, and (3) separation of Re and Os for isotopic analysis. The practice has revealed ineffectiveness of the standard approaches for decomposition of a probe and its equilibration with an isotope spike. It requires special methodological approaches for osmium separation by distillation and rhenium by ion-exchange chromatography (Shirey and Walker 1995; Markey et al. 1998). The Re–Os dating can be performed also with preliminary neutron irradiation of samples, similar to the $^{40}Ar/^{39}Ar$ dating (Yin et al. 1993).

For isotopic analysis of Os, ten masses are measured simultaneously: ^{183}W (for monitoring of isobar imposing $^{184}W \rightarrow {}^{184}Os$), ^{185}Re (for monitoring of isobar imposing $^{187}Re \rightarrow {}^{187}Os$), ^{184}Os, ^{186}Os, ^{187}Os, ^{188}Os, ^{189}Os, ^{190}Os, ^{192}Os И ^{194}Pt (for monitoring of isobar imposing $^{192}Pt \rightarrow {}^{192}Os$). Precise measurements of isotope ratios are performed by the TIMS and ICP–MS methods. The necessary accuracy is reached by using multi-collector devices. In the former method, osmium is measured as negative ions OS_3^- (Creaser et al. 1991; Völkening et al. 1991; Kostoyanov and Pushkarev 1996). The Re concentration is measured by isotope dilution. In the latter method, the positively charged Os ions are analyzed. The substance is input into plasma by laser ablation of a sample, as water solution or fugitive tetra oxide of osmium. The high accuracy makes the multi-collector ICP–MS method more preferable for Re–Os isotope analysis than the N–TIMS method (Schoenberg et al. 2000).

3.6 Isotope Systems of Extinct Nuclides

Extinct radionuclides were generated before formation of the Solar system and had a short period of half-decay. Only the daughter nuclides are analyzed. Some extinct nuclides and their application (Chap. 10) are listed in Table 3.4.

The isotope systems of extinct nuclides are applied for definition of a time interval between generations of two or more meteorites. This information is a basis for

Table 3.4 Systems of extinct nuclides

Extinct nuclide	Daughter isotope	Scheme of decay	$T_{1/2}$, 10^9 years	Application for dating
^{146}Sm	^{143}Nd	α	103 ± 5	Meteorites, rocks of the early Earth
^{244}Pu	^{131}Xe, ^{132}Xe, ^{134}Xe, ^{136}Xe	SF	80 ± 0.9	Meteorites, atmosphere
	^{208}Pb etc.	9α, $6\beta^-$, SF		Not used
^{129}I	^{129}Xe	β^-	15.7 ± 0.4	Atmosphere
^{182}Hf	^{182}W	$2\beta^-$	9 ± 2*	Meteorites
^{53}Mn	^{53}Cr	e.c.	3.74 ± 0.04	Meteorites
^{26}Al	^{26}Mg	e.c., β^+	0.717 ± 0.024	Solar system

Periods of half-decay are shown after Golashvili et al. (2002). Schemes of radioactive transformations: e.c. – electronic capture, SF – spontaneous fission.
*precise measurements of Vöckenhuber et al. (2004) yielded value 8.90 ± 0.09 Ma

knowledge about early evolution of planets of the terrestrial group. The isotope systems with final products of decay such as xenon are used for age definition of the Earth. The heavy gas xenon has been kept within the atmosphere by the gravitational field of the Earth since the period of its early history, when in process of planetary accretion the Earth achieved critical mass a little bit more than the mass of the modern Moon. The isotope systems of extinct nuclides with the periods of half-decay of the order 100 Ma can be used potentially for dating of the oldest rocks and minerals (Dickin 1997).

3.7 Isotope System of ^{14}C

The isotope system ^{14}C is used for dating of organic substances with age not more than 60 Ka. The system was described in monographs (Geyh and Schleicher 1990; Dickin 1997 etc.). Here we present the basic information, which is useful for correct interpretations of ^{14}C dating results.

The isotope ^{14}C belongs to cosmogenic nuclides formed due to radioactive transformation of nuclides under influence of cosmic rays in the terrestrial atmosphere. A parent isotope ^{14}N generates daughter ^{14}C on the n, p reaction. Rapid oxidation of ^{14}C results in formation of molecules $^{14}CO_2$ and ^{14}CO, which quickly mix with the atmosphere and water. The molecules $^{14}CO_2$ get into the fabric of living organisms during breath or photosynthesis. Thus, the ^{14}C activity in living fabrics remains practically constant and depends on ^{14}C activity in the atmosphere adjusted for mass fractionation at photosynthesis. The death of an organism means cessation of the exchange processes and closing of the system. The specific activity of ^{14}C (A_0) begins to be reduced as a result of radioactive decay:

$$A = A_0 \exp(-\lambda t) \tag{3.23}$$

Accepting A_0 as a constant value for plants or other material appropriate ^{14}C activity in the atmosphere, the radiocarbon age is calculated from the measured (residual) activity of a sample.

It has been noticed that the rate of ^{14}C generation in the atmosphere depends on variations of a magnetic field in the Earth and activity of the Sun. Rate of a carbon exchange between large reservoirs, such as ocean, soil, and atmosphere, also varies in time. Methane with the "old" carbon that releases due to drainage of swamps reduces ^{14}C activity in the atmosphere.

The problem of irregular temporal variations of ^{14}C activity is solved by means of radiocarbon dating of annual rings of trees, based on results of dendrochronology. The calibration scale for the Northern Hemisphere (IntCal04) was compiled from the dendrochronological data in a range of the past 12.4 Ka (Reimer et al. 2004). The time interval of 26.0–12.4 Ka was calibrated on data obtained for corals that have zones of seasonal growth, similar to trees. A reservoir effect is taken into account. The older samples are not subjected to recalculation of radiocarbon ages. The calibration scale for ^{14}C data on the Southern Hemisphere (SHCal04) was compiled for the past 11 Ka (McCormack et al. 2004). The ^{14}C activity in the atmosphere depends on latitude, but its variations are close to accuracy of measurements.

With involvement of the calibration scales, the method ^{14}C becomes one of the geochronological methods based on relative temporal variations of isotope parameters (chemostratigraphic methods such as $\delta^{18}O$, $\delta^{13}C$, $^{87}Sr/^{86}Sr$ on carbonaceous rocks).

Determination of ^{14}C age consists of (1) measurement and calculation of conventional radiocarbon age on the Eq. (3.23) and (2) its change to the calendar age by means of the calibration curve. An example of recalculation is given in Fig. 3.4a. A laboratory radiocarbon age is measured with an error 1σ and a calibration is done with an error 2σ. The calendar age is designated by probability intervals. One conventional radiocarbon date can correspond to several intervals of calendar ages (Fig. 3.4b). The calibration is referred to 1950 A.D.

Geochronologists use the value of the period of half-decay ^{14}C 5568 $\pm$ 30 years determined by the pioneers of the radiocarbon method (Libby et al. 1949). This value is lower than the true one by approximately 3%. The values for the period of half-decay of ^{14}C in handbooks on physics vary from 5700 $\pm$ 30 to 5730 $\pm$ 40 years. Recalculation procedure to calendar age is done using the half-decay ^{14}C value 5568 $\pm$ 30 years corrected for the true one. Ages older than 26 Ka are calculated using the half-decay value of ^{14}C 5730 $\pm$ 40 years.

Radioactive decay $^{14}C \rightarrow {}^{14}N$ is accompanied with release of β^- particles, allowing measurements of ^{14}C activity in a sample by scintillation or proportional counters of β-activity. The sample is cleaned and translated into benzol (for a scintillator) or into gas (for a proportional counter).

Photosynthesis results in isotope fractionation. Plants absorb preferably light carbon isotopes, so the measured values of ^{14}C activity in an organic sample systematically differ from ^{14}C activity in the atmosphere (or water reservoir, in case of water plants) at the moment of closing of the isotope system in the sample (i.e. death of a plant or an animal). In terms of mechanism of photosynthesis all plants are subdivided into C_3 and C_4 types. The former type includes trees, seaweeds, autotrophic bacterium as well as a majority of cultivated species (rice, cereals etc.). The latter type exhibits grass of tropical regions. The C_3 and C_4 plants provide mass fractionating $\delta^{13}C = [^{13}C/^{12}C - 1] \times 1000 = -25‰$ and $-13‰$ respectively.

Some species of tropical plants (cactuses, pineapples) show mechanisms of photosynthesis both C_3 and C_4 plants and, hence, intermediate values of mass fractionating. Fractionation of ^{14}C, connected with $\delta^{13}C$ by ln ($^{12}C/^{14}C$)/ln ($^{12}C/^{13}C$) ratio ca. 1.9, allows us to compare the measured values of activity ^{14}C in plants with activity ^{14}C in the environmental reservoirs (atmosphere or water) (Stern and Vogel 1971). Analysis of recent publications shows that biological effects of carbon mass fractionating are strongly variable and not well understood yet (Morgun et al. 2008).

Carbon isotopes are measured also by a tandem mass spectrometer with the accelerator. This technique allows the direct account of mass fractionating in a sample by measurements of all three isotopes of carbon. Analysis is performed with a small quantity of carbon material.

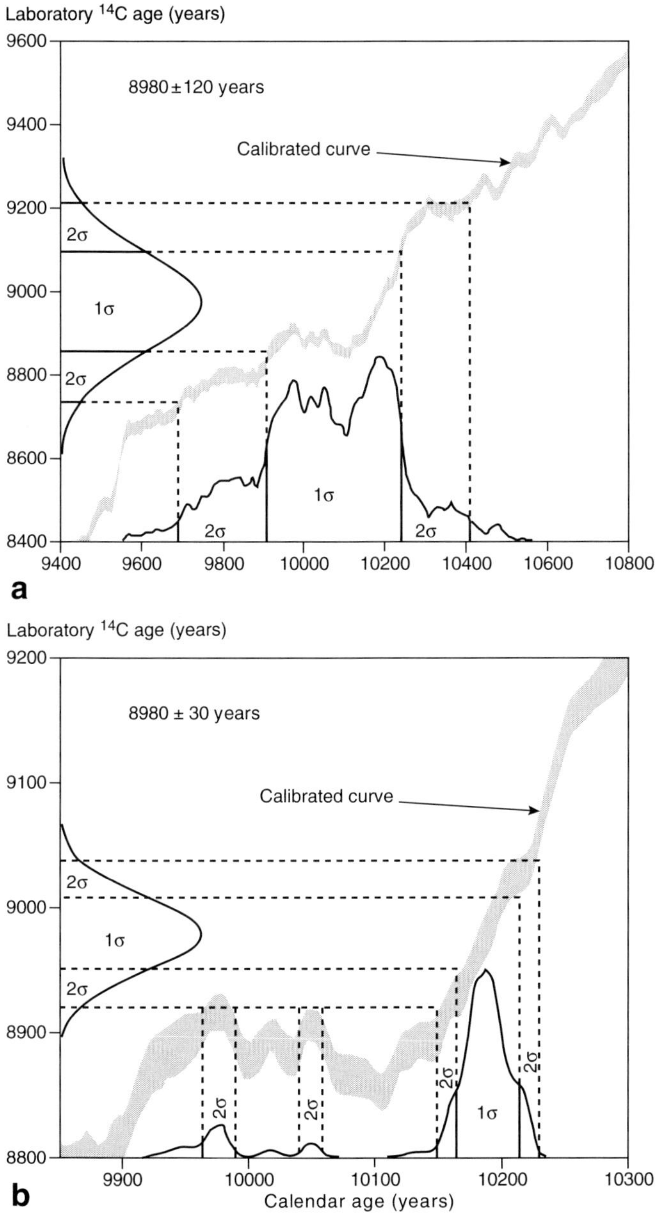

Fig. 3.4 Comparisons of probabilities distribution of calendar (calibrated) ages for the measured ^{14}C age of 8980 ± 120 years of burned wood, found below lavas of the volcano Syni (Udokan range) (**a**), and for the hypothetic "more precise" ^{14}C age of 8980 ± 30 years (**b**). The ^{14}C age of the burned wood was recalculated assuming a half-life of ^{14}C 5568 years introduced by Libby using the calibration curve IntCal04 (Reimer et al. 2004). The age of wood from the volcano Syni was reported by Devirts et al. (1981)

3.8 Summary: Reliability of the Modern Isotopic Methods

This overview has been intended as a statement of modern approaches to precise age determinations. We have examined Th, U–Pb, K–Ar, Rb–Sr, Sm–Nd, Re–Os, and ^{14}C systems, used for dating of geologic processes on the Earth and also (except ^{14}C) on the Moon and planets of the terrestrial group. The currently used exact decay constants for ^{235}U and ^{238}U and interpretation of data by the Concordia–Discordia model promote the highest precision for age determinations by the U–Pb method. Other isotope methods provide dating of geologic processes, generally consistent with the U–Pb data. The developed $^{40}Ar/^{39}Ar$ technique could yield reliable age determinations, comparable with the U-Pb dating results, but application of this method requires additional substantiation of constants for radioactive transformation of ^{40}K.

Chapter 4
Atmospheric Argon

4.1 Origin of ^{40}Ar

The so called Maxwell's demon is considered as a paradox of statistical thermodynamics. Velocities of molecules are distributed in a gas according to a certain rule. At every temperature, the gas is composed of fast (or hot) and slow (or cold) particles. In this medium, a closed vessel is placed. In the wall of this vessel, there is a gate (or valve), which is operated by the "demon". When observing a hot particle approach the gate, this "demon" opens the valve and lets the hot particle inside the vessel. For a cold particle, he keeps the valve closed. Thus, ostensibly contrary to the First Law of thermodynamics, it seems to be possible to capture hot gas without performing work.

Something similar occurs in the terrestrial atmosphere. Fast particles of He reach cosmic speeds and leave the gravitational field of the Earth, while heavy Ar particles have been caught "for ever" by the gravitation of the Earth and from the very beginning of its accretion as a planet have been kept in the atmosphere. At every temperature, the speed of gas particles obeys the Maxwell distribution law:

$$f_c = \frac{4}{\sqrt{\pi}} c^2 e^{-c^2}; \quad c = \frac{v}{\alpha}; \quad \alpha = \left(\frac{2kT}{m}\right)^{\frac{1}{2}}$$

where v – speed of particles, k – Boltzmann's constant, m – mass of particles, T – Kelvin temperature. To abandon the atmosphere, particles should have speed (at an appropriate length of free run), exceeding the first space value (i.e., $v \geq 9.7$ km/c).

This boundary condition for He is: $c \cong 6$, $e^{-c^2} \approx 2.3 \times 10^{-16}$, and integral: $\int_{c=6}^{\infty} f_c dc$ has an order of 10^{-16}. The value, though small, nevertheless is final and, acting over millions years, causes dispersion of terrestrial He in space. Thus, there is a constant outflow of helium from the atmosphere.

In case of Ar, a parameter c is approximately equal to 30, whereas e^{-c^2} and $\int_{c=30}^{\infty} f_c dc$ should be equal to zero. Particles of Ar cannot overcome Earth's gravitational field and thus remain caught from the moment of the gravitational field

DOI 10.1007/978-90-481-2999-7_4,

formation. Hence, today's concentration of Ar in the atmosphere is:

$$Ar_{atm} = \int_{t=0}^{4.56\times10^9} \frac{dAr}{dt}dt$$

Concentration is a state function of materials, i.e. it does not depend on how the materials were formed. We stand on the ground of the exact theory of diffusion and believe that the atmosphere has been enriched by argon continuously over a great number of years.

In terms of Ar, the system of the Earth remained closed in geologic time, while its atmosphere was supplied with argon from fumaroles, volcanoes, geysers, and circulating natural waters. The ratio of ^{40}Ar in the atmosphere and ^{40}K in the terrestrial sources determine the planetary age of the Earth – ca. 4.6 Ga (Brandt 1965; Schwartzman 1973).

If we accept that all atmospheric argon – 40 is radiogenic (i.e. it is lost by terrestrial rocks) and the quantity of argon – 36 has remained approximately constant, it is easy to find an expression for evolution of isotopic composition of the atmospheric argon in time:

$$\frac{^{40}Ar}{^{39}Ar}(t) = 295.5\frac{1-e^{-\lambda t}}{1-e^{-\lambda\times4.56\cdot10^9}} \quad \text{and} \quad \frac{d}{dt}\left(\frac{^{40}Ar}{^{39}Ar}\right) = \frac{295.5\lambda e^{-\lambda t}}{1-e^{-\lambda\times4.56\cdot10^9}}$$

The growth curve of the isotope ratio is presented in Fig. 4.1. During the initial stage of Earth's history, the atmosphere was quickly enriched by ^{40}Ar. Now the rate of enrichment is slow. During the Phanerozoic period, the ratio ^{40}Ar/^{36}Ar increased from 287.0 up to 295.5.

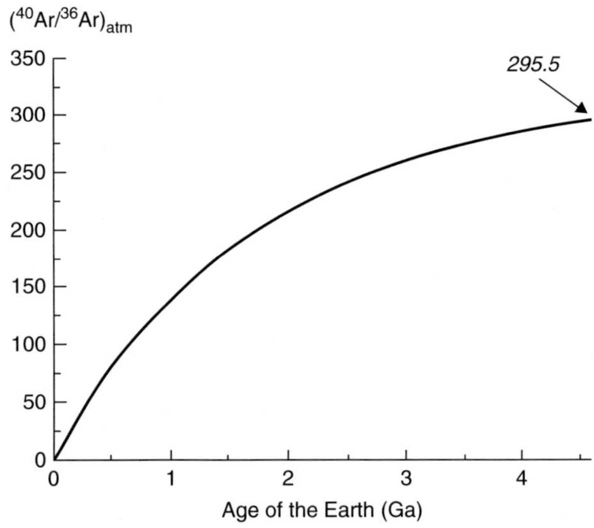

Fig. 4.1 A hypothetic evolution of ^{40}Ar/^{36}Ar ratio in the Earth's atmosphere at sea level

4.2 Variations of $^{40}Ar/^{36}Ar$ and $^{36}Ar/^{38}Ar$

Volcanic rocks often contain contaminating argon with an $^{40}Ar/^{36}Ar$ ratio differing from the air value 295.5 at the surface of the Earth (Ozima and Podosek 2002). In depleted basalts from mid-ocean ridges, this ratio increases up to 25 000 and in carbonatites up to 42 400 (Sarda et al. 1985; Sasada et al. 1997). The increase of the ratio reflects conservation of gas inclusions within minerals and volcanic glass. In lavas of historic eruptions, values of $^{40}Ar/^{36}Ar$ as low as 283.5 have been measured by the conventional K–Ar method (Dalrymple 1969). The presence of contaminating argon components in volcanic rocks has been confirmed by the $^{40}Ar/^{39}Ar$ technique (Heizler and Harrison 1988; Rasskazov et al. 2003e). In the measurements performed in the latter work, formation of ^{36}Ar during irradiation from Cl was prevented by a cadmium cover (Chap. 3). The obtained $^{40}Ar/^{36}Ar$ values for one of two contaminating argon components as low as 233–237 were likely natural. The relative decrease of this ratio could result from isotope fractionation under thermo-gradient conditions in a lava flow or shallow magmatic chamber—enrichment by light masses of hot portions of the cooling system and by heavy masses of the colder ones (Krummenacher 1970; Kaneoka 1980).

The conventional K–Ar method is applied for dating of terrestrial rocks by subtracting from the total quantity of ^{40}Ar in a probe an air admixture that assumes $(^{40}Ar/^{36}Ar)_{air} = 295.5$. To solve the problem of mass fractionation, dating is performed with corrections of results by simultaneous measurements of ^{40}Ar, ^{38}Ar, and ^{36}Ar without using spikes (Matsumoto and Kobayashi 1995).

We suggest that mass fractionation is manifested not only in lava flows and magmatic chambers, but also in a scale of the terrestrial atmosphere due to variations of temperature and pressure. The dependence of temperature on altitude is taken from http://www.grc.nasa.gov/WWW/K-12/airplane/atmosmet.html and approximated by a sectionally-continuous function (Fig. 4.2). Pressure is calculated in percents as a function of altitude (Fig. 4.3).

As a function of altitude and temperature, atmospheric pressure is calculated as:

$$P = P_0 e^{-\frac{B}{1000}\frac{G\cdot m}{R\rho^2}\int_0^h \frac{dh}{T(h)(1+\frac{h}{\rho})^2}} \tag{4.1}$$

where P – partial pressure (%), B – mass number of the isotope, G – gravitational constant, R – gas constant (1.956 cal/grad·mole), m – mass of the Earth, ρ – radius of the Earth, T – Kelvin temperature, h – distance to the surface of the Earth.

This expression generalizes the Boltzmann equation:

$$P = P_0(e^{-kh})^B \tag{4.2}$$

where $k = g/1000RT$; $g = Gm/\rho^2$ – acceleration of the force of gravity for the Earth ($k = 1.380\,66\cdot10^{-23}$ J/K).

The obtained numerical solutions of (4.1–4.2) are presented on the ordinate axis by isotope ratios $^{40}Ar/^{36}Ar$ and $^{36}Ar/^{38}Ar$ relative to their values in air (i.e. in the atmosphere at sea level) and plotted versus the altitude. The curvature of both lines

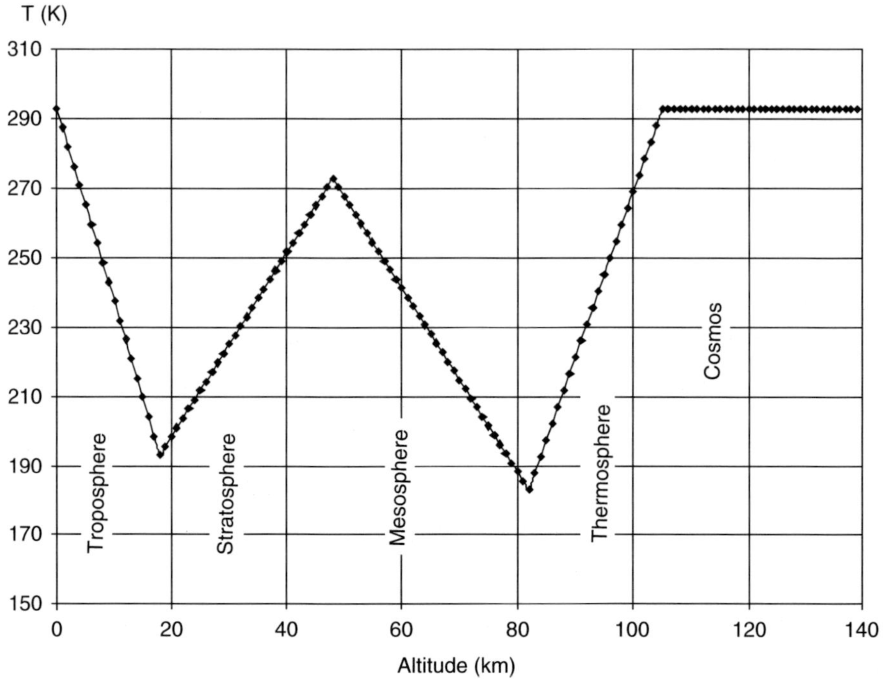

Fig. 4.2 The dependence of temperature in the atmosphere on altitude

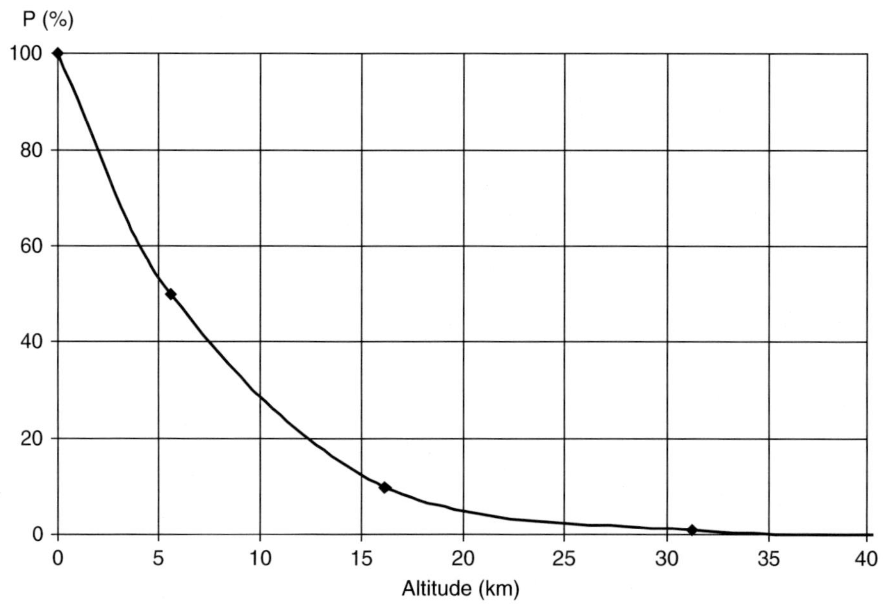

Fig. 4.3 The dependence of pressure in the atmosphere on altitude

is due to a logarithmic scale of the abscissa. The curves intersect the ordinate at an altitude value 1 km and yield trends by points with an interval 1 km. The ends of the curves enter the lower part of the stratosphere (Fig. 4.4).

To the best of our knowledge, direct measurements of isotopic compositions of atmospheric gases as a function of altitude have not been performed yet. Turbulent processes in the troposphere were presumed to stipulate mixing and distortions of the regularities established by the Boltzmann's formula. However, the difference of physical parameters relative to altitude is obvious and stable. For instance, pressure at an altitude of 16.2 km forms stably enough only 10% of the pressure that occurs at the ocean level and the temperature has an order of −50°C.

The atmospheric pressure is a sum of partial pressures of gases (and their isotopes) forming the atmosphere. Is it plausible that with increasing altitude all the partial pressures vary congruently? Are parameters determining the turbulence (such as Reynolds number, friction velocity, pulse flow) direct functions of density, which is varying from component to component? It seems rather unlikely.

The ^{14}C method is based on an empiric adjustment of measured residue activity of probes of organic material to results of dendrochronology and coral dating. For development of the direct measurement method, knowledge of altitude variations of carbon isotopes was not required and homogenizing of the isotopic composition due to a turbulent mixing was assumed. For the southern and northern hemispheres of the Earth, different calibration curves were constructed for a conversion of laboratory values of ^{14}C ages into calendar data (McCormack et al. 2004; Reimer et al. 2004), which take into account lateral isotopic heterogeneity of the atmosphere.

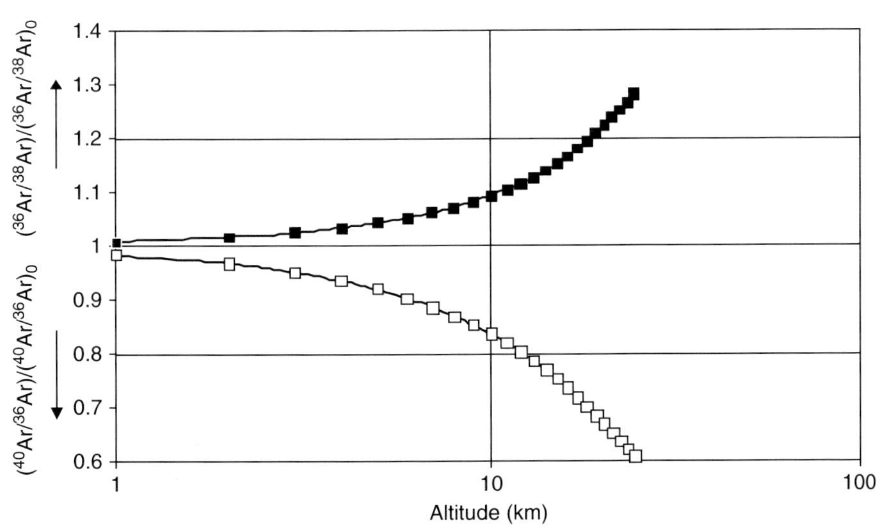

Fig. 4.4 Variations of isotopic ratios of argon in the atmosphere as a function of altitude within the troposphere with transition to the lower stratosphere. Ratios in air: $(^{40}Ar/^{36}Ar)_0 = 295.5$; $(^{36}Ar/^{38}Ar)_0 = 5.320$ (Ozima and Podosek 2002). Explanations in the text

To check results of calculations, performed here for argon isotopes on the basis of the Boltzmann equation, the isotopic ratios should be measured in atmospheric gases in situ, i.e. at high altitudes.

The predicted variations of $^{40}Ar/^{36}Ar$ and $^{38}Ar/^{36}Ar$ ratios can influence the precision of the K–Ar geochronometry of volcanic rocks. Our theoretical calculations show that, if volcanic eruptions occurred, for instance, in the Pliocene or Quaternary on high elevations of Tibet, Himalaya or Kunlun (altitudes 6–8 km), the trapped atmospheric argon must have $^{40}Ar/^{36}Ar$ ratio as low as 255–265.

4.3 Summary: Isotopically Inhomogeneous Atmosphere

The atmospheric argon composition is a time integrated characteristic of radiogenic ^{40}Ar accumulation during the whole history of the Earth. Calculations using the Boltzmann formula show that the atmosphere is isotopically inhomogeneous. The ratios $^{40}Ar/^{36}Ar$ and $^{38}Ar/^{36}Ar$, determined in air at sea level (295.5 and 5.320), likely have, respectively, the lower and higher values at the upper altitudes of the troposphere and stratosphere due to isotope fractionation under varying temperature and pressure.

Chapter 5
Distribution of Radiogenic Argon Within a Mineral

During development, the potassium–argon method was subjected to proofs of radiogenic argon safety in geologic processes. Minerals were heated to determine kinetic parameters of diffusion in them. Then, using the Arrhenius equation, the diffusion parameters were extrapolated to low temperatures of geologic sites. Thus, the suitability of many rock-forming minerals for age determinations was established.

Through theory and experience, a significant role of radiogenic argon diffusion was stressed as compared with valence daughter isotope substances, such as strontium, lead or neodymium. Kinetic equations, comprising radioactive accumulation and diffusion loss of argon, were suggested with the purpose of rehabilitation of the potassium–argon method. The equations were based on diffusion parameters of argon: activation energy (E) and frequency factor (D_0), which were used in kinetic equations with transition to the possible true age. The diffusion parameters were considered as fundamental constants describing a certain mineral type. In physics and engineering, for example, the knowledge of specific electroconductivity or thermal capacity of metals allows calculations of resistivity and heat capacity for bodies with any weight and configuration.

Unfortunately, extremely labor-intensive kinetic experiments in laboratories brought no desirable results for determination of the diffusion parameters. Values of kinetic parameters for one mineral species were widely scattered and made senseless any age corrections.

It is evident that diffusion properties of enclosed particles are controlled by parameters of the host crystal lattice. Due to X-ray analysis, structural parameters of minerals are quite characteristic for the given species and can serve as a diagnostic attribute. Hence, the diversity of diffusion properties in minerals is not strictly regulated by structural parameters of crystal lattices. The assumption arises that the explanation lies in distribution of radiogenic argon within mineral grains. Consider this point in detail.

S.V. Rasskazov et al., *Radiogenic Isotopes in Geologic Processes*,
DOI 10.1007/978-90-481-2999-7_5, © Springer Science+Business Media B.V. 2010

5.1 Closing Isotopic System: α-Factor

Our approach is based on the formal theory of diffusion. Theoretical results were many times proved by laboratory heating experiments. We reject any objections against applicability of the classical diffusion theory to natural mineral associations. If reactions had occurred in gas or an indefinitely dispersed environment, the objections would be valid (in this case, reactions of the first, second etc. orders would be considered). Real chemical reactions take place between grains of final sizes. Reagents are transported from central parts of grains to places of interaction by the mechanism of Brownian movement (diffusion), which is described by the classical theory. Indeed, the classical diffusion theory is completely valid for kinetic equations of chemical reactions (e.g. Haase 1967). It is noteworthy, however, that values of diffusion parameters, obtained in a laboratory, can differ from those that control natural geologic processes.

The differential equations of diffusion are solved by separation of variables. The solution is suggested as the sum of multiplied functions from time and spatial coordinates:

$$^{40}Ar = \sum_{\nu} T_{\nu}(t) \times X_{\nu}(x)$$

In the one-dimensional case:

$$e^{-(2\nu+1)^2\pi^2 Fo} \times X_{\nu}(x) = e^{-(2\nu+1)^2\pi^2 Fo} \times \frac{1}{(2\nu+1)} \sin \frac{(2\nu+1)\pi}{h} x \qquad (5.1)$$

where $Fo = Dt/h^2$ – dimensionless criterion of similarity, called the Fourier number, D – the diffusion coefficient, t– time, h – the characteristic size of a crystal, from which diffusion occurs.

Peculiarities of argon distribution inside a mineral that suffered losses of radiogenic argon have been studied in several papers (Lister and Baldwin 1996; Rasskazov et al. 2000b; Brandt et al. 2002b etc.). If argon is not subjected to diffusion, its concentration within a mineral is constant and does not depend on the coordinates. Respectively, argon accumulates according to the Law of Rutherford–Soddy in a form of the rectangle "a" (Fig. 5.1). At a certain instant, there arises an imposed process such as regional or contact metamorphism or a fluid flow. According to the Arrhenius law, heating sharply increases rate of diffusion in a mineral. Due to argon losses, its distribution becomes semi-sinusoidal, as shown by curves "b", "c", and "d" in Fig. 5.1. If temperature decreases (diffusion is getting lower), argon begins accumulating in the mineral again according to the Law of Rutherford–Soddy. This fraction of argon is distributed homogeneously within the mineral (in a rectangular shape).

Thus, a natural mineral is expected to show both semi-sinusoidal and rectangular distribution of argon fractions. A ratio of the former to the total argon, we designate

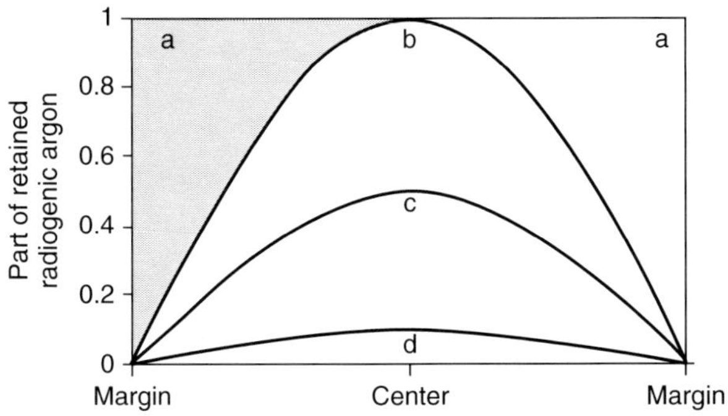

Fig. 5.1 Distribution of radiogenic argon inside a mineral (a one-dimensional case). Originally there was a rectangular distribution *a*, the concentration is constant inside a mineral. Due to diffusion, the distribution readily becomes semi-sinusoidal *b*. At continuing losses the distribution remains semi-sinusoidal, but the amplitude continually decreases: *c*, *d*. It should be noted that the diagram shows the concentration as a function of coordinates. The shadowed area does not mean that "the argon in the corner has disappeared". The semi-sinusoid is only a form of description for the argon concentration

as the α-factor:

$$\alpha = \frac{{}^{40}Ar_{s/\sin}}{{}^{40}Ar_{s/\sin} + {}^{40}Ar_{rect}}$$

where s/sin – semi-sinusoidal, rect – rectangular.

5.2 The α-Factor in the Conventional K–Ar Method

We have developed a method for age determination of a previous thermal event, which affected mineral phases, from a quantitative relation between residual radiogenic argon and argon, accumulated in a closed K–Ar isotope system (Rasskazov et al. 2001; Logatchev et al. 2002; Brandt et al. 2002b, 2006). Radiogenic argon enters a mineral grain according to the solution of the partial differential equation:

$$\frac{{}^{40}Ar}{{}^{40}K} = \frac{8}{\pi^2}\frac{\lambda_k}{\lambda}\sum_{\nu=0}^{\infty}\frac{1}{(2\nu+1)^2}\frac{e^{\left(1-\frac{(2\nu+1)^2\pi^2 D}{h^2\lambda}\right)\lambda\cdot t}-1}{1-\frac{(2\nu+1)^2\pi^2 D}{h^2\lambda}} \tag{5.2}$$

(D – coefficient of diffusion, λ – constant of radioactive transformation for argon-40, h – size of a grain, from which diffusion occurs) (Fig. 5.2).

If $D/h_2 \gg \lambda$, "instantaneous events" (i.e. heating in laboratory conditions) are described by the equation:

$$\frac{Ar}{Ar_0} = \frac{8}{\pi^2}\sum_{\nu=0}^{\infty}\frac{1}{(2\nu+1)}e^{-\frac{(2\nu+1)^2\pi^2 D\cdot t}{h^2}} \tag{5.3}$$

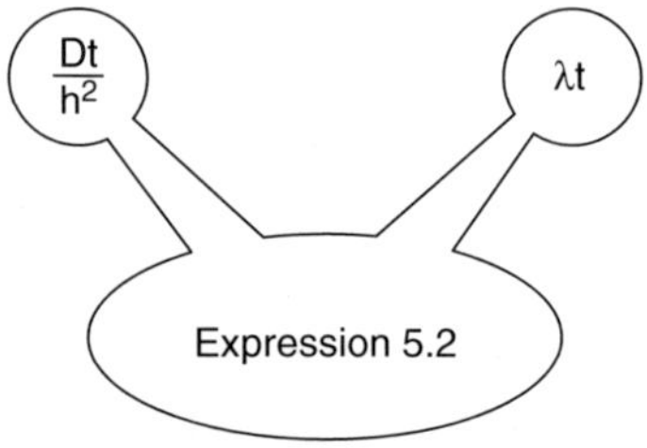

Fig. 5.2 The expression (5.2) means a difference between a quantity depending on radioactive accumulation only (λt) and a quantity depending on kinetics only (Dt/h^2). So, radioactive accumulation and kinetics are symmetric, i.e. have, so to say, equal rights

Space is characterized by the dimensionless Fourier criterion. For the Eq. (5.3), two asymptotic approximations are valid: for small *Fo*:

$$\frac{Ar}{Ar_0} = 1 - 4\sqrt{\frac{Fo}{\pi}} \tag{5.4}$$

and for large *Fo*:

$$\frac{Ar}{Ar_0} = e^{-\pi^2 Fo} \tag{5.5}$$

If radiogenic argon has not been lost from a homogeneous mineral grain, its distribution is rectangular and fits into Eq. (5.4). In case of diffusion losses, the distribution is semi-sinusoidal and satisfies Eq. (5.5).

In laboratory heating experiments, the distribution of argon (Ar_{exp}) obeys the laws of kinetics:

$$Ar_{\text{exp}} = (1-\alpha) \times 4\sqrt{\frac{Fo}{\pi}} + \alpha(1 - e^{-\pi^2 Fo}) \tag{5.6}$$

Equation (5.6) contains unknown parameters *Fo* and α, which are defined from two laboratory heating experiments of various durations. The method of iterations on a computer or a simplified formula can be applied:

$$\alpha = \frac{a_1 - \frac{4}{\sqrt{\pi}}\sqrt{Fo}}{\pi^2 Fo - \frac{4}{\sqrt{\pi}}\sqrt{Fo}} \tag{5.7}$$

where a_1, a_2 – relative quantities of argon, released by a probe after the first and second laboratory heating, respectively; *Fo* – Fourier number of the first heating. The Fourier number of the second heating experiment will be: $Fo_2 = PFo$, where $P = (t_1 + t_2)/t_1$.

5.3 Influence of Argon Distribution on Diffusion Parameters

How will the diffusion parameters E and D vary relative to distribution of argon within a mineral (i.e. relative to the α-factor)?

Consider as an ultimate purpose the definition of E and D_0. Put two equations for two isothermal experiments for argon distribution in two mineral probes with symmetric appearance:

$$E = R \times \ln\left[\left(\frac{Fo_2}{Fo_1}\frac{t_1}{t_2}\right)^{\frac{T_1T_2}{T_2-T_1}}\right] \tag{5.8}$$

and

$$\frac{D_0}{h^2} = \frac{\left[\frac{t_2}{Fo_2}\right]^{\frac{T_2}{T_1-T_2}}}{\left[\frac{t_1}{Fo_1}\right]^{\frac{T_1}{T_1-T_2}}} \tag{5.9}$$

where T_1, T_2 – respective Kelvin temperatures.

Formulas (5.8) and (5.9) are derived from the Arrhenius equation. The results of calculations show strong dependence of E and D_0 on argon distribution (Figs. 5.3, 5.4). Increasing of the α values from 0 to 1 is accompanied with increasing of the E values by a factor of 2 and D_0 values by 13–15 logarithmic orders. These variations correspond approximately to natural dispersion of these parameters.

While Fig. 5.4 gives dependence on the α-factor of size variations of a numerical kind, the numerical influence of the exponential multiplier in the Arrhenius equation $D = D_0 \times e^{-E/RT}$ is not manifested directly. The latter is estimated in a diagram in

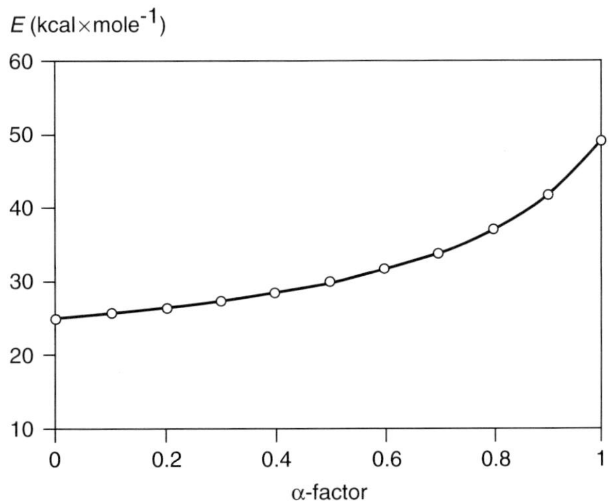

Fig. 5.3 Co-variations of the activation energy E with the α-factor. The dependence is shown in linear coordinates

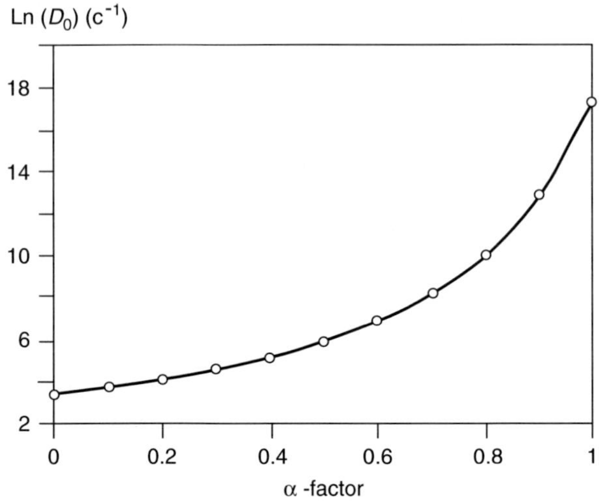

Fig. 5.4 Co-variations of the frequency factor D_0 with the α-factor. The dependence is shown in logarithmic scale

Fig. 5.5. Taking, for example, temperature 500 K, we obtain a change of the E values, expressed in variations of exponential multiplier of more than four logarithmic orders, due to influence of the α-factor.

The compensation rule of Hart (1981) says that increasing E, resulting in reduction of D, is compensated by growth of D_0 in natural minerals according to the formula:

$$\frac{D}{h^2} = \frac{D_0}{h^2} \times e^{-\frac{E}{RT}}$$

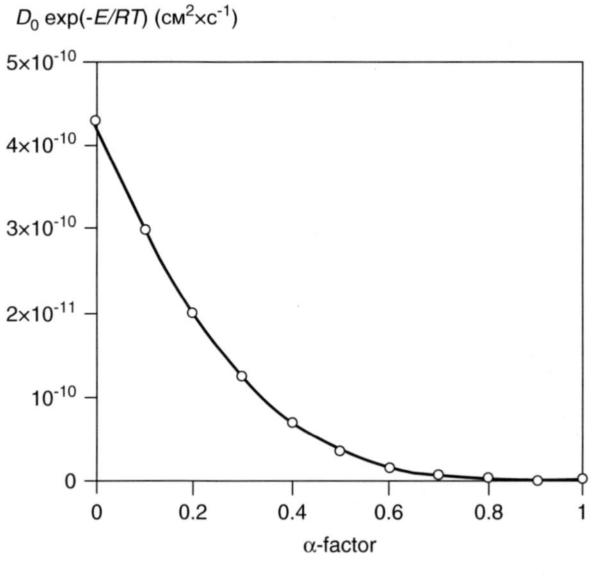

Fig. 5.5 Co-variations of the exponential factor in the Arrhenius law $D = D_0 exp$ $(-E/RT)$ with the α-factor

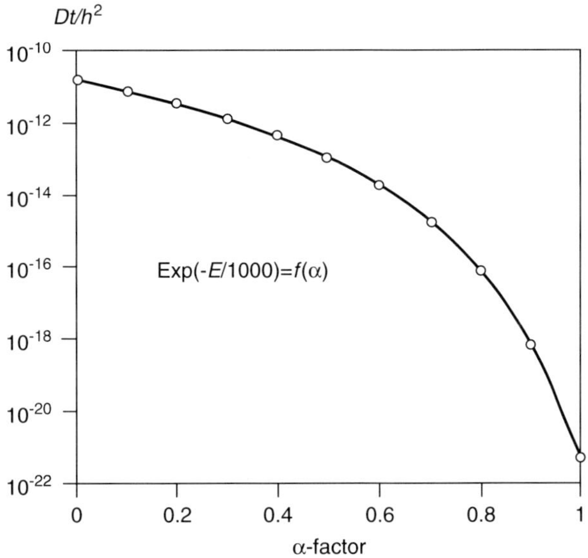

Fig. 5.6 Co-variations of the $Fo = Dt/h^2$ with the α-factor

Hart has deduced this rule empirically. We confirm it theoretically by multiplying values of ordinates in Figs. 5.3, 5.4, and 5.5. In our case, compensation is incomplete, though it reduces dependence. At a range of α-factor from 0 to 1, *Fo* varies within 10 logarithmic orders (Fig. 5.6).

5.4 Application for Dating of Imposed Geologic Processes

Irradiation of probes in the Ar–Ar technique often results in partial losses of ^{39}Ar from crystalline structures in low-temperature steps of heating (e.g. Brandt et al. 1997). Therefore, irradiated probes cannot be used properly for determination of the α-factor, which is estimated from the total quantity of argon components. We report here results of α-factor determinations by the conventional K–Ar method.

To demonstrate argon distribution in natural minerals, we have selected representative data obtained for (1) a closed K–Ar isotopic system with a rectangular fraction of argon ($\alpha = 0$), (2) relatively small argon losses, when the rectangular fraction of argon still dominates ($0 < \alpha < 0.5$), and (3) more sufficient argon losses, when a portion of the semi-sinusoidal argon exceeds the rectangular one ($0.5 < \alpha < 1.0$).

Isotopes ^{40}Ar and ^{36}Ar in probes were measured using a mass spectrometer MI-1201 modified for simultaneous receiving of the two masses. Atmospheric argon was employed as a spike in the isotope dilution procedure similar to the Cassignol–Gillot technique (Gillot and Cornette 1986). Details of measurements as well as accuracy and precision estimates were described by Rasskazov et al. (2000b). Potassium concentrations in the studied aliquots were determined by flame photometry. Ages were calculated using constants determined by Steiger and Jäger (1977).

5.4.1 Rectangular Distribution of Argon ($\alpha = 0$)

The northern part of the Riphean Tuva-Mongolian massif, adjacent to the south-western margin of the Siberian craton, was overlapped by the Oka foredeep tectonically active in the Vendian through Middle Paleozoic. The area was subjected to alkaline, basic, and granite magmatism all through the Paleozoic with final emplacement of the Samsal Complex dated at 305–261 Ma by the Rb–Sr method. The Shagaita-Gol diatreme formed at the western exocontact zone of the largest (10 × 15 km) Samsal granite body of the complex (Rasskazov et al. 2003d).

Breccia of the diatreme is cut by dikes of coarse-grained amazonite granite. Amazonite grains as large as 3–4 mm were selected for kinetic experiments. A three-step heating procedure was done using three different aliquots. Time of heating and relative quantity of argon released a_1 were obtained at the first step:

$$a_1 = (1 - \alpha)\Phi(Fo_1) + \alpha(1 - e^{-\pi^2 Fo_1}) \tag{5.10}$$

where $\Phi(Fo) = 1 - (6/\pi^2)\sum_{n=1}^{\infty}(1/n^2) \times e^{-\pi^2 n^2 Fo}$ for the case of spherical symmetry. Heating of the second aliquot yielded the new relative quantity of argon a_2 with increasing Fourier number *Fo* to *PFo*:

$$a_2 = (1 - \alpha)\Phi(PFo_1) + \alpha(1 - e^{-\pi^2 PFo_1}) \tag{5.11}$$

At last the third aliquot was melted completely with release of the whole quantity of argon. The equations were used for calculating both *Fo* and α by iteration. The obtained values were equal to zero, demonstrating no argon losses from the amazonite crystals.

The complete melting released $72.68 \cdot 10^{-3}$ nmm^3/g of radiogenic ^{40}Ar with a small quantity of Ar_{air} (7.7%). This value and the measured K content of 6.14% yielded an age of 281 ± 9 Ma, which is consistent within expected error with Rb–Sr model ages of large biotite crystals from amazonite-bearing pegmatite veins located within the main Samsal granite body. The value was accepted as a true crystallization age of the amazonite granite dike.

5.4.2 Dominating Rectangular Argon Fraction ($0 < \alpha < 0.5$)

The Sludyanka Complex is situated within the Central Asian fold belt, on the south-western shore of Lake Baikal (Korzhinskii 1947; Vasil'ev et al. 1981). A deposit of large phlogopite crystals in pegmatite veins was formed after granulite-facies metamorphism of 480–490 Ma (Kotov et al. 1997). Their crystallization age was defined by the Rb–Sr method on a phlogopite–calcite–apatite paragenetic assemblage at 460 ± 7 Ma (Brandt et al. 2006).

Numerous K–Ar-isotopic measurements on the phlogopite crystals from many veins, performed in 1960–1980, showed a wide interval of apparent ages in a range of the Early Paleozoic. ^{40}Ar/^{39}Ar dating of three phlogopite crystals, sampled from different veins, yielded plateau age spectra of 480 ± 4 Ma and 450 ± 8 Ma, and one staircase age spectrum with increasing apparent ages from 415 to 450 Ma (Ivanenko et al. 1990). The former plateau value contradicted the Rb–Sr isochron age estimate for the phlogopite deposit and had no satisfactory explanation. The latter plateau value, younger than the Rb–Sr age, demonstrated partial reset of the K–Ar system. These processes were more pronounced in results of measurements on a hyalophane (Ba-feldspar, sample 88a/71, vein No 501, horizon 29 m), which yielded a $^{40}Ar/^{39}Ar$ subplateau age of 271 ± 2 Ma (Fig. 5.7). The hyalophane was paragenetic to the phlogopite-bearing assemblage. Hence, the K–Ar isotope system in some of the primary minerals from the phlogopite-bearing veins, such as hyalophane and phlogopite, was reset long after the emplacement of "post-phlogopite" pegmatites. This is expressed by the later relatively low-temperature hydrothermal calcite–sulphide (pyrite, pyrrhotite) and calcite–quartz–carbonate mineralization.

For determination of the α-factor, a large recrystallization-free phlogopite crystal 25 × 30 cm (sample 814–Φ) was selected. Its two aliquots (9 g each) from the central unaltered part were used; one was melted completely, another one was treated by a three-step heating procedure. At the first step of the experiment, the probe was heated to temperature $T \sim 590°C$ during time interval $t_1 = 7$ min. The extracted $^{40}Ar_1$ was measured. At the second step, the probe was again heated to the same temperature during the same period of time. The extracted $^{40}Ar_2$ was again measured. At the third stage, the probe was melted to complete argon release to define the rest of the $^{40}Ar_3$.

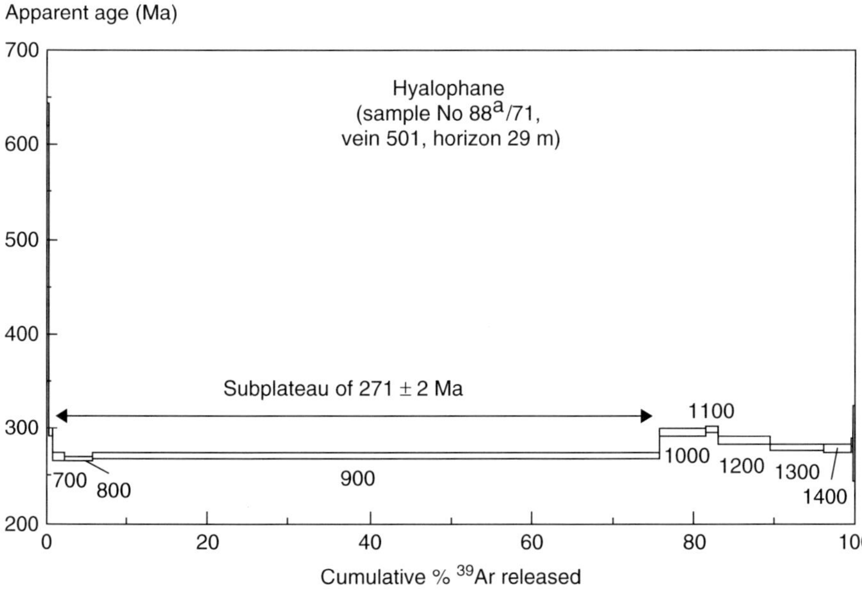

Fig. 5.7 $^{40}Ar/^{39}Ar$ stepwise-heating spectrum for the hyalophane crystal. Closure of the K–Ar isotope system occurred at 271 ± 2 Ma ago. The subplateau age is recalculated from data by Ivanenko et al. (1990). Temperatures (°C) of individual steps are shown

The first aliquot yielded the total ^{40}Ar and K concentration of 0.129 ± 0.0036 nmm^3/g and 8.189%, respectively. The calculated apparent age of 365.6 ± 10.4 Ma was obviously much younger than the crystallisation age of the host phlogopite vein. The three steps of the second aliquot showed $^{40}Ar_1 = 3.596 \pm 0.12 \times 10^{-3}$ nmm^3/g, $^{40}Ar_2 = 1.82 \pm 0.07 \times 10^{-3}$ nmm^3/g, and $^{40}Ar_3 = 0.129 \pm 0.005$ nmm^3/g. The obtained values were:

$$a_1 = \frac{Ar_1^{40}}{Ar_3^{40}} = 2.788 \times 10^{-2}; \quad a_2 = \frac{Ar_1^{40} + Ar_2^{40}}{Ar_3^{40}} = 3.985 \times 10^{-2}$$

The programmed iteration procedure with the $W(\alpha)$-function on a computer resulted in $\alpha = 0.27$ and $Fo_1 = 2.717 \times 10^{-4}$, $a_1 = 2.788 \times 10^{-2}$ and $a_2 = 3.983 \times 10^{-2}$. For checking these data, the values obtained were inserted into the set of Eqs. (5.10) and (5.11). The number $P = 2$ was employed in an approximation of the former equation with calculation of a_1:

$$(1 - 0.27) \times 4\sqrt{\frac{2.717 \times 10^{-4}}{\pi}} + 0.27 \times (1 - e^{-\pi^2 \times 2.717 \times 10^{-4}}) = 2.789 \times 10^{-2}$$

From the latter equation, a_2 is calculated as:

$$(1 - 0.27) \times 4\sqrt{\frac{2.717 \times 10^{-4} \times 2}{\pi}} + (1 - e^{-2\pi^2 2.717 \times 10^{-4}}) \times 0.27 = 3.980 \times 10^{-2}$$

Both values satisfactorily coincided with those obtained by iteration.

From the determined value of the α-factor, the rectangular argon fraction is calculated as $\alpha \times 0.129$ nmm^3/g = 94.17 nmm^3/g. The age of the fraction at 274 Ma shows a period in the Sludyanka complex without argon losses. This age is consistent with dating of minor mineralization in the vein. The closing a K–Ar isotope system in the phlogopite was marked by a paragenetic phlogopite-bearing vein mineral (hyalophane), crystallized simultaneously with resetting Rb–Sr isotope system and yielded ^{40}Ar/^{39}Ar subplateau age of 271 ± 2 Ma (Fig. 5.7).

During the time span between crystallization of the phlogopite 814–Φ at ca. 460 Ma and K–Ar isotope system closure at ca. 274 Ma, approximately 50% of the radiogenic argon fraction was lost by the phlogopite crystal at low temperature processes (in a range of 50–300°C) without recrystallization. We suggest that the process of argon loss was controlled by an exchange mechanism of OH-groups between the surrounding fluid and hydrocrystal structure. The argon particles seem to have been captured and carried to the mineral surface by vacancies formed during displacements of OH-groups (Brandt and Voronovskii 1967).

5.4.3 Dominating Semi-Sinusoidal Argon Fraction ($0.5 < \alpha \leq 1$)

Weathering processes are pronounced in rocks exposed to deep drainage of ground water in a warm humid climate. The rocks are affected by processes that result in diffusion losses of radiogenic argon until overtaken by climate change or cessation of weathering processes due to uplift and dissection of the region. Favorable climatic conditions provided intensively developed weathering processes of South Siberia in the Late Jurassic and also at the Cretaceous-Paleocene and Paleocene-Eocene boundaries (Logatchev et al. 2002).

Suitable material for isotopic dating of a weathering crust was found in the Zadoi alkaline ultramafic massif, located in the Archean Sharyzhalgai block—the uplifted southwestern margin of the Siberian craton basement. A magmatic sequence exhibited four phases: (1) perovskite- and ilmenite-bearing clinopyroxenites, (2) iolites, (3) nepheline syenites, and (4) carbonatites (Rasskazov et al. 2007c; Chap. 11). The final carbonatite veins intruded into perovskite-bearing pyroxenites affected by cataclasis. In exocontact zones of the veins, metasomatic processes resulted in crystallization of isometric mica grains as large as 2 cm.

Weathering processes have transformed these rocks into friable brown material, in which the large mica crystals look like distinct inclusions. Due to weathering, a biotite has been transformed into a vermiculite. Micas from the unaltered clinopyroxenites and the weathering crust contain 7–8 and 5.6 wt% of potassium, respectively. The measured abundance of radiogenic argon of the vermiculite (0.1616 nmm^3/g) yields an apparent K–Ar age of 625 ± 18 Ma, which has no geologic sense either for former metasomatism or imposed weathering processes.

By laboratory kinetic experiments, 0.012 928 nmm^3/g of radiogenic argon has been determined as accumulated in a closed K–Ar isotope system ($\alpha = 0.92$). The calculated age of 59 ± 5 Ma corresponds to cessation of weathering processes. Within the range of expected error, this age coincides with a climatic optimum of ca. 55 Ma determined from data on $\delta^{18}O$ of oceanic carbonates (Berger and Meyer 1987).

We speculate that the tectonically stable environment in the Sharyzhalgai block, as well as in adjacent areas of the Baikal region, was favorable for formation of the weathering crust until ca. 59 Ma but drastically ceased due to tectonic reactivation and uplifting of the Sharyzhalgai block. Afterwards, the mica from the weathering crust preserved all radiogenic argon.

5.5 Summary: Low-Temperature Argon Losses

The classical theory of diffusion demonstrates that experimentally determined values of the frequency factor (D_0) and activation energy (E) for radiogenic argon in a mineral grain depend on the character of its distribution. The characteristic diffusion parameters of a mineral E and D_0 should satisfy uniform distribution of radiogenic argon (i.e. $\alpha = 0$). Only in this case, the parameters E and D_0 can take the meaning of physical constants peculiar to certain minerals species.

The spatial distribution of radiogenic argon seems to be an important feature, because the two parameters of Arrhenius law – the activation energy and the frequency factor – are determined for mineral species with a great degree of uncertainty. Geochronologists may employ the α-factor to answer the question: has the given mineral suffered from radiogenic argon losses or not? There exists a possibility to perform detailed constructions of argon distributions in grains of various mineral species, which would be of interest to crystal chemistry.

To demonstrate reliability of the α-factor determination, we have done routine laboratory experiments with aggregates of mineral grains and have adequately described spatial distribution of radiogenic argon by means of classical diffusion theory. In this case, we have not employed the temperature-dependence of diffusion, expressed by the Arrhenius law. We have applied the theoretical approach for determinations of timing effectively closed K–Ar systems through calculating relative proportions of the rectangular and semi-sinusoidal argon fractions.

The completely preserved (rectangular) argon in the amazonite from the Shagaita-Gol diatreme likely reflects a capability of the mineral phase to keep the whole argon fraction. Additional work is yet to be done for determination of its activation energy of diffusion. The calculated K–Ar age is consistent with dating of amazonite-bearing pegmatite veins by the Rb–Sr method.

In two other cases, marking the last hydrothermal processes in the Sludyanka Complex and formation of a weathering crust in the Zadoi massif of the Sharyzhalgai block, we have found partial losses of radiogenic argon from micas. The obtained value for the α-factor (0.27) in the Sludyanka phlogopite 814–Φ demonstrates that the K–Ar age of 274 Ma, determined by the rectangular argon, corresponds to a hydrothermal event responsible for calcite–sulphide and/or calcite–quartz–carbonate mineralization. Independently, this very hydrothermal event appears to be indicated by an $^{40}Ar/^{39}Ar$ subplateau age of 271 ± 2 Ma, determined on hyalophane from the phlogopite-bearing vein. Similarly, the obtained value for the α-factor (0.92) of the vermiculite from the weathering crust, developed on the Late Precambrian Zadoi alkaline ultrabasic massif, makes possible calculations of a geologically real event at ca. 59 Ma, which corresponds to initiation of the Early Tertiary tectonic motions in South Siberia (Chap. 12).

We have found argon losses from minerals at relatively low temperatures (50–300°C). The three examined cases express an ultimate transition of mineral crystals to conditions where diffusion of argon becomes negligible.

Chapter 6
Thermal Spectra of Argon Isotopes

Obtaining spectra of argon release from minerals and rocks by stepwise heating is at present an indispensable part of age determination by means of the $^{40}Ar/^{39}Ar$ method. A probe is subjected to heating in a vacuum vessel at temperatures increasing stepwise, for instance 700, 900, 1100 K etc. The duration of heat treating at every step is usually constant, for instance, 10 min. Argon released at every step is separated and measured.

Solid state physics knows only one way of mass transfer in crystalline structures (including a release of argon)—a random walk of particles. This forms the mechanism of diffusion as well. In an ideal crystalline lattice, random walks can not be performed, just as moves can not be performed on a checker-board with all positions occupied. With rising temperature in crystals, there arise vacant sites, which "capture" inclusions in the lattice and realize elemental jumps. There exists only a single method to describe these movements quantitatively—Fick differential equations and the temperature law of Arrhenius. A system composed of several phases would require a sum of individual solutions of these equations.

Spectra of isotopes in the $^{40}Ar/^{39}Ar$ method allow us to conjugate ^{40}Ar, being in a definite energetic state (or in a definite temperature dependence), with the potassium (permutated into ^{39}Ar), which has been parent to this fraction of argon. This approach defines fractions of argon losses, "domains" with no argon distortions or containing excessive (inherited) radiogenic argon, absorbed by a mineral during crystallization as well as an identity of contaminating (admixed) and atmospheric (air) argon.

6.1 Presentation of Argon–Argon Spectra on Diagrams

For argon–argon spectra, two types of coordinates are used: apparent (measured) ages versus cumulative ^{39}Ar and $^{36}Ar/^{40}Ar$ versus $^{39}Ar/^{40}Ar$.

The former plot is often applied as a basic method of interpretation. In an ideal case, when no argon losses and no anomalies occur, radiogenic and nucleogenic components of argon are released congruently; a diagram of this type becomes a straight line parallel to the abscissa axis. A parallel portion of such a diagram is

S.V. Rasskazov et al., *Radiogenic Isotopes in Geologic Processes*,
DOI 10.1007/978-90-481-2999-7_6, © Springer Science+Business Media B.V. 2010

called "a plateau". Often it is believed that the plateau expresses the genuine age of the probe.

The latter plot, to present data as an argon–argon isochron, was suggested by Merrihue and Turner (1966) at the beginning of the development of argon–argon age determinations. In the diagram constructions, radiogenic ($^{40}Ar_{rad}$), nucleogenic (^{39}Ar), and contaminating (^{36}Ar, ^{40}Ar) components were involved. In case of a contaminating fraction such as air argon, the problem seemed to be simple. The measured ^{36}Ar multiplied by 295.5—the $^{40}Ar/^{36}Ar$ ratio in the terrestrial atmosphere at the sea level—should be subtracted from the bulk argon. Probes could have a cosmic (cosmic dust, tektites, meteorites, comets) or extraterrestrial (Moon, Mars) origin, however, with an isotopic $^{40}Ar/^{36}Ar$ ratio strongly differing from that of the terrestrial atmosphere. Suggesting $^{40}Ar/^{36}Ar = \text{const}$ and $^{40}Ar/^{39}Ar = \text{const}$ for all thermal fractions of the stepwise heating procedure, the isotopic composition of the contaminating argon could be determined:

$$(^{40}Ar/^{36}Ar)_{tot} = (^{40}Ar/^{36}Ar)_{cont} + (^{40}Ar/^{39}Ar)_{rad} \times (^{39}Ar/^{36}Ar) \qquad (*)$$

where the lower cases indexes *tot, con*, and *rad* are total, contaminating, and radiogenic, respectively.

After further improvements of measuring procedures, a diagram with inverse coordinates $x = {}^{39}Ar/^{40}Ar; y = {}^{36}Ar/^{40}Ar$ was suggested to avoid infinite coordinates when $^{36}Ar \to 0$ in probes enriched with $^{40}Ar_{rad}$ (Turner 1971; Roddick et al. 1980). The argon–argon diagram has become a standard method of interpreting stepwise heating experiments. Rectilinear isochrons with their linear regressions are widely used for determinations of excessive argon. There exist many descriptions of this isochron and recommendations for using a "plateau" to obtain reliable results. Here we present a strict mathematical analysis of these procedures.

The expression (∗) is linear, if $^{40}Ar_{rad}/^{39}Ar = \text{const}$ or all temperature steps give an *identical age*. A linearity of the expression:

$$(^{36}Ar_{cont}/^{40}Ar_{tot}) = f(^{39}Ar/^{40}Ar_{tot}) \qquad (**)$$

is not evident. In some cases it may be so, in others not. Therefore, there arises a question. If expression (∗∗) is not linear, can we accept linear regressions and extrapolations for age interpretation?

The standard mathematical apparatus of partial differential equations, described in such textbooks as (Webster and Szegoe 1934; Frank and Mises 1937; Tikhonov and Samarskii 1953; Madelung 1961) and taught by most mathematical, physical and chemical faculties of universities, are used here for a formal quantitative interpretation of released argon. Problems connected with construction of Ar–Ar isochrons in routine mass-spectrometric determinations of isotopes, for instance, background corrections or presence of several contaminating components, were objects of special publications (Roddick 1978; Heizler and Harrison 1988).

6.2 Histograms of Argon Release

This way of analyzing is important in studies of released argon isotopes in terms of the energy of activation E and diffusion parameter D/h^2 (Brandt et al. 2003, 2004).

To begin with, we consider a single mineral, the argon in which is in a single energetic position. Let the time of heating at the i-th step be Δt_i. In this case ^{40}Ar will be released at every step relative to the total $^{40}Ar_0$ in the mineral:

$$^{40}Ar/^{40}Ar_0 = 1 - \frac{8}{\kappa^2}\sum_{\nu=0}^{\infty}\frac{1}{(2\nu+1)^2}e^{-(2\nu+1)^2\pi^2\sum_{k=0}^{i}\Delta Fo_k}$$

where $\Delta Fo_i = (D_i/h^2)\Delta t_i$ and $D = D_0 e^{(-E/RT)}$

Then the i-th fraction of the spectrum is obtained as a difference:

$$(^{40}Ar/^{40}Ar_0)_i - (^{40}Ar/^{40}Ar_0)_{i-1}$$

These calculations are easily programmed. Parameters controlling the spectrum are the activation energy E and frequency factor D_0/h^2. The impact of these parameters on the spectra of an argon release is demonstrated in Fig. 6.1.

A neutron irradiation induces a certain ambiguity in interpretation, as a bombardment of the crystalline lattice by neutrons causes a possibility that the argon isotopes

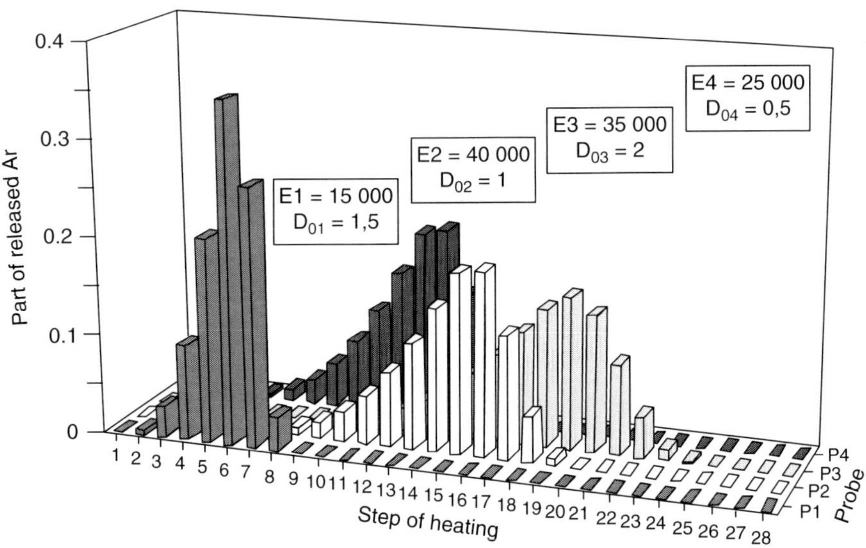

Fig. 6.1 Examples of argon release spectra. Increase of activation energy E (cal/mole) causes a shift towards elevated temperatures. Increase of D_0/h^2 broadens the spectrum. On the abscissa, numbers of heating steps are shown. The first step corresponds to 300°C and the twentieth to 2000°C. Calculations were performed using Eq. (6.6)

gain different activation energies (Brandt et al. 2004). In conventional K–Ar dating, the factor of distortion of kinetic parameters during bombardment does not exist, but age is calculated using the whole argon released and the problem of possible distortion of kinetic parameters due to natural geologic processes remains. There exist two approaches to solution of the problem in this case; the K–Ar system is not distorted, (1) if ages of a whole rock and their mineral phases are identical or (2) mineral grains show homogeneous distribution of radiogenic argon, determined by kinetical experiment (Chap. 5).

A comparative experimental study of aliquots of a poly-mineral basaltic andesite SH–00–58, subjected and not subjected to neutron activation, was performed with the aim to determine the mechanism of formation of temperature spectra and kinetic parameters of argon released. The basaltic andesite was sampled from the Klerk peninsula, Southeastern Primorye in Southeastern Russia (Chap. 12). An examination of thin sections and X-ray phase analysis showed that a sanidine was the only potassium-bearing phase in the sample. Other rock-forming minerals (olivine, ore phase, plagioclase, clinopyroxene) did not contain potassium.

For the irradiated sample, experimental data were obtained *in high vacuo* on a Noblegas–5400 mass-spectrometer in a routine procedure of stepwise heating (analyst A.V. Travin). Eight steps were measured in the temperature interval 600–1200°C. Experiments on heating of aliquots, not subjected to neutron irradiation, were conducted in the atmosphere at different temperatures and time intervals. Argon released at every step was measured by isotope dilution on a mass spectrometer MI–1201 by means of a procedure described in (Rasskazov et al. 2000b).

If kinetic parameters of ^{39}Ar and ^{40}Ar do not differ, the fractions of released argon at every step should be equal. However, three low-temperature fractions in the temperature interval 600–800°C have shown that the quantity of ^{40}Ar exceeded that of ^{39}Ar and in the range 900–1200°C, on the contrary, the lower quantity (Fig. 6.2). Nucleogenic argon is tied with a crystalline lattice that is weaker than radiogenic. As a result, a part of its low-temperature fraction was lost during the neutron irradiation. The difference between spectra of ^{40}Ar and ^{39}Ar in the basaltic andesite SH–00–58 is due to effects of the recoil of ^{39}Ar. The calculated age for three of the four steps will be overestimated. The irradiated sample has not lost, however, ^{39}Ar in the temperature interval 800–1200°C. The four high-temperature fractions are used for age estimates. They represent 76.3% of ^{39}Ar released and form a plateau with an age 37.3 ± 1.3 Ma. For the four steps used, χ^2 has a rather small value (0.49), certifying a high statistical authenticity of the age calculated (Chap. 1).

Experimental points plotted in the coordinates $1/T$, ln (D/h^2) (T– Kelvin temperature, D – coefficient of diffusion, h – grain size, from which diffusion occurs) are located with a satisfactory precision near a straight line. Therefore, the following temperature dependence is accepted:

$$\frac{D}{h^2} = \frac{D_0}{h^2} e^{-\frac{E}{RT}}$$

where E – activation energy of argon diffusion, D_0/h^2 – frequency factor, meaning D at $T \to \infty$, R – gas constant. The values obtained are: E = 84 672 cal/mole;

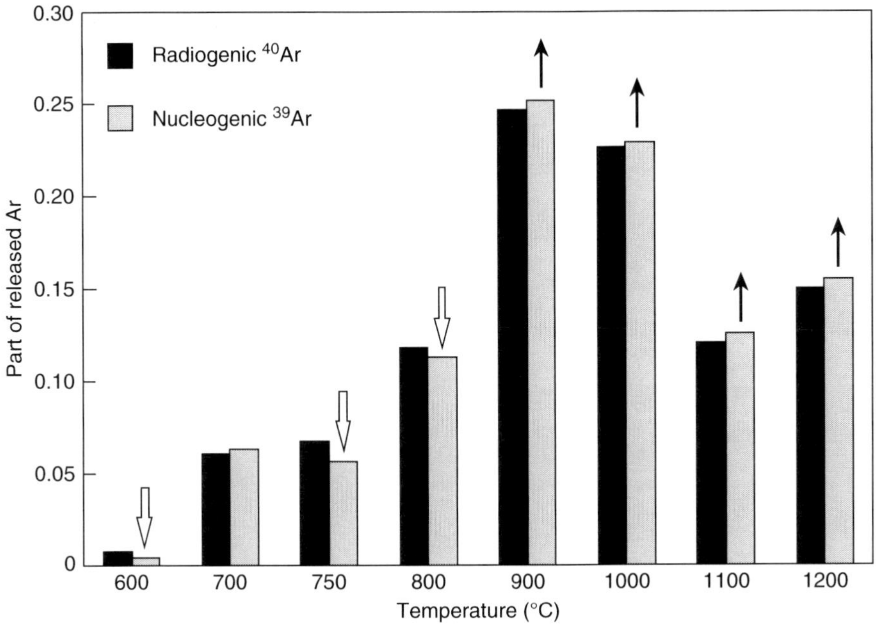

Fig. 6.2 Histogram of measured release spectra of radiogenic ^{40}Ar and nucleogenic ^{39}Ar from the basaltic andesite SH–00–58 subjected to neutron activation. *Open arrows* directed downwards show low-temperature steps with partial losses of ^{39}Ar, *black arrows* directed upward show high-temperature steps, for which, due to losses on low-temperature steps, overestimated concentrations of ^{39}Ar were employed as compared with concentrations of ^{40}Ar recalculated to 100%

$(D_0/h^2) = 2.05{\cdot}10^{12}\, s^{-1}$. The high energy of activation is explained by concentration of argon in the disordered sanidine phase. The high frequency factor obeys the compensation rule of Hart (1981).

Time dependence of argon concentrations has been obtained as well. In preheated aliquots, the observed linear relation between $\sqrt{t}$ and $Ar_{released}$ is inferred from Fick second law.

The stepwise heating procedure was computed theoretically with approximations of argon release spectra for the probe, not subjected to irradiation, under different conditions. Such an experimentally obtained spectrum of radiogenic argon release, compared with a spectrum computed for parameters $E = 84\,672$ cal/mole; $(D_0/h^2) = 2.05{\cdot}10^{12}\,\mathrm{s}^{-1}$, is shown in Fig. 6.3.

The probe exhibits a mixture of coarse and fine grains. The variable is not only temperature, but D_0/h^2 as well. The fact that the process of release is controlled by one parameter can not be generalized for all basalts. If the basalt had two phases with equal potassium contents, the Arrhenius diagram could yield two E-values.

Experimental data on annealing of a poly-mineral probe of the basaltic andesite SH–00–58 have shown that kinetics of argon release is controlled by the Arrhenius and Fick laws. The obtained parameters correspond to a satisfactory preservation of radiogenic argon in the probe with its concentration in a sanidine.

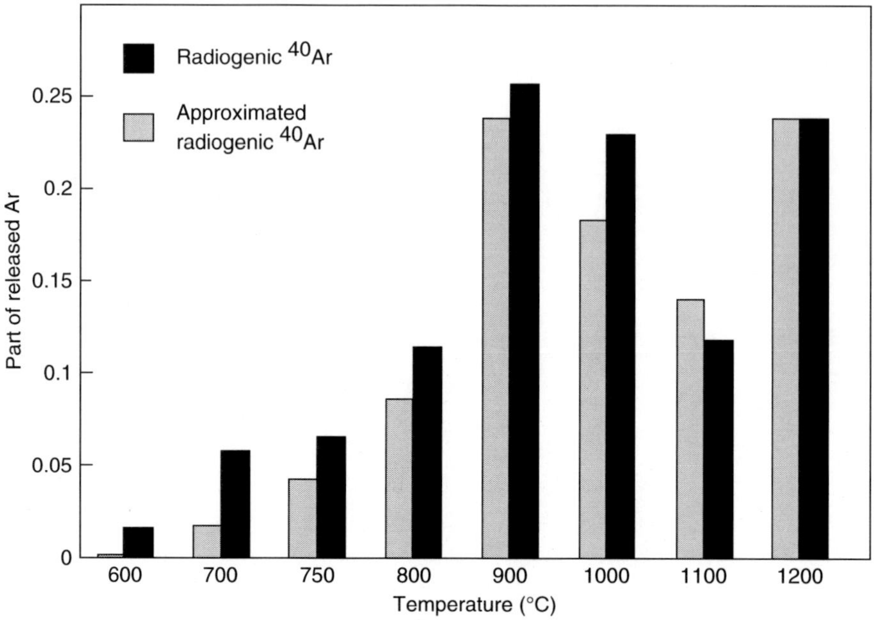

Fig. 6.3 Radiogenic ^{40}Ar release spectra of a probe of the basaltic andesite SH–00–58, not subjected to neutron irradiation. Experimental—(*black columns*) and approximated—(*gray columns*)

6.3 Deduction of an Algebraic Expression for an Isochron Line in Inverse Ar–Ar Coordinates

In constructions of an argon–argon isochron, three components are involved: radiogenic $^{40}Ar_{rad}$, nucleogenic ^{39}Ar, and contaminating argon—a mixture of $^{36}Ar_{cont}$ and $^{40}Ar_{cont}$. If the contaminating argon is air, the ratio $^{40}Ar/^{36}Ar$ equals 295.5. The argon–argon isochron is defined by variation of $^{36}Ar/^{40}Ar$ as a function of $^{39}Ar/^{40}Ar$. For deduction of an algebraic expression $y = f(x)$, let:

$$y = {}^{36}Ar/{}^{40}Ar \tag{6.1a}$$

$$x = {}^{39}Ar/{}^{40}Ar \tag{6.1b}$$

$$Ar_{tot} = {}^{40}Ar_{rad} + {}^{40}Ar_{cont} \tag{6.1c}$$

Further, we suggest that:

$$k = {}^{36}Ar/{}^{40}Ar_{cont}; \tag{6.2a}$$

$$l = {}^{39}Ar/{}^{40}Ar_{rad} \tag{6.2b}$$

are constant during the whole procedure of stepwise heating. If these conditions are not met, the isochron becomes curved. The expression (6.2a) means that the isotopic composition of contaminating argon must remain constant for all steps of annealing. The expression (6.2b) requires a constant $^{39}Ar/^{40}Ar_{rad}$. On the diagram "age versus temperature fractions of ^{39}Ar released", the steps form a straight line parallel to the axis of abscissa. From Eqs. (6.2a, b):

$$y = \frac{k}{1 + {}^{40}Ar_{rad}/{}^{40}Ar_{cont}} \tag{6.3a}$$

$$x = \frac{l}{1 + {}^{40}Ar_{cont}/{}^{40}Ar_{rad}} \tag{6.3b}$$

Substitute the value $^{40}Ar_{rad}$ from (6.3b) into (6.3a) to obtain:

$$y = \frac{k}{1 + \frac{x}{l - x}} \quad \text{or finally} \quad \frac{y}{k} + \frac{x}{l} = 1 \tag{6.4}$$

Expression (6.4) corresponds to a straight line intercepting the axis of ordinates at a value k and with the axis of abscissa at a value l.

6.4 Kinetics of Stepwise Heating

Generally, as already mentioned, the radiogenic argon in crystalline structures obeys Arrhenius and Fick laws. Fick law determines the dependence of argon concentrations on time and coordinates:

$$\frac{\partial {}^{40}Ar}{\partial t} = D\frac{\partial^2\, {}^{40}Ar}{\partial x^2} \tag{6.5}$$

where t – time, x– one-dimensional space coordinate, D – diffusion coefficient. The Arrhenius law determines the temperature dependence of the diffusion coefficient:

$$\frac{D}{h^2} = \frac{D_0}{h^2} e^{-\frac{E}{RT}} \tag{6.6}$$

where h– characteristic size of grains, from which diffusion occurs, D_0 – frequency factor, E– activation energy of diffusion, R – the gas constant, T – Kelvin temperature.

For the case of stepwise heating, the relation with stepwise constant diffusion is valid (Brandt et al. 1996):

$$\Delta c_i = \frac{8}{\pi^2} \sum_{\nu=0}^{\infty} \frac{1}{(2\nu + 1)^2} e^{-(2\nu+1)^2 \rho^2 \sum\limits_{i=0}^{\infty} \Delta Fo_i} (1 - e^{-(2\nu+1)\pi^2 \Delta Fo_i}) \tag{6.7}$$

where Δc_i – fraction of $(^{36}Ar, {}^{40}Ar)_{cont}$, $^{40}Ar_{rad}$ or ^{39}Ar released during i-th step of a probe heating, Fo – a dimensionless Fourier number, $Fo = Dt/h^2$. This formula is easily programmed and tabulated.

It should be noted that originally the migration of argon particles in crystalline lattices was interpreted as a reaction of the first order, such as simple annihilation of a migrating particle in situ. The size of the host mineral grains was not accounted for. The ordinary differential equation for this case is:

$$\frac{d\Delta}{dt} + K\Delta = \lambda P_0 \text{ with } P = P_0 e^{-\lambda t}$$

The solution of this equation is:

$$\frac{\Delta}{P} = \frac{e^{(\lambda - k)t} - 1}{1 - \frac{K}{\lambda}}$$

where$K$ – reaction rate, Δ – daughter substance, P – parent substance, λ – constant of radioactive transformation. Details of the first order reaction are reported by Damask and Dins (1966).

6.5 Separating Air and Radiogenic Argon

Division of radiogenic and air argon, contained as a mixture in every mineral or probe, is an important procedure in routine measurements. Air argon penetrates into probes, which have been exposed to the atmosphere, in quantities frequently exceeding the radiogenic argon by several orders. Forced by a huge gradient of pressure, air argon penetrates into intercrystal boundaries, crystal defects, and dislocations. Air argon causes a masking effect, increasing an error of measurement. Laboratory thermal training of a probe is used to diminish interfering effect of air argon.

A formal approach to separation of radiogenic and air argon in probes by a thermal procedure has been developed by Rasskazov et al. (2000b). To preserve in the probe a k-th fraction of radiogenic argon and extract an l-th fraction of air argon, the probe should be heated at a temperature (°C):

$$T = \frac{E - e}{R} \cdot \frac{1}{\ln\left(\frac{fo}{Fo} \cdot \frac{D_0/H^2}{d_0/h^2}\right)} \tag{6.8a}$$

during a time interval (in seconds):

$$t = \frac{(D_0/H^2 Fo)^{\frac{e}{E-e}}}{(d_0/h^2 fo)^{\frac{E}{E-e}}} \tag{6.8b}$$

where the capital letters D, H, Fo refer to radiogenic argon and small letters d, e, fo to air argon; $Fo = (\pi k^2/16); fo = -(1/\pi^2) \ln(\pi^2 l/8)$.

The separation, satisfying to 99% of radiogenic argon preserved and 99% of air argon released, is achieved by heating a probe to a temperature calculated from the Eq. (6.9a) during the time interval derived by Eq. (6.8b). These calculations show

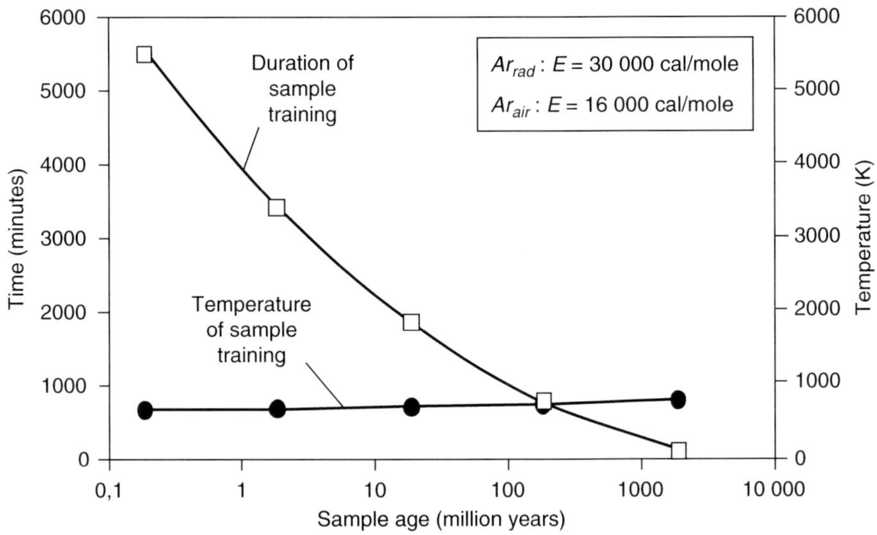

Fig. 6.4 Time and temperature of thermal treatments of a probe, after which the initial content of air argon of 1 nmm^3 becomes equal to the content of radiogenic argon in the probe

that, at $E = 25\ 000$ cal/mole for radiogenic argon and $e = 15\ 000$ cal/mole for air argon, the probe is to be trained for 16 days at the temperature 373°C. This result is unequivocal and cannot be reproduced by any other mode of training. The training can not be accelerated.

Consider a probe containing originally 1 nmm^3 (mm^3 under normal temperature and pressure) of air argon. By thermal training it is required to lower its content and to make it equal to radiogenic argon. Time and temperature of training are given in the diagram of Fig. 6.4. The older the probe, the less time is required for training, the easier is the measurement. Temperature of training is slightly increasing with age. For young probes the time of training can reach 24 h and more.

6.6 $E_{cont} = E_{rad} = E_{39}$

In stepwise heating, every argon component is characterized kinetically by the frequency factor D_0 and energy of activation E. Differences of these parameters enable us to discriminate different kinds of argon by thermal procedures. If $E - e = 0$, T becomes zero, and t—uncertain. If two or more sorts of argon, contained in a crystal, have identical diffusion parameters, they become undistinguishable and unfit for division by thermal procedures. It is understandable as all substations are released congruently and their relations remain constant.

In coordinates $x = {}^{39}Ar/{}^{40}Ar$ versus $y = {}^{36}Ar/{}^{40}Ar$, i-steps of heating will be undistinguishable and represented by a single point multiple to i. As contaminating, radiogenic, and nucleogenic argon are undistinguishable, this point has no geochronological sense.

6.7 $E_{cont} < E_{rad} = E_{39}$

Assume now, that the contaminating argon is tied with the host crystal structure more poorly than the radiogenic and nucleogenic. For definiteness, consider a crystal of an age:

$$t = 1.804 \cdot 10^9 \ln \left(J \frac{^{40}Ar_{rad}}{^{39}Ar} + 1 \right) = 12.5\,\text{Ma}$$

(for these parameters: $J = 1.32 \cdot 10^{-5}$, $^{40}Ar_{rad}/^{39}\text{Ar} = 5.27 \cdot 10^{-4}$). Accept further $E_{\text{rad}} = E_{39}$ 25 000 cal/mole; $E_{cont} = 23\,000$ cal/mole. The frequency factor D_0/h^2 in all cases is considered as equal to 0.5 s^{-1}.

The calculations are accomplished by the programmed formula (6.7). The results are given in Table 6.1. The further calculations were performed to find ratios of the spectra with different energies of activation. We see that, if the spectra are congruent, their ratios are constant. Shifts along the temperature axis cause variations of these ratios and shift of activation energies for contaminating argon into the range of low values. An isochron in form of a straight line with all points, belonging to it, intercepts the coordinate axes precisely at points of initials given (Fig. 6.5) And inversely, if we receive an isochron as a straight line, with all points laying on it, then the activation energies for radiogenic and nucleogenic argon are identical, and the activation energy for contaminating argon is somewhat lower. The intercepts of the isochron with axes of coordinates determine factors k and l in expression (6.4) This is a classical case of an inverse argon–argon isochron, in terms of which researchers aspire to interpret experimental data.

Table 6.1 Calculation of the isochron shown in Fig. 6.5

T (K)	$^{40}Ar_{\text{rad}}$	$^{39}Ar\ 10^4$	$^{40}Ar_{\text{cont}}$	$^{36}Ar_{\text{cont}}$	$^{40}Ar_{\text{tot}}$	$^{36}Ar/^{40}Ar_{\text{tot}}$	$^{39}Ar/^{40}Ar_{\text{tot}}$
573	$8.43{\cdot}10^{-7}$	0.0016	$1.8{\cdot}10^{-6}$	$6.1{\cdot}10^{-9}$	$2.65{\cdot}10^{-6}$	0.002 305	604.7915
673	$3.06{\cdot}10^{-6}$	0.0058	$6.41{\cdot}10^{-6}$	$2.17{\cdot}10^{-8}$	$9.46{\cdot}10^{-6}$	0.002 291	612.7891
773	$9.38{\cdot}10^{-6}$	0.0178	$1.73{\cdot}10^{-5}$	$5.85{\cdot}10^{-8}$	$2.67{\cdot}10^{-5}$	0.002 194	667.5117
873	$2.15{\cdot}10^{-5}$	0.0408	$3.67{\cdot}10^{-5}$	$1.24{\cdot}10^{-7}$	$5.82{\cdot}10^{-5}$	0.002 133	701.2623
973	$4.06{\cdot}10^{-5}$	0.077	$6.57{\cdot}10^{-5}$	$2.22{\cdot}10^{-7}$	0.000 106	0.002 092	724.7523
1073	$6.78{\cdot}10^{-5}$	0.1286	0.000 104	$3.53{\cdot}10^{-7}$	0.000 172	0.002 052	747.161
1173	0.000 103	0.195	0.000 152	$5.14{\cdot}10^{-7}$	0.000 255	0.002 019	765.4509
1273	0.000 145	0.2752	0.000 206	$6.96{\cdot}10^{-7}$	0.000 351	0.001 984	785.0288
1373	0.000 192	0.364	0.000 23	$7.77{\cdot}10^{-7}$	0.000 421	0.001 844	863.5935
1473	0.000 216	0.4092	0.000 164	$5.56{\cdot}10^{-7}$	0.000 38	0.001 463	1077.234
1573	0.000 166	0.315	$6.05{\cdot}10^{-5}$	$2.05{\cdot}10^{-7}$	0.000 226	0.000 904	1390.83
1673	$7.3{\cdot}10^{-5}$	0.1386	$9.26{\cdot}10^{-6}$	$3.13{\cdot}10^{-8}$	$8.23{\cdot}10^{-5}$	0.000 381	1683.964
1773	$1.53{\cdot}10^{-5}$	0.02990	$4.5{\cdot}10^{-7}$	$1.52{\cdot}10^{-9}$	$1.57{\cdot}10^{-5}$	$9.68{\cdot}10^{-5}$	1843.242
1873	$1.25{\cdot}10^{-6}$	0.002 36	$4.38{\cdot}10^{-9}$	$1.48{\cdot}10^{-11}$	$1.25{\cdot}10^{-6}$	$1.19{\cdot}10^{-5}$	1890.886
1973	$3.01{\cdot}10^{-8}$	$5.7{\cdot}10^{-5}$	$6.06{\cdot}10^{-11}$	$2.05{\cdot}10^{-13}$	$3.02{\cdot}10^{-8}$	$6.8{\cdot}10^{-6}$	1893.719

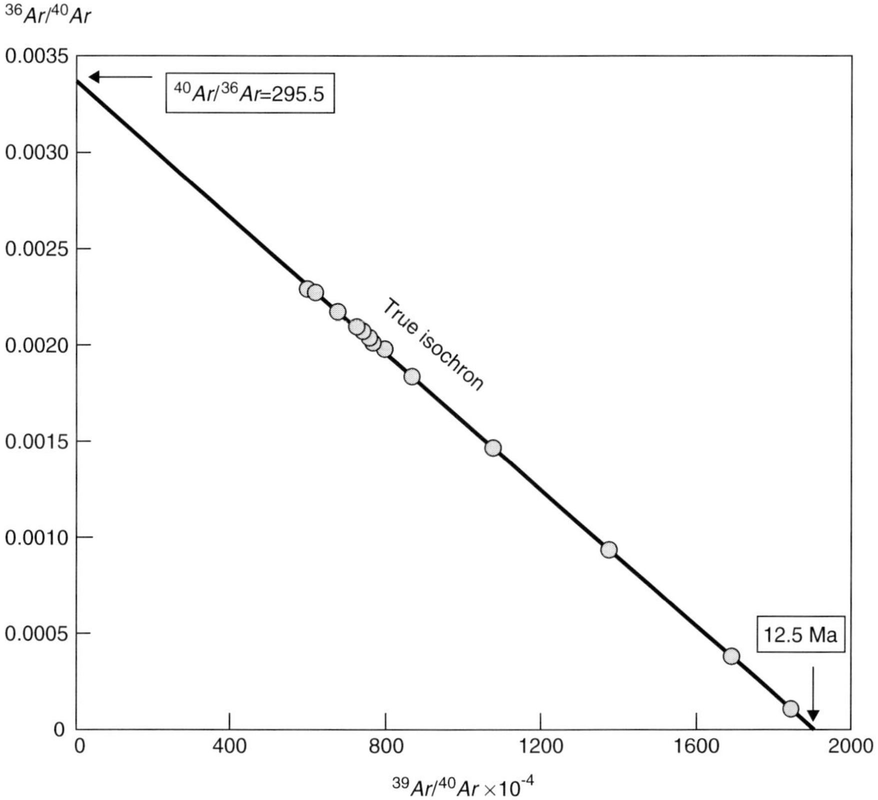

Fig. 6.5 An argon–argon isotopic diagram in case of the equal activation energies for radiogenic and nucleogenic argon components with the somewhat lower activation energy of air argon. The isochron obtained is an ideal straight line that intersects axes of coordinates at the isotopic ratio of the contaminating (air) argon and the age

6.8 $E_{cont} < E_{39} < E_{rad}$

Equal diffusion parameters of radiogenic and nucleogenic argon was confirmed by many results of practical measurements. It is interesting, nevertheless, to examine consequences resulting from a shift of activation energy for diffusion of radiogenic and nucleogenic argon. This case can occur because the cross-sections of neutron reaction at collisions are final not only for potassium, but also for central elements of the host crystalline structure. The structure, affected by neutron bombardment, can tend to swell with formation of vacant lattice sites of a Frenkel type. Whether the spectrum of nucleogenic argon-39 relative to the spectrum of radiogenic argon-40 is shifted, has been an object of long-term discussion (Villa 1997). Does the neutron bombardment influence the host crystal structure? Is the radioactive recoil of the ^{39}K (n, p) ^{39}Ar transition sufficient enough to damage the ties of the formed particle with

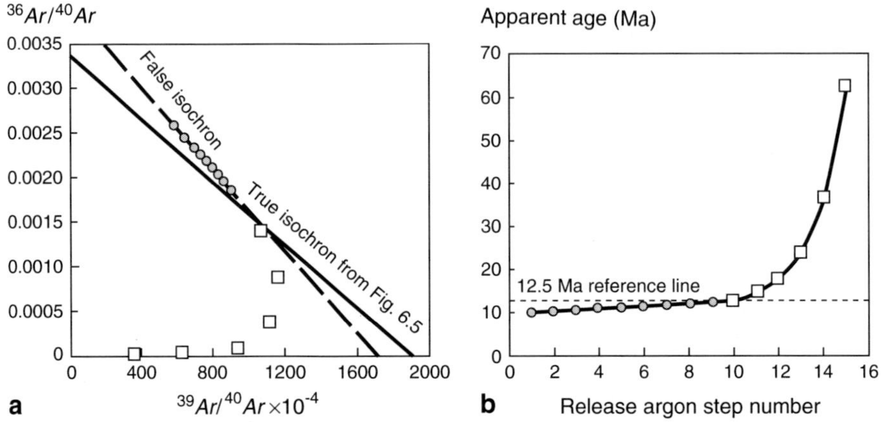

Fig. 6.6 An *Ar–Ar* isotope diagram (**a**) and age-spectrum of fractions released at different temperatures (**b**) for a case when the activation energy of nucleogenic argon is lower than that of radiogenic and contaminating. The locus of points is not a straight line. Some points, denoted by *gray circles*, form a fictive isochron. Its intercepts with the coordinate axes give distorted values of age and isotopic composition of contaminating argon. Points not fitting to the fictive isochron, are denoted by *squares*

the host structure? A valence charged atom of potassium has turned to a neutral atom of argon! The answer "does not influence" is vital for the method.

Consider whether the shape of the line that approximates heating steps in inverse Ar–Ar coordinates can be indicative for shifted spectra or not. In Fig. 6.6, the spectrum is given in the case when, during the irradiation of a probe, radioactive recoil and impact of neutrons on central elements of a host lattice have inflicted damages and the energy of activation of nucleogenic argon has become less than that of radiogenic. It has caused a false slope of the isochron and distorted abundance of initial air argon. Does a way exist to distinguish true and fictitious cases? As far as we know—it does not. The parameters of contaminating argon are unknown. This problem is certainly solvable, but requires some experimental and theoretical work.

The input values are $E_{\mathrm{cont}} = 23\,000$ cal/mole, $E_{39} = 24\,500$ cal/mole, $E_{\mathrm{rad}} = 25\,000$ cal/mole, and $D_0/h^2 = 0.5\,\mathrm{s}^{-1}$ (Fig. 6.6). Calculations similar to those demonstrated in Table 6.1 show that even at such small differences of activation energies, the points do not lie on a straight line, but rather on a complex curve, shifted from the true isochron. The conditions (6.2a,b) are not met. Occurrences of intercepts of this curve with the coordinate axes are far from nominal.

The experimenter is inclined to attribute the apparent "dispersion" of points to experimental error, "heterogeneity of crystal nuclei" etc., and thus to choose as apparently quasi-correct a rectilinear part of the curve and get a regression line. The latter gives an intercept with the axis $^{36}Ar/^{40}Ar = 0.0042$, instead of 0.003 384, i.e. the true input value of the calculation, and an intercept with the $^{39}Ar/^{40}Ar$-axis = 1400, instead of 1900. An erroneous conclusion may be reached of an abnormal isotope abundance of contaminating argon, differing from air, and of an overestimated age.

On the diagram of fractional release of argon at step heating (Fig. 6.6b), one can see "a plateau", located below the true age of a probe, and an ascending sequence of high temperature steps.

6.9 Loss of Radiogenic Argon in the Geologic Past

From the character of expression (6.7) and from its expansion into a Taylor's series, it may be seen that:

$$f(\Delta Fo) \neq f(Fo_0 + \Delta Fo)$$

where the Fourier number Fo_0 characterizes losses of argon by a probe and ΔFo is obtained from laboratory heating. A curve of argon release becomes lower as the step number increases.

Assume that a probe has lost in the past a half of its radiogenic argon. Such loss corresponds to a Fourier number $Fo_i = 0.049$. Perform calculations similar to that shown in Table 6.1, but to each value of Fourier's step add 0.049. The appropriate sequence of points (Fig. 6.7) forms a curve declined from the true isochron.

The curve can be erroneously interpreted as "dispersion of points" or "defects in crystal nuclei". In interpretation, the experimenter can try to reject points, accept points 7–11 as the "true", and obtain a line of regression $({}^{36}Ar/{}^{40}Ar_{tot}) = -1.76 \cdot 10^{-6}({}^{39}Ar/{}^{40}Ar_{tot}) + 0.0035$ intercepting the ${}^{39}Ar/{}^{40}Ar_{tot}$-axis at 1550, instead of 1900, and the axis ${}^{36}Ar/{}^{40}Ar$-axis at 0.0035, instead of 0.003 384. The distorted values make no sense.

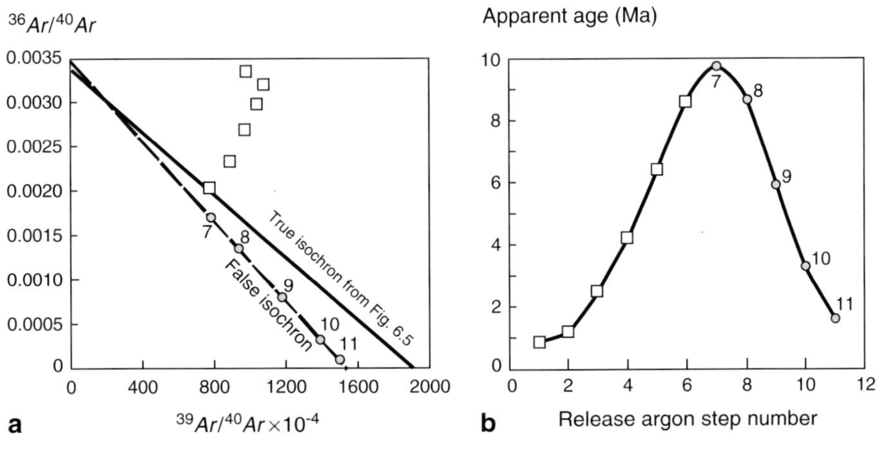

Fig. 6.7 An argon–argon isotope diagram (**a**) and age spectrum for release fractions at different temperatures (**b**) for the case of 50% loss of radiogenic argon in the geologic past. Symbols as on Fig. 6.6

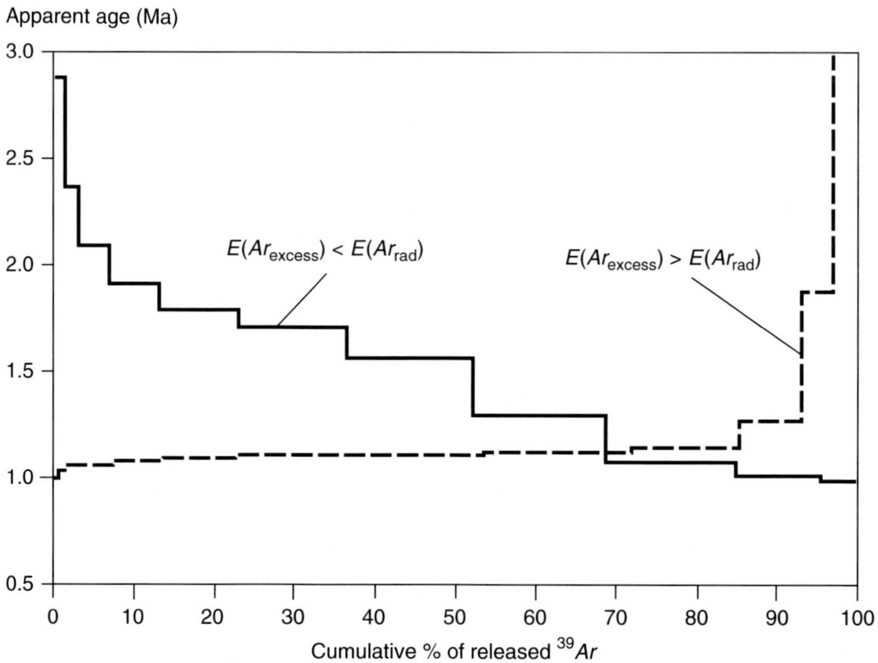

Fig. 6.8 Examples of release spectra for excessive argon ($Ar_{excess} = 0.25\,Ar_{tot}$). The true age (age of the plateau) corresponds to an ordinate equal unity. The *heavy line* refers to excessive argon loosely tied with the crystalline lattice with $E = 20\,000$ cal/mole, E of radiogenic argon—25 000 cal/mole. The *broken line* refers to strongly bonded excessive argon with $E = 30\,000$ cal/mole; E of radiogenic argon—25 000 cal/mole

6.10 Excessive Argon-40

Radiogenic argon, lost by P–D systems and "picked up" by newly forming structures which may or may not contain stoichiometric potassium, is identified, for example, in low-pressure pyroxenes and olivines.

How does excessive argon affect an age spectrum? The energy of activation for excessive argon can be higher or lower than the energy of activation for radiogenic argon (Fig. 6.8). For an "older" spectrum with excessive argon (Fig. 6.9a), a choice of a plateau appears difficult (Fig. 6.9b). Excessive and radiogenic parts of argon are not distinguishable, if their energies of activation are equal.

6.11 Argon of Poly-mineral Aggregates

In studies of natural objects, researchers seldom deal with one ideal crystal. As a rule, even the carefully selected mono-mineral fraction is really a mixture of crystals somewhat differing in chemical compositions (for example, in

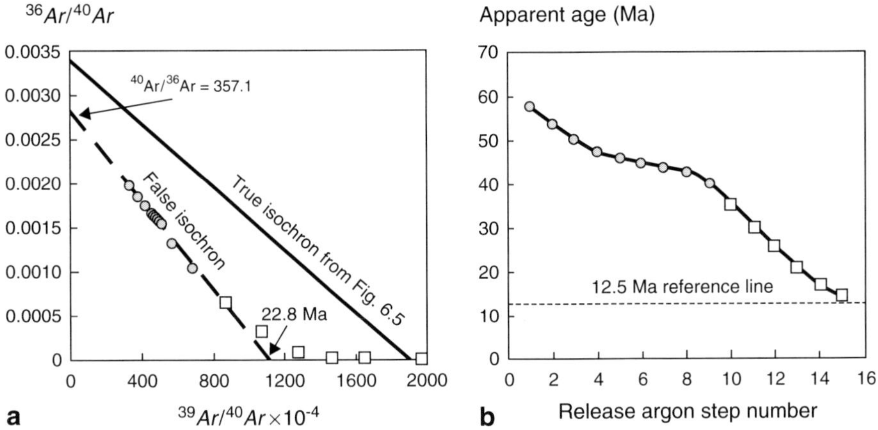

Fig. 6.9 An argon–argon isotope diagram (**a**) and spectrum of fractions released at different temperatures (**b**), when the activation energy of excessive argon is less than that of radiogenic argon. Symbols as on Fig. 6.6

concentrations of potassium). Crystallizing mineral phases may trap different quantities of contaminating argon. Such probes should be considered as mixtures of ideal crystals.

Obviously, a series of points will be expressed on the Ar–Ar isochron diagram, if energies of activation for radiogenic, nucleogenic, and contaminating argon are the same in each ideal crystal. Respectively, the linear relation of intercepts with the axis of abscissa will show the age of the poly-mineral unit and the intercept with the axis of ordinates isotopic composition of the contaminating argon, if the crystals have equal ages and isotopic compositions of contaminating argon in them are identical. It is clear, that significant linear relation will remain valid in case of the somewhat lower energy of activation for contaminating argon than for radiogenic and nucleogenic. A nonlinearity of points in isochron coordinates, inferred from formal consideration of an ideal crystal with unequal energies of activation for nucleogenic and radiogenic argon, results in a mixture of ideal crystals as well. The same conclusion for a nonlinear relation of points remains valid for a crystal aggregate that has lost radiogenic argon in the geologic past by diffusion.

Some researchers argue that text-book diffusion is too small to cause appreciable effects in geochronology. Numerous papers have demonstrated, however, the measured quantity of migrating particles in laboratory heating experiments. The particles are released down to their complete loss. Thus, the reference to "text-book diffusion" on closer examination has appeared as deprived of sense.

The present-day theory of migrating radiogenic particles in minerals is based on formal consideration of diffusion processes. Interpretation of experimental results in terms of differential equations of diffusion and Arrhenius law requires determination of two important parameters: the energy of activation and the frequency factor, which allow extrapolation of particle migration from elevated to lower temperature conditions and estimation of particle safety in geologic time.

6.12 Argon in Two-Component Crystal Structure

Practical application of the $^{40}Ar/^{39}Ar$ method demonstrates a control of argon kinetics in real crystals by a set of activation energies. Crystals are not ideal. Their elementary cells might be destroyed or deformed. Is it possible to define deviation of a real crystalline structure from an ideal monocrystalline one in terms of the migration theory for a structure exhibiting a combination of two ideal components with differing energies of activation?

In an ideal crystal, the coefficient of diffusion D of radiogenic argon is connected to Kelvin temperature $T$by the Arrhenius law:

$$D = D_0 e^{-\frac{E}{RT}} \tag{6.9}$$

where D_0, E are diffusion parameters.

Let A and A_f be, respectively, a quantity of argon released from a probe by heating during a time t at a temperature T and a quantity of argon released at a greater temperature and longer heating. Then:

$$F(Fo) = 1 - \frac{8}{\pi^2}\sum_{\nu=0}^{\infty}\frac{1}{(2\nu+1)^2}e^{-(2\nu+1)^2\pi^2 Fo} = \frac{A}{A_f} = a \tag{6.10}$$

where $Fo = (Dt/h^2)$ and D are calculated using the formula (6.9).

If a crystal is not ideal, there appear two pairs of diffusion parameters; one pair D_1, E_1 characterizes the greater part of argon (1 - $\nu)A_f$ and another pair D_2, E_2 the smaller part νA_f. In this case, the formula (6.10) is not valid for release of argon and a more complicated calculation is required. Hence, a strict correspondence of a unique pair of diffusion parameters D_0 and E to Fick law should be required.

Consider a probe treated at temperatures T_1, T_2, and T_3during time intervals t_1, t_2, and t_3 with argon released A_1, A_2, and A_3, respectively. Then:

$$\begin{aligned} F\left(t_1\frac{D_0}{h^2}e^{-\frac{E}{RT_1}}\right) &= \frac{A_1}{A_f} = a_1 \\ F\left(t_2\frac{D_0}{h^2}e^{-\frac{E}{RT_2}}\right) &= \frac{A_2}{A_f} = a_2 \\ F\left(t_3\frac{D_0}{h^2}e^{-\frac{E}{RT_3}}\right) &= \frac{A_3}{A_f} = a_3 \end{aligned} \tag{6.11}$$

Further:

$$\frac{D_0}{h^2}e^{-\frac{E}{RT_1}} = \frac{\Phi(a_1)}{t_1};\quad \frac{D_0}{h^2}e^{-\frac{E}{RT_2}} = \frac{\Phi(a_2)}{t_2};\quad \frac{D_0}{h^2}e^{-\frac{E}{RT_3}} = \frac{\Phi(a_3)}{t_3} \tag{6.12}$$

where $\Phi(a)$ is the inverse function of the function $F(Fo)$.

Take the logarithm of the system (6.12):

$$\ln\frac{D_0}{h^2}-\frac{E}{RT_1}=\ln\left[\frac{\Phi(a_1)}{t_1}\right]$$

$$\ln\frac{D_0}{h^2}-\frac{E}{RT_2}=\ln\left[\frac{\Phi(a_2)}{t_2}\right]$$

$$\ln\frac{D_0}{h^2}-\frac{E}{RT_{31}}=\ln\left[\frac{\Phi(a_3)}{t_3}\right] \quad (6.13)$$

A linear system of three equations with two unknowns $\ln(D_0/h^2)$ and E is received. Hence, columns:

$$\begin{pmatrix}1\\1\\1\end{pmatrix};\quad \begin{pmatrix}-\dfrac{1}{RT_1}\\ -\dfrac{1}{RT_2}\\ -\dfrac{1}{RT_3}\end{pmatrix};\quad \begin{pmatrix}\ln\left[\dfrac{\Phi(a_1)}{t_1}\right]\\ \ln\left[\dfrac{\Phi(a_2)}{t_2}\right]\\ \ln\left[\dfrac{\Phi(a_3)}{t_3}\right]\end{pmatrix}$$

are lineally dependent, as well as the determinant:

$$\begin{vmatrix}1 & -\dfrac{1}{RT_1} & \ln\left[\dfrac{\Phi(a_1)}{t_1}\right]\\ 1 & -\dfrac{1}{RT_2} & \ln\left[\dfrac{\Phi(a_2)}{t_2}\right]\\ 1 & -\dfrac{1}{RT_3} & \ln\left[\dfrac{\Phi(a_3)}{t_3}\right]\end{vmatrix}=0$$

Decompose the determinant on the last column and obtain (as multiplied by R):

$$\left(\frac{1}{T_2}-\frac{1}{T_3}\right)\ln\left[\frac{\Phi(a_1)}{t_1}\right]+\left(\frac{1}{T_3}-\frac{1}{T_1}\right)\ln\left[\frac{\Phi(a_2)}{t_2}\right]+\left(\frac{1}{T_1}-\frac{1}{T_2}\right)\ln\left[\frac{\Phi(a_3)}{t_3}\right]=0 \quad (6.14)$$

If an experiment yields data of the left-hand part of the equation differing from zero for a single pair of diffusion parameters D_0 and E, the Fick law is invalid. The calculated value of the left-hand part of the equation can serve as a measure of a deviation from a case of an ideal crystal satisfying the Fick law.

Let $(D_1/h^2) = 10\,\text{s}^{-1}$; $E_1 = 30\,000$ cal/mole; $(D_2/h^2) = 10\,\text{s}^{-1}$; $E_2 = 25\,000$ cal/mole; $t_1 = 3600$ s; $t_2 = 2000$ s; $t_3 = 1500$ s; $T_1 = 1000$ K; $T_2 = 1100$ K; $T_3 = 1300$ K. Attribute $(1-\nu)\cdot A_f$ of argon the index (1) and $\nu\cdot A_f$ of argon the index (2). The law of argon release takes the form:

$$(1-\nu)F(Fo_1)+\nu F(Fo_2)=\frac{A}{A_f}=a \quad (6.15)$$

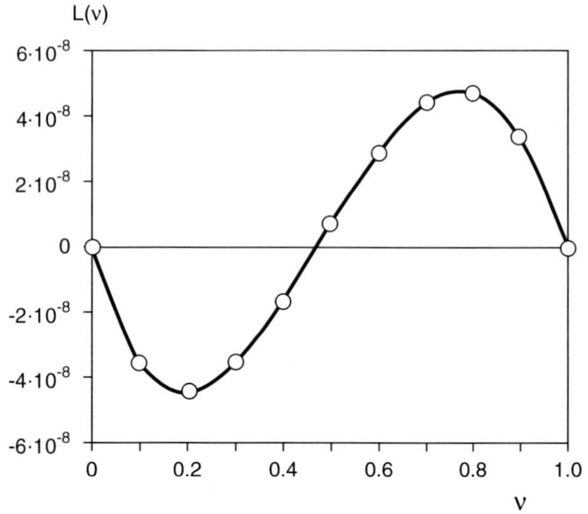

Fig. 6.10 Graph of $L(\nu)$ (see text). The function values less and more than 0 correspond to predomination of the phases 1 and 2, respectively. At $\nu = 0.470\,62$, a mixture of the two phases structurally behaves as a single phase

Let $L(\nu)$ be the left-hand part of the Eq. (6.14) that yields an alternating function from ν with zero at a point $\nu = 0.470\,62$ (Fig. 6.10).

From two heating experiments, D_0/h^2 and E can be determined as:

$$E = R \times \ln\left[\left(\frac{Fo_2}{Fo_1}\frac{t_1}{t_2}\right)^{\frac{T_1 T_2}{T_2 - T_1}}\right]; \quad \frac{D_0}{h^2} = \frac{\left[\dfrac{t_2}{Fo_2}\right]^{\frac{T_2}{T_1 - T_2}}}{\left[\dfrac{t_1}{Fo_1}\right]^{\frac{T_1}{T_1 - T_2}}}$$

We obtain $D_0 = 3.304\ \text{s}^{-1}$; $E = 27\,594$ cal/mole.

Deduction of the inverse function $\Phi(a)$ is a complicated mathematical problem, because the direct function is expressed by infinite series of transcendent functions. The solution is obtained by iteration on a computer after development of an appropriate algorithm. The necessary condition (6.14) is satisfied, the structure behaves as a monocrystal with the parameters D_0 and E. However deviations of $L(\nu)$ from zero show that this condition is not sufficient for a proof of monocrystallinity. Multiphase and single-phase structures might be undistinguishable despite different kinetic parameters. These results demonstrate that the radiation damage of probes in the argon–argon method may be more efficient than it is believed at present.

Is it possible to determine diffusion parameters (and consequently an age) taking into account losses in natural probes? It is, but after accepting a thermal model.

6.13 Irreversible Processes of Argon Release from Mineral Phases

The migration of radiogenic particles is accompanied by rearrangement in the host crystalline structure. Is laboratory or imposed natural heating reversible? What processes do occur in crystals during stepwise heating applied in the $^{40}Ar/^{39}Ar$ method?

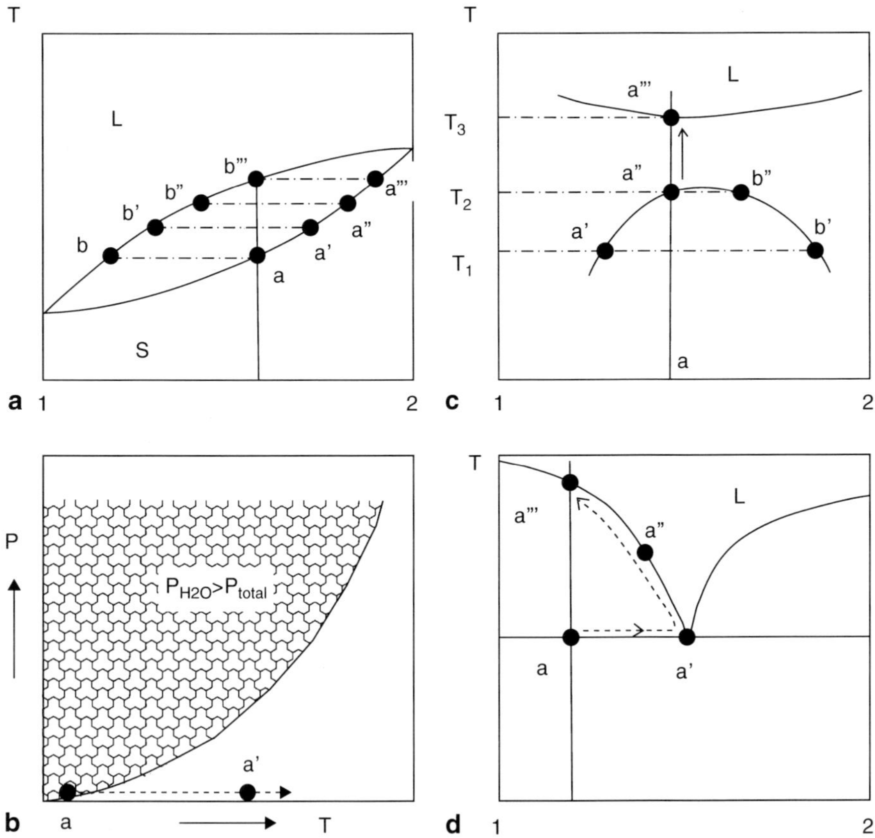

Fig. 6.11 Diagrams illustrating processes in crystals during stepwise heating applied in argon–argon geochronometry. **a** System with unlimited miscibility of two phases. During heating, the temperature reaches the solidus line at points a, a′, a″, or a‴ and melting begins. The system evolves irreversibly. **b** System with constitutional water. Its stability area is marked. If the temperature at low pressures is elevated, the mineral loses water. The process of drying is irreversible. **c** System with a solvus. This system is characteristic for alkali feldspars. At temperatures lower than solidus, order–disorder reactions take place in laboratory heating experiments. At higher temperatures, metastable phases appear. Here heating is irreversible as well. **d** System with an eutectic. A melt of eutectic composition appears after increasing temperature to the solidus value. This composition becomes enriched in component 1 and moves irreversibly along *a*′, *a*″ up to total melting point *a*‴

Consider evolution of mineral systems with (1) unlimited miscibility of two phases, (2) constitutional water, (3) solvus, and (4) eutectic (Fig. 6.11).

1. In the system with unlimited miscibility of two phases, radiogenic argon is retained below the solidus line and totally released above the liquidus line. Natural processes in solid solutions are governed by equilibrium reached kinetically during infinite time, unlike short-term heating experiments resulting in formation of melt patches. The latter are transformed to glass inclusions by

fast cooling. In this respect, the process of short-term heating of the system is irreversible.

Consider the evolution of a composition with a given proportion of components 1 and 2 marked by a vertical line in Fig. 6.11a. Heat up the probe. Assume that the argon system remains unchanged below the solidus line. A liquid phase with the composition b appears at the point a with initiation of argon release. The further heating shifts the composition of the solidus phase to the right from the point a to the points a′, a″ etc. simultaneously with displacement of composition of the liquidus phase from the point b to points b′, b″ etc. A combination of points a‴ and b‴ marks the total release of radiogenic argon. Such release of argon occurs, for example, in plagioclase. Respectively, argon spectra of lower temperature steps shows a phase with lower Ca/K, whereas progressive annealing results in increasing Ca/K.

2. Minerals with structural water (micas, amphiboles, leucites etc.) are stable inside the marked area of the diagram in Fig. 6.11b under significant P_{H_2O} (pressure of water vapors). At laboratory heating in high vacuum or at one atmosphere, the figurative point shifts to the right from a to a′ i.e. outside the area of stability. As a result, the annealed mineral dries up. After return back to lower temperature condition, the mineral differs from the former phase by partial dehydration.

   At some elevated temperature, structural water and other volatile components of a hydrocrystal are released with endo- or exothermal effects. Complications in the character of argon release are reflected in changes of apparent ages. For instance, the $^{40}Ar/^{39}Ar$ plateau age of 14 Ma was obtained both for a leucite and groundmass from the olivine melaleucitite U–94–53 in the temperature interval of 500–740°C, whereas apparent ages 2–3 times older than the plateau were measured for the leucite at higher temperature fractions of 800–1200°C. The plateau was accepted as the true age governed by diffusion losses of argon. Respectively, the steps of the leucite with overestimated ages were considered to indicate destruction of the hydrocrystalline structure (Rasskazov et al. 2000b).

   Similar phase transitions occur in other minerals. During ascension on temperature steps, new diffusion parameters begin to operate, although the experimenter believes in their constancy for all argon of a spectrum. The low-temperature part of the spectrum in biotite (temperature less than 600°C) can reflect losses by diffusion, but high temperature steps result in disintegration of structure due to oxidation of iron and dehydration (Lo et al. 2000). It is noteworthy that diffusion parameters vary at elevated temperatures not only because of dehydration. The structural transition of nepheline to carnegieite, for example, explains the specific spectrum of argon release from nepheline (Chap. 11).

3. In the solvus system (Fig. 6.11c), order–disorder reactions take place due to varying temperatures. For example, as a melt cools, one feldspar phase crystallizes firstly and normal diffusion in a homogeneous crystal structure occurs. A quick cooling of the melt results in crystallization of disordered plagioclase, sanidine or anorthoclase. A slow cooling leads to disintegration of solid solutions into two feldspar phases. In up-hill diffusion, potassium migrates towards a phase enriched by potassium and sodium towards a phase enriched by sodium. A zero diffusion

(spinodal) takes place somewhere below the solvus. If a feldspar has crystallized through slow cooling of a melt, laboratory annealing of the feldspar releases argon firstly at the temperature T_1 from solvus phases a′ and b′, then at temperature T_2 from phases a″ and b″ with transition to high-temperature argon, derived from homogeneous feldspar, up to complete melting at temperature T_3. It is clear that a microcline-perthite from granites produces the low-temperature fractions of argon spectrum, whereas homogeneous feldspar, if available, yields high-temperature steps.

Feldspar temperature spectra of argon isotopes were used to discriminate "domains", each of which was described by specific diffusion parameters. Such "domains" were argued to reflect the crystal structure of feldspars and contain thermochronologic information (Lovera et al. 1989, 1991). The sizes of domains were calculated to be comparable sometimes with lengths of displacement of ^{39}Ar during recoil of 78 ± 20 nm (Villa 1997).

4. In the eutectic system (Fig. 6.11d), proportions of components 1 and 2 depend on the eutectic temperature. Laboratory annealing of the composition a results in appearance of a melt with eutectic composition a′ and in afterward shift of melt sequence along a curve to the point a″ until total melting at the point a‴.

All diagrams considered represent equilibrium systems. Laboratory argon release spectra might exhibit disequilibrium irreversible processes, however. Strictly speaking, a spectrum of argon release does not exist in nature (i.e. during the geologic life of a mineral) and occurs only as an artefact in a laboratory.

6.14 Summary: Peculiarities of Ar–Ar Spectra

The $^{40}Ar/^{39}Ar$ method involves three substations: (1) radiogenic ^{40}Ar, (2) nucleogenic ^{39}Ar, and (3) contaminating ^{40}Ar and ^{36}Ar. The energy of activation of diffusion of each of these constituents depends on the strength of their bonds with a host crystal lattice. In this chapter, we have examined temperature argon spectra to solve the problem of separation of contaminating and radiogenic argon, i.e. the major problem of age measurements.

Such separation has been suggested by formal consideration of the Ar–Ar-isochron diagram. The algebraic expression for an isochron, yielding $^{36}Ar/^{40}Ar$ values as function of $^{39}Ar/^{40}Ar$ for thermal fractions of stepwise heating of an ideal crystal is linear only if both the isotopic composition of contaminating argon and $^{39}Ar/^{40}Ar_{rad}$ remain constant. On the diagram of stepwise release of ^{39}Ar, these constraints correspond to the plateau age.

This analysis of the argon–argon isochron diagram is based on the classical theory of mass-transfer, classical equations of theoretical physics in partial derivatives. Geochronological information cannot be determined for an ideal crystal if the energy of activation for all three argon substances is identical, because in this case the argon constituents are inseparable from each other by thermal procedures and are released congruently. The isotope ratios are constant and define a single figurative point of a multiplicity i (i – number of heating step). An ideal crystal yields a direct line appropriate to conditions of isochron, if the energies of activation for radiogenic and nucleogenic argon components are identical and the energy of activation for contaminating argon is lower. And inversely, if the dependence on isochron coordinates is represented by a straight line, the energies of activation for radiogenic and nucleogenic argon components are equal and the energy of activation for contaminating argon is lower. True linearity can be obtained also in the isochron coordinates through analysis of poly-mineral mixes due to non-ideality of natural crystals, even in the case when all energies of activation for contaminating, radiogenic, and nucleogenic argon are identical.

Late Tertiary and Quaternary rocks and minerals contain relatively small quantities of radiogenic argon, released at varying temperatures in mixtures with contaminating argon. The latter quite often prevails, so the data points of the measured temperature steps fall close to intersection of the isochron line with an axis of ordinates. The older samples are enriched by radiogenic argon with the subordinate role of contaminating argon, and the points are plotted near intersection of the isochron line with an axis of abscissa.

Atmospheric argon plays frequently a role of a contaminating admixture. Unlike radiogenic and nucleogenic argon components, produced directly inside a host crystalline structure, the atmospheric argon component is impregnated into structural weaknesses under a huge gradient of pressure. Respectively, the energy of activation for the contaminated admixture should be lower. At measurements by the $^{40}Ar/^{39}Ar$ method, some part of atmospheric argon is usually removed from a probe by its preliminary low temperature training in high vacuum. Besides the impregnated atmospheric argon, the contaminating admixture might be exhibited by excessive or inherited argon captured during crystallization. Analysis of Ar–Ar diagrams of minerals and rocks with complicated thermal history presumes the existence of several components of captured argon with different isotopic compositions (Roddick et al. 1980; Heizler and Harrison 1988; Widdowson et al. 2000; Rasskazov et al. 2003e). Such cases require special studies of kinetical differences between these argon components.

If energies of activation for radiogenic and nucleogenic argon differ, inverse isochron coordinates yield complex curves with pseudoisochrones. The interpretation of these curves in terms of linear regressions and extrapolations might result in erroneous conclusions about isotopic compositions of contaminating argon (i.e. about presence of excessive or inherited argon). The conclusion about excessive argon should be proved by direct observations of gas-liquid inclusions in an analyzed mineral. The conclusion about inherited argon can be admitted only in case of strict correspondence to the requirements of linearity of all sets of data points obtained by stepwise heating. The discrepancy between the data points and a linear trend, especially in low- or high-temperature steps, likely demonstrates relative shifts of the activation energies for radiogenic and nucleogenic argon components, excessive argon or distortion of the isotope system in the geologic past.

The formal theory will be developed to solve a reversal task, i.e. determination of a character of activation energies for analyzed substations of argon components from the form of experimental curves in inverse isochron coordinates. For this kind of kinetical studies, besides the two widely accepted types of diagrams, histograms of argon components are also indicative.

Reliable interpretation of argon–argon spectra requires development of a physico-chemical approach to analysis of stepwise heating experiments. Such studies should identify phases with kinetically differing parameters. We have demonstrated that the condition of a mono-crystalline phase (6.14) that was postulated as a basis of the $^{40}Ar/^{39}Ar$ method can be expressed by parameters D_0 and E. However, deviations of $L(\nu)$ from zero show that this condition is not sufficient for a proof of mono-crystalline structure. Some multi-phase structures are kinetically undistinguishable from single-phase counterparts. But numerous examples for crystalline phases reveal also obvious distinctions of their kinetic parameters often expressed by distorted argon–argon spectra.

Chapter 7
Radiogenic Argon in a Cooling Dike

High precision geochronological studies require understanding of relations between apparent (measured) K–Ar ages and various episodes in a cooling history of magmatic bodies: intrusion of liquid magma, chemical closure of a magmatic system (decreasing temperature below a solidus), and closure of a parent–daughter isotope system. There arises a number of problems connected not only with distortions of true ages due to repeated thermal or tectonic influence on rocks after crystallization, but also due to radiogenic argon accumulation in the process of post-crystallization sub-solidus decrease of temperatures. For definition of an instant for liquid intrusion, chemical closing of a system should be taken into account in connection with influence of the latent heat of crystallization. The account of the latter is feasible, though it complicates calculations. So far, it is important to show the principal approach to time definition of intruding magma.

7.1 Argon Kinetics

We start with an overview of formal relations between accumulation and diffusion of radiogenic argon. This introduction is necessary, because the appropriate formulas are dispersed in a variety of publications, some of which are not accessible.

The second Fick law gives dependence of concentration on time and coordinates of a diffusional substation:

$$\frac{\partial^{40}Ar}{\partial t} = D\Delta^{40}Ar \tag{7.1}$$

where D – coefficient of diffusion, Δ – Laplace operator.

This equation fits for interpretation of laboratory experiments on argon kinetics, the duration of which is negligibly small relative to the geologic time. If the geologic time interval is considered, the function of a source is taken into account:

$$\frac{\partial^{40}Ar}{\partial t} = D\Delta^{40}Ar + \lambda_K{}^{40}K_0 e^{-\lambda t} \tag{7.2}$$

where λ_K – constant for electron capture by a nucleus of potassium-40, λ – complete constant of ^{40}K transformation, $^{40}K_0$ – initial concentration of ^{40}K in a probe.

S.V. Rasskazov et al., *Radiogenic Isotopes in Geologic Processes*,
DOI 10.1007/978-90-481-2999-7_7, © Springer Science+Business Media B.V. 2010

This is a non-uniform partial differential equation. A diffusion coefficient varies with temperature according to the Arrhenius law:

$$D = D_0 e^{-\frac{E}{RT}} \tag{7.3}$$

where D_0 – frequency factor, e.g. a diffusion coefficient at indefinitely high temperature, E – energy of activation, T – Kelvin temperature. In general, a diffusion coefficient and other parameters (for example, melting temperature), depend on pressure as well. But this aspect is beyond the current study. At last, the radioactive transformation obeys the Law of Radioactivity of Rutherford–Soddy:

$$\frac{d^{40}K}{dt} = -\lambda^{40}K; \quad \frac{^{40}Ar}{^{40}K_0} = -\frac{\lambda_k}{\lambda}(1 - e^{-\lambda t}); \quad t = \frac{1}{\lambda}\ln\left(\frac{\lambda}{\lambda_k}\frac{^{40}Ar}{^{40}K} + 1\right) \tag{7.4}$$

Solutions of Eqs. (7.1) and (7.2) have been obtained by the method of separation of variables and Laplace transformation (Tables 7.1, 7.2, 7.3, and 7.4). The solutions

Table 7.1 List of diffusion equations for a flat-parallel body

Homogeneous equation (7.1), D = const	$\bar{c}(t) = \frac{8}{\pi^2}\sum_{\nu=0}^{\infty}\frac{1}{(2\nu+1)^2} \times e^{-(2\nu+1)^2\pi^2 Fo}$ (7.5) (Amirkhanov et al. 1960)
Approximation	$\bar{c}(t) = \frac{8}{\pi^2}e^{-\pi^2 Fo}$ (7.6) (Brandt et al. 2000)
Alternative solution of the Eq. (7.1) by the Laplace transformation	$\bar{c}(t) = 1 - 4\sqrt{\frac{Fo}{\pi}} + 16\sqrt{Fo}\sum_{n=1}^{\infty}(-1)^{n+1} \times ierfc\frac{n}{2\sqrt{Fo}}$ (7.7) (Brandt et al. 1996)
First approximation	$\bar{c}(t) = 1 - 4\sqrt{\frac{Fo}{\pi}}$ (7.8) (Brandt et al. 2000)
Inhomogeneous equation, D = const.	$\frac{^{40}Ar}{^{40}K_0} = \frac{8}{\pi^2}\frac{\lambda_e}{\lambda}\sum_{\nu=0}^{\infty}\frac{1}{(2\nu+1)^2} \times \frac{e^{-\frac{(2\nu+1)^2\pi^2 D}{h^2}t} - e^{-\lambda t}}{1 - \frac{(2\nu+1)^2\pi^2 D}{h^2\lambda}}$ (7.9) (Amirkhanov et al. 1960)
The same, relative to present-day potassium	$\frac{^{40}Ar}{^{40}K_0} = \frac{8}{\pi^2}\frac{\lambda_e}{\lambda}\sum_{\nu=0}^{\infty}\frac{1}{(2\nu+1)^2} \times \frac{e^{\left(1-\frac{(2\nu+1)^2\pi^2 D}{h^2}\right)\lambda t} - 1}{1 - \frac{(2\nu+1)^2\pi^2 D}{h^2\lambda}}$ (7.10) (Amirkhanov et al. 1960)
Combination of (7.7) and (7.8) provides precision ~1%	$\bar{c}(t) = e^{-4\sqrt{\frac{Fo}{\pi}}\frac{0.09}{Fo+0.09} + \frac{Fo}{Fo+0.09}\left(\ln\frac{8}{\pi^2} - \pi^2 Fo\right)}$ (7.11) (Brandt et al. 2000)
Combination of (7.9) and (7.11) provides precision ~1%, $D = var.$	$\frac{^{40}Ar}{^{40}K} = \lambda_k\int_0^{\infty} e^{-\lambda\cdot\tau} e^{-4\sqrt{\frac{\Phi o}{\pi}}\frac{0.09}{\Phi o+0.09} + \frac{\Phi o}{\Phi o+0.09}\left(\ln\frac{8}{\pi^2} + \pi^2\Phi o\right)} d\tau$ (7.12) (Brandt et al. 2000)

Table 7.2 List of diffusion equations for a cylindrical body

Homogeneous equation (7.1), D = const	$$\bar{c}(t) = 4\sum_{m=1}^{\infty}\frac{1}{(\mu_0^m)^2}e^{-(\mu_0^m)^2 Fo}$$ μ_0^m – routes of the equation $J_0(z) = 0$ (7.13) (Amirkhanov et al. 1960)
Approximation	$$\bar{c}(t) = \frac{4}{(\mu_0^1)^2}e^{-(\mu_0^1)^2 Fo} \quad (7.14)$$ (Brandt et al. 2000)
Inhomogeneous equation, D = const.	$$\frac{^{40}Ar}{^{40}K_0} = 4\frac{\lambda_e}{\lambda}\sum_{m=1}^{\infty}\frac{1}{(\mu_0^m)^2}\times\frac{e^{-(\mu_0^m)^2 Fo} - e^{-\lambda t}}{1-(\mu_0^m)^2\frac{Fo}{\lambda t}} \quad (7.15)$$ (Amirkhanov et al. 1960)
The same, relative to present-day potassium	$$\frac{^{40}Ar}{^{40}K_0} = 4\frac{\lambda_e}{\lambda}\sum_{m=1}^{\infty}\frac{1}{(\mu_0^m)^2}\times\frac{e^{\left(1-\frac{(\mu_0^m)^2 D}{a^2\lambda}\right)\lambda t} - 1}{1-\frac{(\mu_0^m)^2 D}{a^2\lambda}} \quad (7.16)$$ (Brandt et al. 2000)

(7.5)–(7.6), (7.13)–(7.14), (7.17)–(7.18) and (7.12), (7.21)–(7.22) concern constant diffusion coefficients (isothermal and isobaric processes) and those of an arbitrary function of time, respectively. These equations demonstrate that the average argon concentration is neither a function of time, nor a coefficient of diffusion or grain size,

Table 7.3 List of diffusion equations for a spherical body

Homogeneous equation (7.1), D = const	$$\bar{c}(t) = \frac{6}{\pi^2}\sum_{n=1}^{\infty}\frac{1}{n^2}e^{-n^2\pi^2 Fo} \quad (7.17)$$ (Allan 1956)
Approximation	$$\bar{c}(t) = \frac{6}{\pi^2}e^{-\pi^2 Fo} \quad (7.18)$$ (Brandt et al. 2000)
Inhomogeneous equation, D = const.	$$\frac{^{40}Ar}{^{40}K} = \frac{6}{\pi^2}\frac{\lambda_e}{\lambda}\sum_{n=1}^{\infty}\frac{1}{n^2}\times\frac{e^{-n^2\pi^2 Fo} - e^{-\lambda t}}{1-n^2\pi^2\frac{Fo}{\lambda t}} \quad (7.19)$$ (Allan 1956)
The same, relative to present-day potassium	$$\frac{^{40}Ar}{^{40}K} = \frac{6}{\pi^2}\frac{\lambda_e}{\lambda}\sum_{n=1}^{\infty}\frac{1}{n^2}\times\frac{e^{-\left(1-n^2\pi^2\frac{Fo}{\lambda t}\right)\lambda t} - 1}{1-n^2\pi^2\frac{Fo}{\lambda t}} \quad (7.20)$$ (Allan 1956)

Table 7.4 List of diffusion equations for flat-parallel, cylindrical, and spherical bodies

Solution for any $D(t)$	$$\frac{^{40}Ar}{^{40}K} = \frac{8}{\pi^2}\frac{\lambda_{.k}}{\lambda}\sum_{\nu=0}^{\infty}\frac{1}{(2\nu+1)^2} \times \int_0^t e^{\left(1-\frac{(2\nu+1)^2\pi^2}{h^2}\frac{1}{(t-\tau)}\int_r^t \frac{D(t)}{\lambda}dt\right)\lambda(t-\tau)}d\tau \quad (7.21)$$ (Brandt et al. 2000)
The same, in final differences (stepwise approximation)	$$\frac{^{40}Ar}{^{40}K} = \frac{8}{\pi^2}\frac{\lambda_{.e}}{\lambda}\sum_{\nu=0}^{\infty}\frac{1}{(2\nu+1)^2} \times \sum_{k=1}^{n}\frac{e^{\lambda(t_k-t_{k-1})(1-\beta_\nu D_k)}}{1-\beta_\nu D_k} \times \prod_{i=k+1}^{n} e^{\lambda(t_i-t_{i-1})\left(1-\beta_\nu\frac{D_i}{\lambda}\right)}, \quad where\ \beta_\nu = \frac{(2\nu+1)^2\pi^2}{h^2} \quad (7.22)$$ (Brandt et al. 2000)

but is a function of their combination—Fourier criterion:

$$Fo = \frac{Dt}{h^2} \tag{7.23}$$

where h – the size of a grain, from which constant diffusion occurs.

Some cases of variable diffusion fit to an integrated Fourier criterion:

$$Fo_i = \frac{1}{h^2}\int_0^t D(\tau)d\tau \tag{7.24}$$

Other cases satisfy a "lagging behind" Fourier criterion:

$$\Phi o = \frac{1}{h^2}\int_\tau^t D(\tau)d\tau \tag{7.25}$$

7.2 K–Ar Isotopic Balance in a Cooling Dike

The expressions (7.20–7.22) enable quantitative estimates of ^{40}Ar accumulation in any thermal history without simplifications and artificial hypotheses (for example, assumption that the inverse value for Kelvin temperature in a dike grows with time proportionally).

Let basalt magma be intruded at a time instant $t = 0$ in form of a flat-parallel dike of width 2R = 1000 m (Fig. 7.1). For simplicity, assume that host rocks are

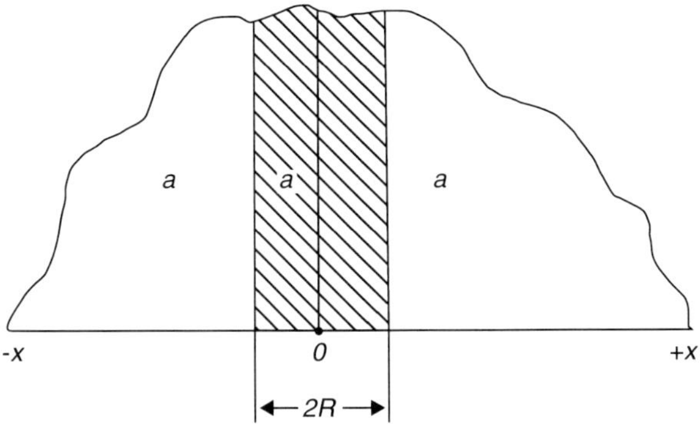

Fig. 7.1 Schematic drawing for a statement of a problem on calculating model of radiogenic argon distribution in a cooling dike. A basaltic dike of a width $2R$ was intruded into basaltic host rocks with a thermal diffusivity $a = 10^{-6}$ m²/s at $T = 1300$ K. The temperature of the center of the dike is computed

also of basaltic composition. Temperature conductivity of the basalt $a = 10^{-6}{\cdot}c^{-1}$ ($a = \lambda/c\gamma$, λ – heat conductivity, c – heat capacity, γ – density).

Temperature of the dike centre will be computed as (Lykov 1964):

$$\frac{T(0,t) - T_e}{T(0,0) - T_e} = erf\frac{R}{2\sqrt{at}} \tag{7.26}$$

where T_e – initial temperature of host rocks. The argument *erf* can be interpreted as a criterion of similarity. If the thickness of the dike increases by a factor of 10, the time of cooling will increase by a factor of 100.

In our case, the temperature and diffusion coefficients of argon in the Eq. (7.3) are functions of time. Hence, we should use formula (7.21). However, integration of the equation in quadratures represents some difficulties, therefore, we use the formula of step heating (7.22).

A calculation procedure is given in Table 7.5. Twenty steps are put in column 1. Then intervals of existence for the $erf\left(R/2\sqrt{at}\right)$ function (from 0 to 1) are arbitrarily attributed to 20 values in column 2 with transition to values in column 3. From these values, timing of cooling is calculated in column 4. Duration of each step, mean temperatures of each step, and reverse value of temperature are specified in columns 5, 6, and 7, respectively. Mean values of a diffusion coefficient for radiogenic argon, appropriate to these temperatures, are given in column 8. The parameters suggested here for basalts ($D_0/h^2 = 3.15 \cdot 10^6 year^{-1}$ and $E = 30\,000$ cal/mole) are inferred from interpretation of stepwise heating experiments. The relative increase of the parameter $D/h^2\lambda$ for the given interval is shown in column 9.

Then an expression (7.22) is used to get the sum of multiplying expressions (7.13) and (7.5) for any interval of time. Values of $^{40}Ar/^{40}K_0$ (i.e. argon formed with simultaneous diffusion calculated according to expression (7.13) on the program "A") are placed in column 10. Respectively, residual values of this ratio from previous steps, taking into account diffusion in the i-th interval (calculated according to expression (7.5) on the programs "B"), are given in column 11. The sums of values in columns 10 and 11, determined from step 1, are represented in column 12. At last, apparent

Table 7.5 Calculations of argon in a cooling dike using the Eq. (7.22) (Table 7.4)

	erf x	x	t, year·10^{-6}	Δt, year·10^{-6}	T_K	$1/T_{1/K}\cdot 10^3$	$(D/h^2)_{mean}$, year^{-1}	$(D/h^2\lambda)$, mean	Progr. "A" Ar/K_0	Progr. "B" C/C_0	$^{40}Ar/K$	$t_{appar.}$, Ma	$(D/h^2\lambda)$, mean	Progr. "A"	Progr. "B"	$^{40}Ar/K$	$t_{appar.}$, Ma
1	2	3	4	5	6	7	8	9	10	11	12	13	14	15	16	17	18
20			∞		300	3.33											
				~50			$1.93\cdot10^{-15}$	$3.54\cdot10^{-6}$	$2.53\cdot10^{3}$	1.000			$3.54\cdot10^{-4}$	$2.52\cdot10^{-3}$	0.995		
19	0.01128	0.01	9.8		311	3.21					$9.88\cdot10^{-4}$	19.62				$9\cdot10^{-4}$	19.6
				14.84			$1.19\cdot10^{-14}$	$2.15\cdot10^{-5}$	$7.43\cdot10^{-4}$	1.000			$2.15\cdot10^{-3}$	$7.4\cdot10^{-4}$	0.994		
18	0.0226	0.02	4.96		323	3.1					$2.45\cdot10^{-4}$	4.88				2.31^{-4}	4.75
				2.756			$6.2\cdot10^{-14}$	$1.14\cdot10^{-4}$	$1.38\cdot10^{-4}$	1.000			$1.14\cdot10^{-2}$	$1.37\cdot10^{-4}$	0.993		
17	0.0340	0.03	2.204		334	2.99					$1.07\cdot10^{-4}$	2.13				$9.49\cdot10^{-5}$	1.95
				0.964			$2.51\cdot10^{-13}$	$4.66\cdot10^{-4}$	$4.81\cdot10^{-5}$	1.000			$4.66\cdot10^{-2}$	$4.75\cdot10^{-5}$	0.992		
16	0.045	0.04	1.24		345	2.9					$5.84\cdot10^{-3}$	1.17		$1.85\cdot10^{-5}$		$4.78\cdot10^{-5}$	0.98
				0.273			$6.29\cdot10^{-13}$	$1.15\cdot10^{-3}$	$1.36\cdot10^{-5}$	0.999			0.115		0.993		
15	0.05	0.045	0.967		350	2.85					$4.48\cdot10^{-5}$	0.895		$3.14\cdot10^{-5}$		$3.59\cdot10^{-5}$	0.73
				07.11			$8.19\cdot10^{-11}$	0.15	$3.51\cdot10^{-5}$	0.995			15		0.932		
14	0.1	0.088	0.256		400	2.5					$9.76\cdot10^{-6}$	0.195		$4.23\cdot10^{-6}$		$4.44\cdot10^{-6}$	0.0913
				0.147			$5.53\cdot10^{-9}$	10.1	$7.03\cdot10^{-6}$	0.968			10^{3}		0.668		
13	0.15	0.155	0.109		450	2.22					$2.82\cdot10^{-6}$	$5.63\cdot10^{-12}$		$3.2\cdot10^{-7}$		$3.21\cdot10^{-7}$	0.0659
				0.047			$1.53\cdot10^{-7}$	238	$2.07\cdot10^{-6}$	0.881			$2.38\cdot10^{4}$		$4.83\cdot10^{-2}$	–	–
12	0.2	0.18	0.062		500	2.0					$8.53\cdot10^{-7}$	$1.7\cdot10^{-2}$		$1.78\cdot10^{-8}$		–	–
				0.023			$2.34\cdot10^{-6}$	$4.29\cdot10^{3}$	$7.44\cdot10^{-7}$	0.624			$4.29\cdot10^{5}$		$9.33\cdot10^{-10}$		
11	0.25	0.225	0.039		550	1.82					$1.75\cdot10^{-7}$	$3.5\cdot10^{-3}$		$1.81\cdot10^{-9}$		–	–
				0.013			$3.3\cdot10^{-5}$	$4.22\cdot10^{4}$	$1.72\cdot10^{-7}$	0.122			$4.22\cdot10^{6}$		–		
10	0.3	0.275	0.0262		600	1.67					$2.51\cdot10^{-8}$	$5.01\cdot10^{-4}$		$2.51\cdot10^{-10}$		–	–
				0.007			$1.66\cdot10^{-4}$	$3.04\cdot10^{5}$	$2.51\cdot10^{-8}$	$0.9\cdot10^{-4}$			$3.04\cdot10^{7}$		–		
9	0.35	0.32	0.0194		650	1.54					$4.6\cdot10^{-9}$	$1.9\cdot10^{-4}$		–		–	–
				0.0048			$9.06\cdot10^{-4}$	$1.66\cdot10^{6}$	$4.6\cdot10^{-9}$				$1.66\cdot10^{8}$		–		
8	0.4	0.37	0.0146		700	1.43					$1.01\cdot10^{-9}$	$2.02\cdot10^{-5}$		–		–	–
				0.0036			$4.17\cdot10^{-3}$	$7.65\cdot10^{6}$	$1.01\cdot10^{-9}$	0			$7.65\cdot10^{8}$		–		

Table 7.5 (continued)

1	erf x 2	x 3	t, year·10^{-6} 4	Δt, year·10^{-6} 5	T_K 6	$1/T_{1/K}\cdot10^3$ 7	$(D/h^2)_{mean}$, year^{-1} 8	$(D/h^2\lambda)$, mean 9	Progr. "A" Ar/K_0 10	Progr. "B" C/C_0 11	$^{40}Ar/K$ 12	$t_{appar.}$, Ma 13	$(D/h^2\lambda)$, mean 14	Progr. "A" 15	Progr. "B" 16	$^{40}Ar/K$ 17	$t_{appar.}$, Ma 18
7	0.45	0.425	0.011		750	1.33					$2.84\cdot10^{-10}$	$5.07\cdot10^{-6}$		–		–	–
				0.0022			$1.47\cdot10^{-2}$	$2.69\cdot10^7$	$2.84\cdot10^{-10}$	0			$2.69\cdot10^9$		–		
6	0.5	0.475	0.0088		800	1.25					$4.02\cdot10^{-11}$	$8.1\cdot10^{-7}$		–		–	–
				0.0032			0.104	$1.9\cdot10^8$	$4.02\cdot10^{-11}$	0			$1.9\cdot10^9$		–		
5	0.6	0.595	0.0056		900	1.11					$7.28\cdot10^{-12}$	$1.5\cdot10^{-7}$		–		–	–
				0.0019			0.574	$1.05\cdot20^9$	$7.28\cdot10^{-12}$	0			$1.05\cdot10^{11}$		–		
4	0.7	0.73	0.0037		1000	1.0					$1.6\cdot10^{-12}$	$3.7\cdot10^{-8}$		–		–	–
				0.0013			2.606	$4.78\cdot10^9$	$1.6\cdot10^{-12}$	0			$4.78\cdot10^{11}$		–		
3	0.8	0.91	0.0024		1100	0.901					$6.42\cdot10^{-13}$	$1.3\cdot10^{-8}$		–		–	–
				0.00069			6.5	$1.19\cdot10^{10}$	$6.42\cdot10^{-13}$	0			$1.19\cdot10^{12}$		–		
2	0.9	1.16	0.0017		1200	0.853					$2.1\cdot10^{-13}$	10^{-8}		–		–	–
				0.00123			19.8	$3.65\cdot10^{10}$	$2.1\cdot10^{-13}$	0			$5.65\cdot10^{12}$		–		
1	1.0	> 2.0	0.00046		1300	0.769					$1.5\cdot10^{-14}$	$4\cdot10^{-9}$		–		–	–
				469													

ages for the various moments of the geologic time after the beginning of a process (see column 5) are given in column 13.

Calculations are performed for rocks of two textural types: coarse-grained (columns 10–13) and fine-grained (columns 14–18), the sizes (h) are related as 1:10 and diffusion parameters ($D/h^2\lambda$) as 1:100.

In the statement of the problem, we have assumed identical properties of basalts in the dike and host rocks. Moreover, we have postulated a lack of radiogenic argon in the host rocks. The purpose of this study is to demonstrate a formal approach that has been developed in the case of simple relations.

The theory can be further improved by a combination of experimental work and data on natural objects. For example, the older granites as host rocks can contain radiogenic argon with its concentrations much higher than in a cooling basaltic dike. Will the dike rocks be saturated by argon, lost by thermally activated host rocks, or will the released argon be lost in a boundary zone, which might be ideally "transparent" for argon migrated to the atmosphere? These questions still remain without answer.

7.3 Closure of K–Ar Isotope System

Variations of temperature, inverse values of temperature, and apparent ages in a cooling dike are plotted versus true ages in Fig. 7.2. Mutually consistent apparent and true ages are shown by a bisector with the ratio 1:1 at a coordinate corner of panel (c). As shown on panel (b), the inverse temperature of the cooling dike can not be approximated by a straight line. However, this circumstance should not be given large importance, because in reality the latent heat of crystallization, fluids, and chemical reactions might change a curve of cooling. It is essential to demonstrate an approach of calculations on an arbitrarily chosen example.

Hereafter, different conditions based on varied safety values of radiogenic argon will be considered. At high temperature stage, radiogenic argon can not accumulate. As a dike cools down, first a small part of radiogenic argon and then its larger part start accumulating. In the process of cooling, the apparent age asymptotically gets closer to the true one. There is not and cannot be any sudden closure or threshold value of temperature and time, at which diffusion stops and the apparent age becomes equal to the true one. Hence, at least the cooling dike never shows "blocking" temperature in the sense postulated by Dodson (1973).

The curve of an apparent age resembles a function:

$$t_{\text{apparent}} = t_{\text{true}} - a$$

where $a \ll t$, $\lim\ t_{\text{true}} \to \infty$, $(t_{\text{true}} - a)/t_{\text{true}} = 1$

The parameter a can be interpreted as a "backlog in time", which is characteristic for a chosen size of a dike. The backlog should not be considered as the closure (or blocking) temperature, however. It is a slowly converging integral diffusion. Similar backlogs of K–Ar ages relative to "true" Rb–Sr ones were observed by Hanson and Gast (1967) and were mentioned by Wright et al. (1991).

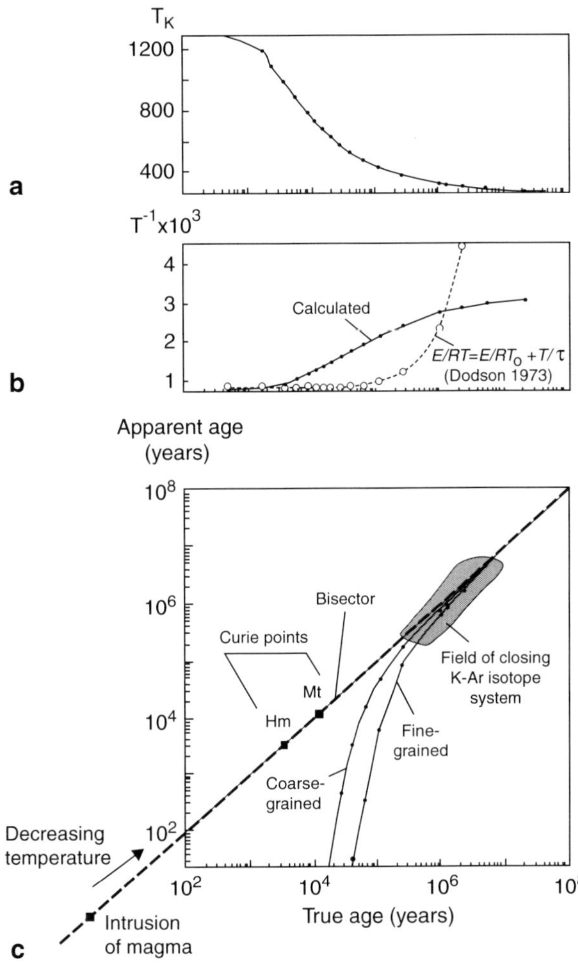

Fig. 7.2 Time dependence of Kelvin temperature (**a**), inverse Kelvin temperature (**b**) and lag of the apparent age from the true one (**c**) in a cooling dike. Panel **a** demonstrates general decreasing temperature in the dike with time. Panel **b** shows that inverse temperature is not approximated by a time dependent linear relation (by a curve of Dodson). Panel **c** indicates lag for a fine-grained mineral fraction greater than for a coarse-grained one. An instant of intrusion and the Curie points for mineral phases: (*Mt* – magnetite, *Hm* – hematite) are shifted in time relative to a field of closing K–Ar isotope system

It is noteworthy that these regularities are valid only for the given model, i.e. for the cooling dike. A peculiarity of the model is the accepted initial point as the moment of liquid magma intrusion, whereas it is more customary that the initial point is chosen for a system as its moment of chemical closure. Other regularities might take place in models of the imposed processes.

According to the expressions (7.23–7.25), a diffusion parameter *Fo* is inversely proportional to the square of the grain size. Apparent ages of fine-grained and coarse-grained fractions differ. We can write:

$$t_{\text{coarse}} = t_{\text{true}} - c,\ t_{\text{fine}} = t_{\text{true}} - f \text{ and } f - c = t_{\text{coarse}} - t_{\text{fine}}$$

Numerically, we obtain approximately: $f = 3{\cdot}10^5$ years and $c = 8{\cdot}10^5$ years; $f - c = 5{\cdot}10^5$. This difference between ages of the fine-grained and coarse-grained fractions will remain constant in time and will have implications in cases of high sensitivity and accuracy of isotope measurements.

The backlog of the apparent ages from the true ones strongly depends on the size of a cooling body, parameters T_0, D_0, h, E, a as well as involvement of fluids. Is it possible to solve an inverse problem, i.e. to restore a thermal history from the data of today's experiment? We have to accept that such a problem belongs to the incorrect type (Tikhonov and Goncharskii 1987), therefore it cannot be solved unequivocally. Within limits of some hypotheses and assumptions and by means of modern development of computer techniques, the problem might be solved by a cut-and-dried method.

A numerical example for a dike (R = 2000 m) that cools down from temperature 1300 to 311 K during $5 \cdot 10^8$ years yields the parameters for a backlog of the apparent age: $f = 8.2 \cdot 10^6$ and $c = 5.2 \cdot 10^6$ years.

7.4 Geologic Application

Study of radioisotope thermochronology of intruded liquid magma is relevant for high precision reconstruction of temporal variations of tectonic stress in the crust. As a rule, magmatic conduits are elongated in a direction of the maximal compression axis acting in a regional stress field (Nakamuro 1977).

Spatial–temporal distribution of dikes shows that the continental rifts have developed at varying stress fields. For instance, in the northeastern part of the Baikal Rift System, dike orientation of the Udokan volcanic field changed simultaneously with transition from olivine tholeiite and alkaline basalt to basanite and leucitite liquids at about 2.6 Ma and accompanied melting of the lower crust on background of dominating mantle-derived magmatic evolution (Rasskazov 1982, 1985; Rasskazov et al. 1997).

Temporal change of tectonic stress in the crust was demonstrated by variations of dike orientations in the southwestern part of the Baikal Rift System as well as in the Rio Grande and East African Rift Systems (Rasskazov 1985, 1993 and reference therein). A similar phenomenon was described in the Japan Arc (Sato 1994; Rasskazov and Taniguchi 2006 and reference therein). A tectonic pulse that lasted during the past 14–15 Ka was comparable with duration of the cooling history of magmatic bodies in the upper crust. This pulse was found to be a common feature for the latest Pleistocene through Holocene volcanoes of Central and East Asia (Chuvashova et al. 2007).

The cooling dike model can be applied also for interpretation of paleomagnetic data on direct or reverse polarity and paleointensity variations in the geomagnetic field. Liquid magma has no spatial magnetic orientation and crystallized rocks can not get it and keep it as long as temperature decreases below a Curie point (thermal energy of magnetic domains in rocks is so great that any primary orientation cannot be kept). On the one hand, the available data assume inversions of a geomagnetic field in the Pliocene and Quaternary lasted up to 200 Ka. On the other hand, a geomagnetic dipole might change rather quickly. For example, its intensity sharply increased during the past 12 Ka. It is clear that the study of these processes in the geologic past puts forward the special requirements for radiogenic isotope thermochronology of cooling magmatic bodies.

We can see that comparisons of Curie points and temperature intervals that correspond to transition of a K–Ar system from radiogenic argon losses to its accumulation according to the Law of Rutherford–Soddy is important for correct interpretation of paleomagnetic data, though the physical sense of magnetic and radioactive processes differ. Unfortunately, we could not find any thermochronologic basis for so-called "blocking temperature" (Dodson 1973), which would have given information on a cooling magmatic body resembling a Curie point. If the magnetic stability is achieved at a higher temperature than an interval of K–Ar isotopic closure of a cooling system, the difference between these temperatures results in additional diffusional loss of radiogenic argon. The true age of the paleomagnetic event in this case will be older than the apparent isotopic estimate. In case of opposite relations between a Curie point and a temperature interval of K–Ar isotopic closure, the age of the paleomagnetic event will be overestimated.

7.5 Summary: Temporally Varying Diffusion

During recent decades, numerous publications have appeared in which geochronometric information about various isotope systems has been given tectonic, thermal and petrologic interpretations (Hanson and Gast 1967; Hart 1964; Harrison and Bé 1983; Treloar et al. 1989; Harris 1997; Zhao et al. 1997). Significant progress in explanations of isotopic data was achieved (Harrison et al. 1991; Lovera et al. 1989, 1991; Richter et al. 1991). In accepted models of geologic processes, researchers were often compelled to resort to simplifications, however. As a result there arose contradictions with the precise theory. An example is the extremely large duration of the periods of thermal activity in a geologic history of rocks according to the concept of "closing temperature" (Dodson 1973; Sharma et al. 1980), in comparison with estimations based on the theory of heat conductivity (Lykov 1964).

Deductions of exact relations were published already in the 1950s–1960s (Allan 1956; Gentner and Kley 1957; Amirkhanov et al. 1961). Partial differential equations were complex and looked like weakly converging infinite rows. This made direct estimations impossible. Now theoretical interpretation of any experimental results can be done in terms of the precise theory.

In this chapter, we have demonstrated calculations of temporally varying diffusion in a cooling dike. In a solidified basaltic body, radiogenic argon is generated due to radioactive transformation of potassium, but is not accumulated at high-temperature stages. As the dike cools down in sub-solidus conditions, losses of daughter substance decrease and apparent K–Ar ages become smoothly and asymptotically closer to the true values. A backlog of the apparent age depends, with other equal conditions, on size of mineral grains in the dike. These radiogenic argon losses can result in essential underestimation of K–Ar ages for Quaternary dikes. In older magmatic bodies, the quantity of initial sub-solidus losses of radiogenic argon is negligible as compared with error of radiogenic argon measurements in a probe.

The model of a cooling dike can be applied for corrections of apparent (measured) K–Ar ages in high-precision geochronometry of liquid magma intrusion for purposes of tectonic stress reconstruction in the crust, or for definition of cooling time below the Curie point in paleomagnetic studies. It is noteworthy also, that for exact timing of intrusion, calculations should take into account an instant of chemical "closure" of magmatic system in connection with influence of the latent heat of crystallization.

Chapter 8
Radiogenic Isotopes in an Exocontact Zone of a Magmatic Body

Recent decades have brought enormous progress in analytical laboratory techniques and publication of many high precision results on radioactive isotope systems. The measured isotope ratios are usually shown in data tables and graphs with standard deviations of 5 or 6 true ciphers. Often, the measured results are considered as an ultimate step for an isotope study. The precise data presume however a satisfactory description of natural geologic processes in terms of precise models and an adequate development of a strict theory based on fundamental differential equations.

Consider, for instance, radiogenic argon accumulation in a mineral as a function of time with a constant diffusion (or other kind of losses) (Amirkhanov et al. 1961; Brandt et al. 1996):

$$\frac{^{40}Ar}{K_0} = \frac{8}{\pi^2}\frac{\lambda_e}{\lambda}\sum_{\nu=0}^{\infty}\frac{1}{(2\nu+1)^2}\frac{e^{-(2\nu+1)^2\pi^2 Fo} - e^{-\lambda t}}{1 - \dfrac{(2\nu+1)^2\pi^2 Fo}{\lambda t}} \tag{8.1}$$

For this and other expressions of this chapter, symbols are listed in Table 8.1.

We can see that the term $e^{-\lambda t}$ expressing radioactive decay and the term $e^{-\pi^2 Fo}$ expressing losses are quite symmetric, so to say, they "enjoy equal rights". But the loss-term is a pronounced function of temperature through the Arrhenius law:

$$D = D_0 e^{-\frac{E}{RT}} \tag{8.2}$$

In this respect, a radioactive system is a clock, as well as a thermometer.

In papers so far published, the impact of secondary heating on radiogenic isotopes in minerals was widely recognised and proven (Hart 1964; Dodson 1973; Brandt 1973; Chernyshev et al. 2000; Brandt et al. 2002a etc.). Hart's work has become classical and although published in 1964 is still often mentioned in modern publications and textbooks (e.g. Dickin 1997). It concerns the behaviour of radiogenic argon and strontium in micas, amphiboles and feldspars from an exocontact zone of the 53 Ma Eldora stock intruded into the 1350–1200 Ma host rocks. The author has succeeded in a study of isotope distribution within some mineral grains by means of painstaking experiments. Precise theories of neither the thermal field, nor of the behaviour of argon at variable temperatures, however, were developed. The Eq. (8.1)

S.V. Rasskazov et al., *Radiogenic Isotopes in Geologic Processes*,
DOI 10.1007/978-90-481-2999-7_8, © Springer Science+Business Media B.V. 2010

Table 8.1 List of symbols used in Chap. 8

^{40}Ar	–	present-day concentration of radiogenic argon
$^{40}K_0$	–	initial concentration of potassium
λ	–	total radioactive decay constant of potassium
λ_e	–	radioactive decay constant of potassium to reach ^{40}Ar
t	–	time
D	–	diffusion coefficient
D_0	–	frequency factor (diffusion coefficient at infinite temperature)
E	–	activation energy of diffusion
R	–	the gas constant
a	–	the temperature diffusivity
a^d	–	the temperature conductivity of an intrusive body
a^m	–	the temperature conductivity of a surrounding medium
λ_h	–	the heat conductivity
λ_h^d	–	the heat conductivity of an intrusive body
λ_h^m	–	the heat conductivity of a surrounding medium
c	–	the heat capacity
γ	–	the density
x, y, z	–	the spatial coordinates
l	–	distance from a contact with a magmatic body
T	–	the Kelvin temperature
T_1^*	–	the temperature of intruded magma
T_2^*	–	the temperature of surrounding medium in an exocontact zone at a time of magma intrusion
T_2	–	the temperature of surrounding medium in an exocontact zone at distance x from a contact with intrusion (one-dimensional case)
T_{max}	–	maximum temperature that can be achieved at a point in surrounding medium of a magmatic body
h	–	width of a mineral grain
d	–	width of a magmatic body
t_{max}	–	time-maximum (time required to heat-up a point in surrounding medium of a magmatic body to a maximum possible value of temperature)
ν	–	velocity of propagation of "temperature pulse" (maximum values of the temperature) in vicinity of a magmatic body
Fo	–	dimensionless Fourier number for diffusion of radiogenic isotopes
ϕo	–	dimensionless Fourier number for thermal conductivity
ρ	–	dimensionless time
ρ_{max}	–	maximum value of dimensionless time, that can be achieved at a point in surrounding medium of a magmatic body
ρ_0	–	an arbitrary constant of time
C	–	heat capacity in arbitrary units

given above can not be applied directly for an analysis of mineral age response to a long-term thermal influence because it involves a constant D, whereas in a system with a long intrinsic thermal history, D varies with time. So, researchers were compelled to use some simplified approximations such as a linear increase of 1/T during cooling (Dodson 1973; Brandt 1973), or to analyse a distribution of thermal field and behaviour of radiogenic isotopes separately. Actually, both processes are so closely bound, that there arises the idea to connect them into a single mathematical expression. This chapter is devoted to this purpose.

8.1 Theoretical Consideration

8.1.1 Thermal Model

A starting point for an investigation of heat transfer is the parabolic partial differential equation (Lykov 1964):

$$\frac{\partial T}{\partial t} = a \frac{\partial^2 T}{\partial x^2} \tag{8.3}$$

Further, we restrict a model to the one-dimensional case, because involvement of more complicated cylindrical, spherical and other cases does not add any new physical revelations, but rather complicates the mathematical symbolism. Solutions for two- and three-dimensional cases, however, are presented in the Appendix.

A solution of a differential equation can be obtained either numerically (Bonneville and Capolsini 1999) or explicit (Courant and Hilbert 1951), in the form of mathematical functions, permitting one to see immediately the trends and features of the process considered. We prefer the latter way.

Consider an intruding flat-parallel body (Fig. 8.1). For the sake of simplicity, let the intrusive and the host rocks be both of a basaltic composition ($a \approx 10^{-6}$ m^2/s). There exists a general solution for this case obtained through Laplace—transform (Lykov and Mikhailov 1963) and applied in an attempt to use radiogenic argon in rocks as a kind of paleogeothermometer (Brandt and Voronovskii 1967):

$$T_r = \frac{T_2 - T_2^*}{T_1^* - T_2^*} = \frac{1}{1 + K_\varepsilon} \left[erfc \frac{l}{2\sqrt{\Phi o}} - (1 - \alpha) \sum_{n=1}^{\infty} \alpha^{n-1} erfc \frac{l + nK_\alpha}{2\sqrt{\Phi o}} \right] \tag{8.4}$$

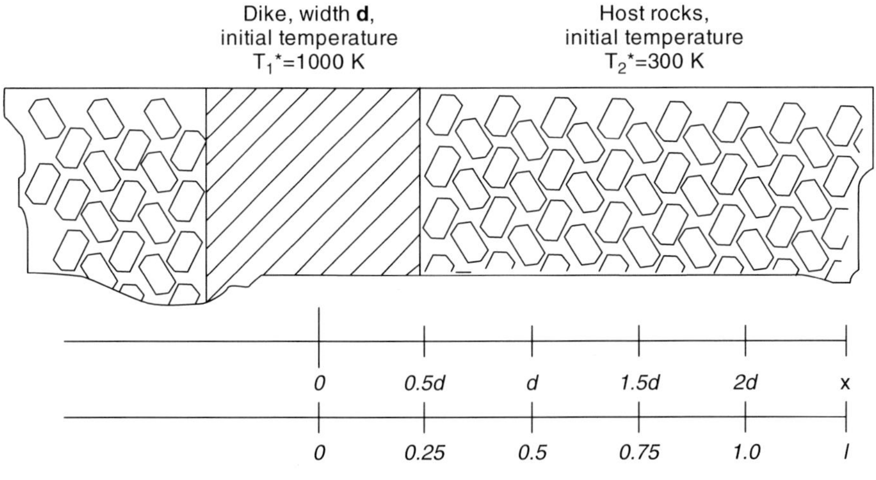

Fig. 8.1 A flat-parallel magmatic body of a width *d* is intruded into host rocks of indefinite dimensions

where

$$K_\varepsilon = (\lambda_h^m/\lambda_h^d)\sqrt{a^d/a^m}, K_\alpha = \sqrt{a^m/a^d}, \quad \text{and} \quad a = \sqrt{a^m/a^d}\,[(1 - K_\varepsilon)/(1 + K_\varepsilon)]$$

This equation, although general, is rather inconvenient for our purposes, containing two transcendental functions *erfc* and their infinite sum, uneasy in algebraic manipulations and in programming. Therefore, we are searching for a solution of (8.3), which would retain all the features of the general solution, but be simpler in manipulation. We depart from the model of a spreading of a delta-function on a straight line, considered in textbooks on differential equations (Courant and Hilbert 1951; Tikhonov and Samarskii 1953). One of the possible solutions of the differential equation of heat conduction above meeting the requirement $T_2(y \to \infty) = T_2^*$ is:

$$T_2 - T_2^* = \frac{C}{\sqrt{4\pi a(\rho + \rho_0)}} e^{-\frac{x^2}{4a(\rho+\rho_0)}} \tag{8.5}$$

where ρ_0 and C are arbitrary constants. We determine these constants from the following conditions:

1. the temperature at $\rho = 0$ and $x = 0$ is $T_1^* - T_2^*$.
2. $\int_{-\infty}^{+\infty} (T - T_2^*)dx = (T_1^* - T_2^*)d$, i.e. the whole quantity of excessive heat is the same as in a magmatic body of a width d at a temperature $T_1^* - T_2^*$ (the thermal capacity we put = 1).

Inserting into Eq. (8.5) $\rho = 0$ and $x = 0$, due to the first condition, we obtain:

$$\frac{C}{\sqrt{4\pi a \rho_0}} = T_1^* - T_2^* \tag{8.6}$$

and from the second condition:

$$\frac{C}{\sqrt{4\pi a \rho_0}} \int_{-\infty}^{+\infty} e^{-\frac{x^2}{4a\rho_0}} dx = (T_1^* - T_2^*)d \tag{8.7}$$

Since the integral equals the denominator in formula (8.6), the expression takes the form:

$$C = (T_1^* - T_2^*)d \tag{8.8}$$

Comparing this expression with the first condition, we obtain as well:

$$\rho_0 = \frac{d^2}{4\pi a} \tag{8.9}$$

Substituting (8.8) and (8.9), produces:

$$T_2 - T_2^* = \frac{(T_1^* - T_2^*)d}{\sqrt{4\pi a\left(\rho + \frac{d^2}{4\pi a}\right)}} e^{-\frac{x^2}{4a\left(\rho + \frac{d^2}{4\pi a}\right)}} \tag{8.10}$$

We introduce new variables of time and coordinates due to formulas:

$$\rho = \frac{a}{d^2}t \quad \text{and} \quad x = 2dl \tag{8.11}$$

With these variables, expression (8.10) takes a form suitable for our purposes:

$$T_r = \frac{T_2 - T_2^*}{T_1 - T_2^*} = \frac{1}{\sqrt{4\pi\left(t + \frac{1}{4\pi}\right)}} e^{-\frac{l^2}{t+\frac{1}{4\pi}}} \tag{8.12}$$

Note that our reduced time ρ equals the dimensionless thermal Fourier number:

$$\rho = \Phi o = \frac{at}{d^2} \tag{8.13}$$

This solution reflects the mean physical features of the process. It may be seen that (8.12) satisfies the fundamental differential equation of heat transfer (8.3), i.e. being inserted into the equation, turns it into an equality. Theoretically, heat propagates with an infinitely high rate (actually with the velocity of sound). There do not exist "retarding potentials", as is the case of hyperbolic wave equations. Nevertheless, temperature variations propagate with finite rates (just as the charge of a capacitor or the current through an inductivity vary with finite rates despite the fact that an electric field propagates with the velocity of light). So, Eq. (8.12) shows that a temperature maximum at points in the vicinity of the intrusive heat-up is reached with a certain time-delay. The longer the delay, the larger their distance from the magmatic body.

Taking a derivative of (8.12) with respect to t (or finally ρ) and setting it equal to zero determines a time-maximum for every point of the environment:

$$t_{\max} = 2l^2 - \frac{1}{4\pi} \tag{8.14}$$

Inserting this expression into (8.12) yields a temperature maximum for every point:

$$T_{r(\max)} = \frac{1}{l\sqrt{8\pi}} e^{-\frac{1}{2}} = \frac{d}{x\sqrt{2\pi}} e^{-\frac{1}{2}} \tag{8.15}$$

Figure 8.2 shows the temperatures as a function of time in various points of the environment of the magmatic body. It is easy to see that the crest of this "temperature pulse" propagates with a velocity:

$$\nu = \frac{dx_{\max}}{d\rho_{\max}} = \frac{a}{\sqrt{2a\rho_{\max} + \frac{d^2}{2\pi}}} \tag{8.16}$$

Actually, the temperature maximum travels with a velocity varying from a maximum at $\rho = 0$ to zero at $\tau \to \infty$. For the case of heat propagation in the vicinity of a flat-parallel magmatic body of a width d = 1 km, with a temperature conductivity $a = 10^{-6}$ m^2/s, characteristic for a basalt, the velocity of the displacement of the temperature maximum varies from 19 cm/year at $\rho = 0$ to zero at large distances

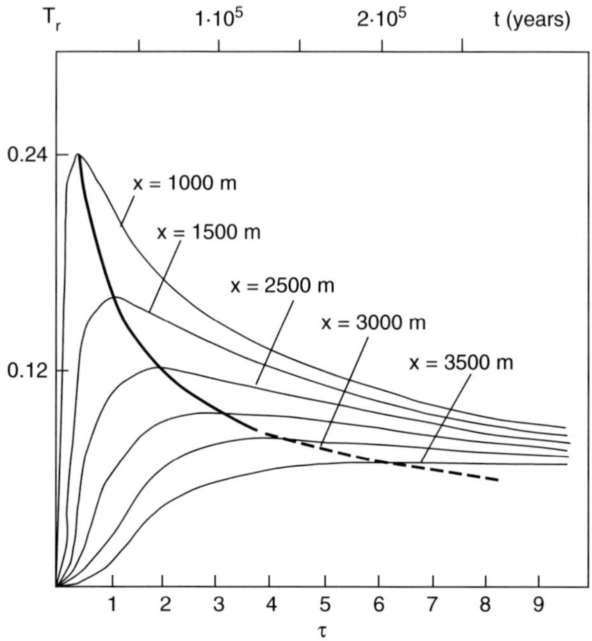

Fig. 8.2 Relative time variations of temperature (T_r) in an exocontact zone ("thermal waves", shown by solid lines) at distances from 500 to 3000 m from the contact of the magmatic body with the surrounding medium (at 1000 to 3500 m from the origin of coordinates). The magmatic body is 1 km wide. In every point of the exocontact zone, a temperature medium occurs with a time-delay. The larger distance between the point and the magmatic body, the greater the delay. Temperature maxima are connected with a heavy line

with an average of a few centimeter per year. Further, it should be noted that "the temperature pulse" in every point of the surrounding space is delayed in time. Hence, considering the behaviour of radiogenic substances (^{40}Ar, ^{87}Sr etc.) in an exocontact zone, not only temperature activation, but its time-delay are to be accounted for. This may stipulate the fact that the apparent age of a mineral in a point of the host rocks can be lower than that of the instant of intrusion. These features are illustrated in Fig. 8.2.

8.1.2 *Radiogenic Argon Retention in a Mineral under Varying Temperatures*

Argon kinetics in the temperature field of the magmatic body is determined by the diffusion equation. The diffusion coefficient D is a function of temperature, which is, in turn, a function of time. Therefore, the diffusion coefficient D becomes a function of time.

Consider a mineral appearing at $t = 0$, having a concentration of ^{40}K at this instant equal to $^{40}K_0$. We suggest that the coefficient D was equal to zero during the whole time interval before t_0 and hence the concentration of ^{40}Ar at t_0 was:

$$^{40}Ar_0 = \frac{\lambda_e}{\lambda}(1 - e^{-\lambda t_0})^{40}K_0 \tag{8.17}$$

At $t > t_0$ the mean concentration of argon is $\overline{^{40}Ar}$. We suggest that in the time interval from t_0 to t, diffusion differed from zero and was a function of time, $D = D(t)$. Here we assume the validity of Arrhenius law (8.2).

During the time span $t_0 - t$, owing to radioactive transformation, there had appeared some argon:

$$\Delta^{40}Ar = \int_{t_0}^{t} \lambda_e {}^{40}K_0 e^{-\lambda\tau} d\tau \tag{8.18}$$

In the same time interval, the mineral had lost through diffusion a quantity of radiogenic argon:

$$\begin{aligned} {}^{40}Ar_0 + \Delta^{40}Ar - \overline{{}^{40}Ar} = {}^{40}Ar_0 \left(1 - \bar{N}\left(\frac{1}{h^2}\int_{t_0}^{t} D(t)dt\right)\right) \\ + \int_{t_0}^{t} \lambda_e {}^{40}K_0 e^{-\lambda\tau}\left(1 - \bar{N}\left(\frac{1}{h^2}\int_{\tau}^{t} D(t)dt\right)\right) d\tau \end{aligned} \tag{8.19}$$

This expression is a generalisation of a formula for Ar-losses obtained by Amirkhanov et al. (1961) for the case of variable diffusion and large time intervals. Here (one-dimensional case):

$$\bar{N} = \frac{8}{\pi^2}\sum_{\nu=0}^{\infty} \frac{1}{(2\nu+1)^2} e^{-(2\nu+1)^2\pi^2 t} \tag{8.20}$$

Transform the Eq. (8.19):

$${}^{40}Ar_0 + \Delta^{40}Ar - \overline{{}^{40}Ar} = {}^{40}Ar_0\mu_1 + \Delta^{40}Ar\mu_2 \tag{8.21}$$

Put ${}^{40}K_0 = {}^{40}Ke^{\lambda t}$ in Eqs. (8.17–8.18) to obtain:

$$\Delta^{40}Ar = \frac{\lambda_e}{\lambda} {}^{40}K(e^{\lambda(t-t_0)} - 1) \tag{8.22}$$

and

$${}^{40}Ar_0 = \frac{\lambda_e}{\lambda} {}^{40}K(e^{\lambda t} - e^{\lambda(t-t_0)}) \tag{8.23}$$

For μ_1 and μ_2, there are the following equations:

$$\mu_1 = 1 - \bar{N}\left(\frac{1}{h^2}\int_{t_0}^{t} D(t)dt\right) \tag{8.24}$$

$$\mu_2 = \frac{\lambda \int_{t_0}^{t} e^{\lambda(t-\tau)}\left(1 - \bar{N}\left(\frac{1}{h^2}\int_{\tau}^{t} D(t)dt\right)\right) d\tau}{e^{\lambda(t-t_0)} - 1} \tag{8.25}$$

In the latter, both the numerator and the denominator become zero at $t = t_0$, which may cause troubles in calculation. Therefore, we take the mean value of the expression under the integral-sign in the numerator. Dividing both the numerator and the denominator by $t - t_0$, we obtain:

$$\mu_2 = \frac{\int_{t_0}^{t} e^{\lambda(t-t_0)} \left(1 - \bar{N}\left(\frac{1}{h^2}\int_{\tau}^{t} D d\tau\right)\right) d\tau}{t - t_0} \times \frac{\lambda(t - t_0)}{e^{\lambda(t-t_0)} - 1} \tag{8.26}$$

where $\lambda(t - t_0)/(e^{\lambda(t-t_0)} - 1) = z/(e^z - 1) = \sum_{k=0}^{\infty}(B_k/k!)z^k$, B_k are Bernoulli numbers and $z = \lambda(t - t_0)$

8.2 Geochronological Application

Incompletely reset K–Ar and Rb–Sr isotope systems are widely known in areas of faulting, magmatism, and hydrothermal activities (Dickin 1997; Dunlap and Fossen 1998; Reddy and Potts 1999 and references therein). In complex conditions, it is often unclear, whether an isotope system distortion has been produced by heat of an magmatic body or by hydrothermal anomaly (Matheney et al. 1990).

A theoretical solution relevant to the case of thermal magmatic overprints means computing an effect of heat produced by an intruded magmatic body of a width d and an initial temperature T_0 on a mineral situated at a distance x at a geologic time instant t since crystallisation of host rocks. The true ages of the intrusive and the host rocks are Δt and $t + \Delta t$, respectively.

In the K–Ar isotope system, an apparent age ($t_{apparent}$) is determined by means of the formula:

$$t_{apparent} = \frac{1}{\lambda} \ln\left(\frac{\lambda}{\lambda_e}\frac{^{40}Ar_{measured}}{^{40}K_{measured}} + 1\right) \tag{8.27}$$

where $^{40}Ar_{measured}$ is the concentration of radiogenic argon in a mineral of any point in the exocontact zone. Rewriting Eq. (8.21), we obtain:

$$^{40}Ar_{measured} = {}^{40}Ar_0\mu_1 + \Delta^{40}Ar\mu_2 \tag{8.28}$$

Similar to Eqs. (8.22) and (8.23), we write:

$$\Delta^{40}Ar = \frac{\lambda_e}{\lambda}{}^{40}K(e^{\lambda\Delta t} - 1) \tag{8.29}$$

and

$$^{40}Ar_0 = \frac{\lambda_e}{\lambda}{}^{40}K(e^{\lambda(t+\Delta t)} - e^{\lambda\Delta t}) \tag{8.30}$$

Further:

$$\mu_1 = 1 - \bar{N}\left(\frac{1}{h^2}\int\limits_{t}^{t+\Delta t} D(t)dt\right) \tag{8.31}$$

$$\mu_2 = \frac{\lambda \int\limits_{t}^{t+\Delta t} e^{\lambda(t+\Delta t-\tau)}\left(1 - \bar{N}\left(\int\limits_{\tau}^{t+\Delta t} D(t)dt\right)\right)d\tau}{e^{\lambda\Delta t} - 1} \tag{8.32}$$

From (8.27) and (8.28), we obtain the formula required:

$$t_{apparent} = \frac{1}{\lambda}\ln\left((e^{\lambda t} - 1)e^{\lambda\Delta t}(1 - \mu_1) + (e^{\lambda\Delta t} - 1)(1 - \mu_2) + 1\right) \tag{8.33}$$

This formula allows us to determine the apparent age of a mineral for any point of the exocontact zone given. The quantities $1 - \mu_1$ and $1 - \mu_2$ are complicated functions of argon diffusion parameters, entering in the Arrhenius law (8.2), as well as of thermal parameters: the width of the magmatic body, the initial temperature of the intruded magma, the temperature conductivity of the medium. They are quantities, synthesising thermal and argon diffusion properties. The formulas (8.31–8.33) are valid for any other radiogenic systems as well. We have put proper values on computer programs, allowing us obtain numerical values for any parameter values. Similar formulas for two-dimensional and three-dimensional cases are given in the Appendix.

Figure 8.3a, b shows, respectively, the exocontact aureole for the Rb–Sr model ages of biotite and K–Ar ages of hornblende obtained by Hart (1964) as compared with theoretical lines computed by means of formulas (8.31–8.33) (solid lines). Both cases were calculated assuming the Eldora stock to be a slab with the dimensions 7.5×15 km. The theoretical lines approximate the experimental data quite satisfactorily. Note that the biotite studied is coarse grained and its Rb–Sr ages are expressed by a smooth theoretical curve. In contrast, the hornblende crystalline particles vary widely in size and result in abrupt changes of apparent K–Ar ages. This is to be accounted for by the factor D_0/h^2 calculated for every point separately.

Another example of a partial isotope reset due to conductive heat impact is host rocks of the Eldjurtinskii granite massif, Caucasus, Russia (Chernyshev et al. 2000). From Rb–Sr isochron dating on whole rock samples and mineral separates, the massif has been intruded into Late Paleozoic host rocks (plagiogranites) during the latest Pliocene at 2.12 ± 0.10 Ma (Kostitsyn and Kremenetsky 1995). At a distance of 200–300 m from the Pliocene intrusion, biotites, muscovites, and potassium feldspars show K–Ar ages of 2.2–2.1 Ma consistent with its timing. Therefore, near the magmatic body, the K–Ar isotope system was reset completely. The heat impact was identified, however, in a lesser degree at a distance of 300–1000 m. Apparent mineral ages increased consequently up to 8.9 Ma. A study of the Rb–Sr isotope system from scarns, sampled within the same exocontact zone, demonstrated garnet–pyroxene isochron ages as young as 5 Ma at a distance of 2–3 m and increasing ages up to 300–260 Ma at greater distances.

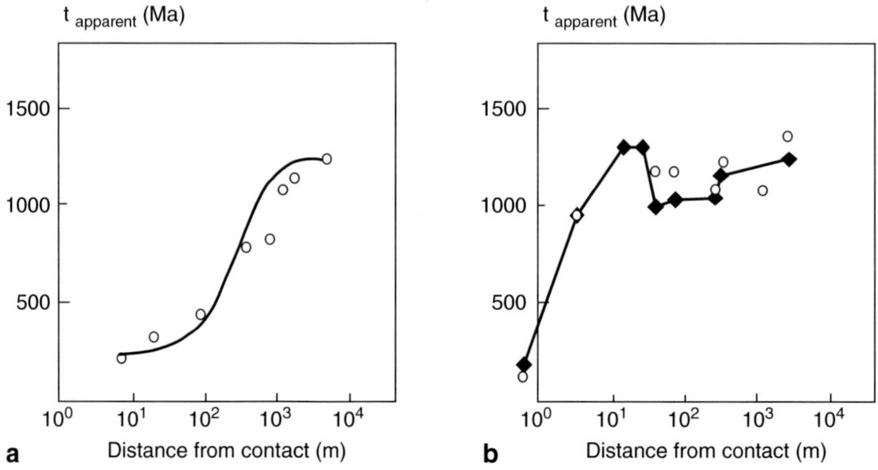

Fig. 8.3 Distortion pf Rb–Sr model ages of biotite (**a**) and K–Ar ages of hornblende (**b**), in the exocontact aureole of the 53 Ma Eldora stock (7.5 × 17.0 km). The host rocks are of about 1250–1200 Ma (Berger 1975). Open circles are the measured values (Hart 1964), solid lines are the theoretical values computed by formulas (8.28–8.30). The angular shape of the line in (**b**) is due to irregular variations of the hornblende grain sizes h reflected in the computed $Fo = Dt/h^2$. Note that the x-axis is in logarithmic scale

8.3 Summary: Theory of Diffusion and Its Application for Study of Exocontact Processes

In this chapter, mathematical expressions were developed to consider accumulation of radiogenic isotopes in an exocontact zone of a magmatic intrusion. It was found that a temperature maximum propagates from a thermal source at a certain rate, decreasing in its absolute value. Hence, there occurs a time-delay of thermal impact on the exocontact rocks. The later the thermal impact, the more distance from the source of heat. It means that apparent ages obtained for the same mineral species sampled in variable distance from such a heat source will differ not only due to less pronounced thermal impact at larger distance, but also because of the time-delay of this impact.

In many natural cases, the time-delay of thermal impact is very short in comparison to ages of studied minerals, and hence cannot be documented by means of present-day analytical techniques. An impact of a long-living thermal field derived from large intrusive massifs of recent ages is expected to affect notably mineral ages in its exocontact zone.

It is noteworthy that the temperature conductivity a is a function of temperature, as well as the coefficient of argon diffusion D. Timing of heating in an exocontact zone of a magmatic body actually seems to be affected by the latent heat of crystallisation as well. A temperature-dependent a will tend to compress the thermal aureole around a magmatic body, whereas the latent heat will tend to expand it, both the effects quantitatively compensating each other. So, in the first-step theory, these effects may be omitted.

Finally, our mathematical expressions approximate fairly well the age–space relation in the exocontact zone of the Eldora stock observed by Hart (1964). That coincidence supports the developed theory.

8.4 Appendix: Two- and Three-Dimensional Cases

Equation of heat-conduction:

$$\frac{\partial T_r}{\partial t} = a\Delta T_r$$

where $\Delta = (\partial/\partial x^2) + (\partial/\partial y^2)$
A not-reduced solution:

$$T = \frac{d_1 d_2 T_0}{4\pi a\sqrt{\rho+\rho_1}\sqrt{\rho+\rho_2}} e^{-\left(\frac{x^2}{4a(\rho+\rho_1)}+\frac{y^2}{4a(\rho+\rho_2)}\right)} \text{ with}$$

$$\rho_1 = \frac{d_1^2}{4\pi a} \text{ and } \rho_2 = \frac{d_2^2}{4\pi a}$$

The reduced formula:

$t_1 = (a/d_1^2)\rho$–reduced time on x-axis $t_2 = (a/d_2^2)\rho$—time reduced on y-axis. $t_2 = (d_1^2/d_2^2)t_1$

Reduced coordinates:

$$\xi = \frac{x}{2d_1} \quad \text{and} \quad \eta = \frac{y}{2d_2}$$

Final equation:

$$\frac{T}{T_0} = \frac{1}{4\pi\sqrt{t_1+\frac{1}{4\pi}}\sqrt{t_2+\frac{1}{4\pi}}} e^{-\left(\frac{\xi^2}{t_1+\frac{1}{4\pi}}+\frac{\eta^2}{t_2+\frac{1}{4\pi}}\right)}$$

Quite similar, the three-dimensional case will be:

$$\frac{T}{T_0} = \frac{1}{(4\pi)^{3/2}\sqrt{t_1+\frac{1}{4\pi}}\sqrt{t_2+\frac{1}{4\pi}}\sqrt{t_3+\frac{1}{4\pi}}} e^{-\left(\frac{\xi^2}{t_1+\frac{1}{4\pi}}+\frac{\eta^2}{t_2+\frac{1}{4\pi}}+\frac{\zeta^2}{t_3+\frac{1}{4\pi}}\right)}$$

with $\xi = (x/2d_1); \eta = (y/2d_2); \zeta = (z/2d_3)$

$$\rho_1 = \frac{d_1^2}{4\pi a}; \qquad \rho_2 = \frac{d_2^2}{4\pi a}; \qquad \rho_3 = \frac{d_3^2}{4\pi a}$$

$$t_1 = \frac{a}{d_1^2}\rho_1; \qquad t_2 = \frac{a}{d_2^2}\rho_2; \qquad t_3 = \frac{a}{d_3^2}\rho_3$$

Chapter 9
Diffusion in a Laplace Regime

Rocks and minerals might occur in conditions of elevated temperatures favorable for accelerated diffusion processes. This results in accumulation of radiogenic isotopes in rocks and K–Ar isotope system in the Laplace diffusion losses. It is important to define boundary conditions of such an isotope system. As diffusion losses increase, the rate of accumulation of daughter substance approaches zero $\left(\partial^{40}Ar/\partial t = 0\right)$ and the Fick equation turns to the Laplace equation. Unlike convective mixing, resulting in complete homogenization of all isotope systems, the Laplace regime may affect some radiogenic isotopes, while the others obey the Law of Radioactivity of Rutherford–Soddy without diffusion losses.

9.1 Deduction of the Laplace Age Equation

From the diffusion theory on radiogenic nuclides in crystal host structures, it is known (Rasskazov et al. 2000b), that the absolute concentration of a substance—inclusion reaches a maximum in time, and then converges to zero. The radioactivity shows itself as a pulse in infinite time, whereas diffusion is "immanent". The locus of a maximum is determined by relation D/λ (D – coefficient of diffusion, λ – decay constant for a parent radioactive substance) (Fig. 9.1).

This takes place if the quantity of daughter substance is compared with the initial quantity of parent substance. If it is compared with its present-day quantity, the daughter/parent ratio of the substance without losses will monotonically increase, but with losses will exhibit a curve, asymptotically approaching a horizontal line (Fig. 9.2). The latter evolution of an isotope system is characterized as the Laplace regime, and the respective apparent age is referred to as the Laplace age.

Find the solution of the non-uniform equation of accumulation–diffusion:

$$\frac{\partial C(x,t)}{\partial t} = D\frac{\partial^2 C}{\partial x^2} + \lambda_k{}^{40}K$$

S.V. Rasskazov et al., *Radiogenic Isotopes in Geologic Processes*,
DOI 10.1007/978-90-481-2999-7_9,

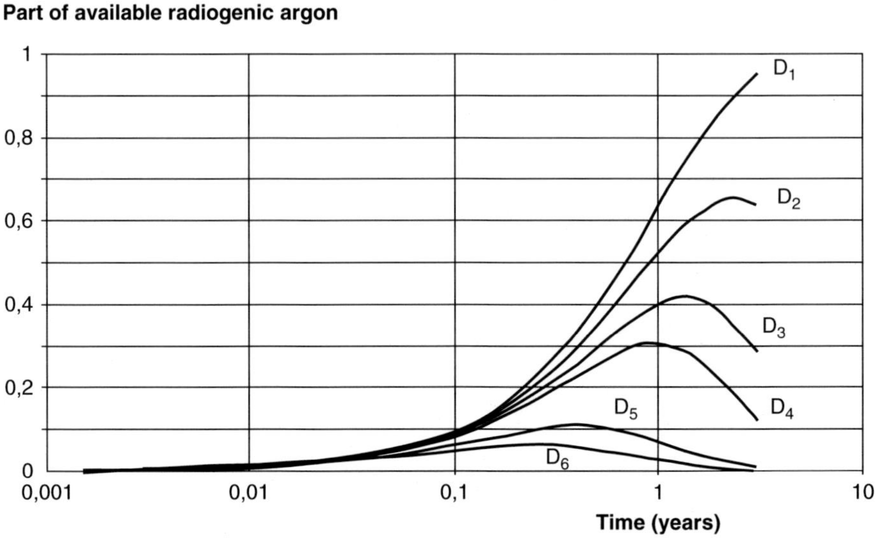

Fig. 9.1 Temporal dependence of available radiogenic argon on diffusion coefficients (D_1–D_6). Absolute concentrations of radiogenic ^{40}Ar have a maximum in time. Radioactivity is a temporary pulse, whereas diffusion acts permanently. The fraction of accumulated argon decreases with increasing diffusion $D_1 < D_2 < D_3 < D_4 < D_5 < D_6$

that corresponds to the Laplace regime. It has a form $C(x,t) = f(x) \cdot e^{-\lambda \cdot t}$, where:

$$f(x) = \frac{\lambda_k}{\lambda} {}^{40}K_0 \left(\frac{\cos\sqrt{\frac{\lambda}{D}}(x - \frac{h}{2})}{\cos\frac{1}{2}\sqrt{\frac{\lambda h^2}{D}}} - 1 \right)$$

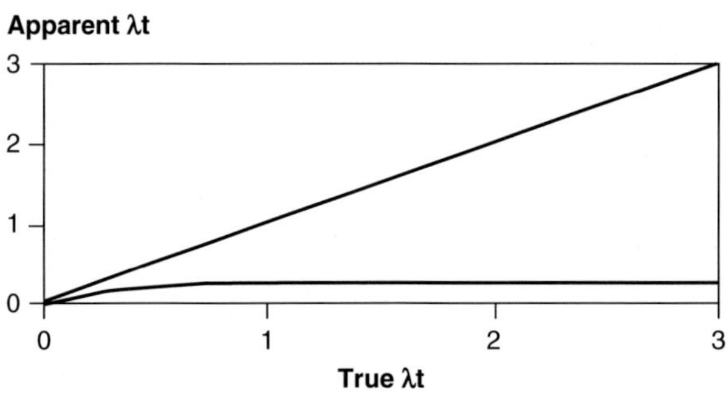

Fig. 9.2 Correlation between apparent and true values of a parameter λt in systems with and without losses of a daughter substance. In a system without losses, the ratio of the concentration of daughter substance to the present-day concentration of parent substance will monotonically increase (*upper curve*) and in a system with losses, it will tend to an asymptotic value (*lower curve*, Laplace regime)

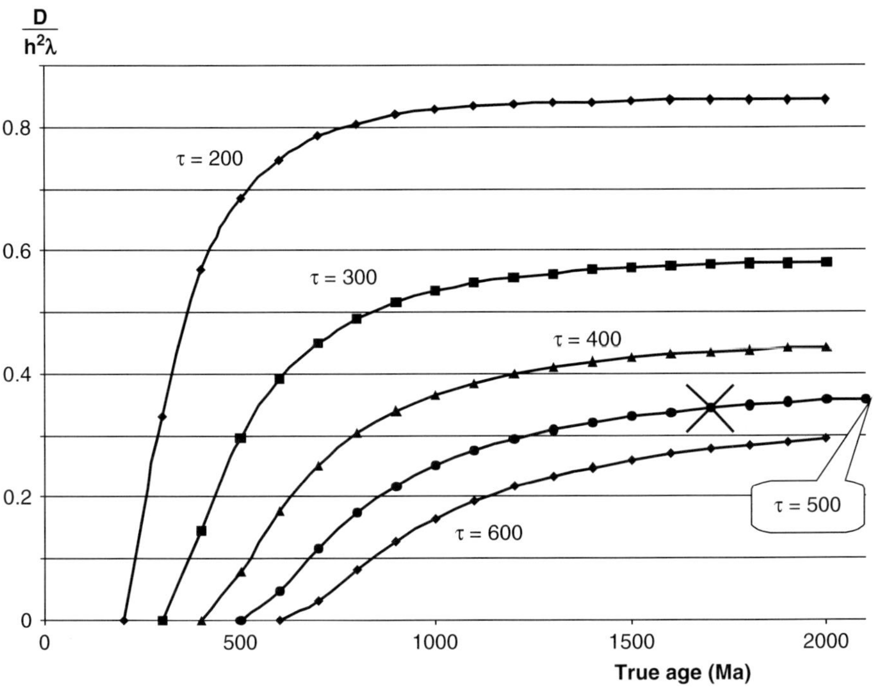

Fig. 9.3 "Isochrons" of Laplace ages in coordinates $D/(h^2\lambda)$ versus true age. On the curve corresponding to an apparent age of 500 Ma and a true age of 1700 Ma, the point is marked by an oblique cross

The average concentration will be:

$$\bar{f} = \frac{\lambda_k}{\lambda}{}^{40}K_0\left(\frac{tgz}{z} - 1\right)$$

where $z = (1/2)\sqrt{(\lambda h^2)/D}$.

To define the Laplace age, use a formula $e^{\lambda t_{Lapl}} = tgz/z$.

Figure 9.3 demonstrates apparent ages as functions of true ages and the diffusion parameter $D/(h^2\lambda)$. The family of curves on the diagram enables us to define the parameter from values of apparent (measured in K–Ar isotope system) and true ages.

9.2 Geologic Examples

Geologic situations of continuous accommodation of rocks in conditions of elevated temperatures are widespread. We have found, for instance, systematically distorted $^{40}Ar/^{39}Ar$ ages of Cretaceous basalts sampled from deep drill holes within the Tatar Strait and its recently uplifted shores, but well-pronounced plateaus for coeval basalts erupted in the adjacent mainland. This distortion may have been caused by elevated temperatures occurring in volcano-sedimentary sequences of actively subsided basins.

The evolution in a Laplace regime is considered from data on K–Ar and Rb–Sr isotope systems for metamorphic and magmatic rocks of the western shore of Lake Baikal. An initial point for evolution of the isotope systems is defined by U–Pb zircon age, obtained for charnokites from the Sharyzhalgai Complex (Aftalion et al. 1991). It has been suggested that conditions similar to those in the middle and upper parts of the crust in the Basin and Range Province of the Western USA were dominant in these rocks at ca. 1880–1860 Ma and were followed by intrusions of granites. Comparative (within error) Rb–Sr-isochron age 1823 $\pm$ 61 Ma with MSWD = 0.82 was obtained for an unaltered gabbronorite from the Ozerskii basic intrusion (Grudinin et al. 2007). Crystal slates and granites from the western shore of South Baikal showed, however, whole rock Rb–Sr isochron ages in an interval of 1700–1630 Ma and K–Ar apparent ages for the same samples in an interval of 500–450 Ma (Brandt et al. 1978).

Mineral phases of granites, sampled 50–100 km to the north of the South Baikal shore line, were subjected to differential losses of radiogenic ^{40}Ar (Didenko et al. 2003). A low-K mineral phase (amphibole) showed six high-temperature steps and exhibited 60% of ^{39}Ar with a plateau age 1853.9 $\pm$ 3.8 Ma, consistent with the U–Pb zircon age. A high-K mineral phase (biotite), however, demonstrated plateau-like steps with an underestimated apparent integrated age of 1741.1 $\pm$ 11.7 Ma (90% of ^{39}Ar). Diffusion losses of argon from feldspars were more sufficient.

The diagram of Fig. 9.4 shows results of age calculations for evolution of an Rb–Sr isotope system in conditions of radioactive accumulation without losses, at least since 1700 Ma, and hypothetical evolution of K–Ar isotope system in the Laplace regime in one- and two-stage models. The former model assumes that the K–Ar age of 500 Ma was derived by the average value of diffusion parameters $(D/h^2\lambda) = 0.343$ since 1700 Ma. The latter model takes into account a change in a mode of evolution in the geologic past. It implies high diffusion of radiogenic argon in the Laplace regime at elevated temperatures before exposure of rocks on a terrestrial surface. Fission track analysis on apatite (Van der Beek et al. 1996) constrains the transition to temperatures less than 120°C in rocks of the Primorskii range (Olkhon island area) in a time interval of 140–120 Ma. This transition resulted in a notable decrease of radiogenic argon diffusion with transition of the K–Ar system into evolution without losses. A flatter slope of the evolution curve of apparent ages in Fig. 9.4 means a diffusion parameter $(D/h^2\lambda) > 0.343$.

The two stage model of the Laplace regime satisfactorily describes evolution of a K–Ar isotope system in rocks composed of amphiboles and micas (i.e. minerals with high energy of activation for radiogenic argon). However, some mineral phases (for example, potassium feldspar with pertites etc.) easily lose argon even at low temperature; therefore rocks with these minerals never reach complete closure of an isotope system.

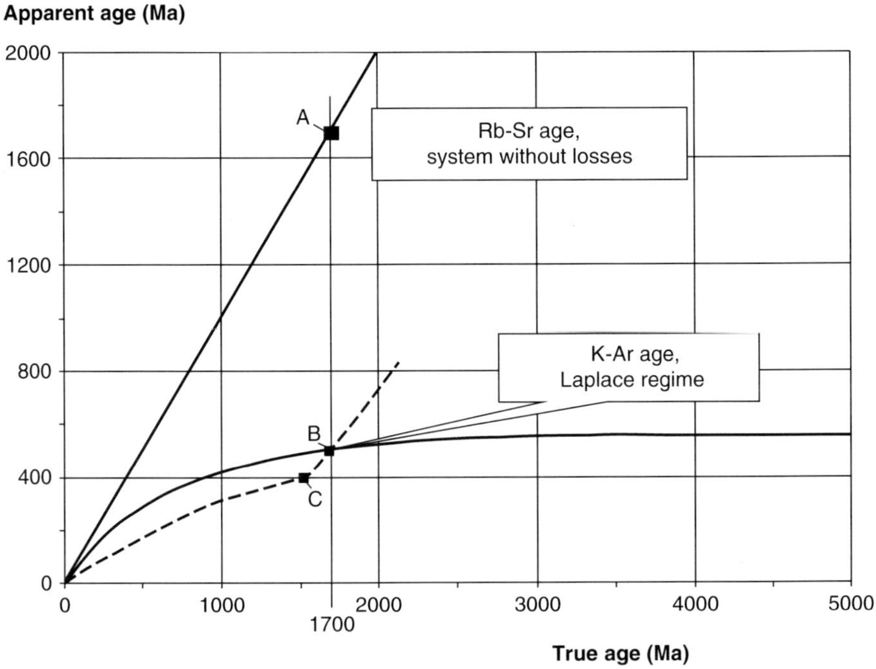

Fig. 9.4 Time evolution of apparent ages for granites and crystalline schists exposed in the western shore of Lake Baikal. The origin of coordinates corresponds to a time instant 1700 Ma and the value 1700 Ma on the axis of abscissa—to the present time. In point A, the apparent and the true time coincide. Point B lies on a hypothetical curve of the Laplace age, values of which have reached today 500 Ma. In the geologic future, this value will increase. A broken line shows a two-stage evolution of a K–Ar system with a transition in point C from a regime with losses to the normal accumulation of radiogenic isotopes (see text)

9.3 Summary: Estimation of the Diffusion Parameter $D/h^2\lambda$

Diffusion coefficients D in minerals and rocks are usually determined in laboratories. Another opportunity for their estimation has been demonstrated here by study of long-lived geologic processes. It has been suggested that the U–Pb and Rb–Sr isotope systems of granites and crystalline schists, currently exposed at the western shore of lake Baikal, show true ages due to evolution in a mode of radiogenic isotope accumulation without losses at least since 1700 Ma, while the K–Ar isotope system yields distorted ages because of radiogenic argon losses in a Laplace regime, probably, changed to radiogenic argon accumulation without losses at 140–110 Ma. Evolution of a K–Ar isotope system in the Laplace regime yields an average value of the diffusion parameter $(D/h^2\lambda) \geq 0.343$.

Understanding behavior of radiogenic isotopes in geologic time requires us to proceed from (1) the Law of Radioactivity of Rutherford–Soddy, determining speed of parent-daughter transition of substances, (2) the second Fick law, determining dependence of particle concentrations on time and coordinates, and (3) Arrhenius law, giving diffusion dependence on temperature. Pressure and temperature affect diffusion in opposite ways. We have not estimated numerically the relations between these factors, but speculate that a situation is quite possible, when the Laplace regime occurs in conditions of elevated temperatures in the upper crust, while the high lithostatic pressure promotes an accumulation of radiogenic isotopes without losses in deeper levels.

Chapter 10
Early Earth

By U–Pb, Rb–Sr, and K–Ar methods, an age interval from 5.0 to 2.8 Ga was estimated for the oldest rocks in the Earth in the late 1950s. Model ages of 5.3–4.5 Ga were calculated from abundances of ^{40}K (Starik 1961) and at ca. 4.6 Ga from mean concentrations of ^{40}K in crust and ^{40}Ar in atmosphere (Brandt 1965; Schwartzman 1973). The latter age was substantiated using more precise dating of meteorites by U–Pb, Lu–Hf, Re–Os, and other methods. Meteorites were unequivocally specified as the oldest subjects of the Solar system (e.g. Shukolyukov and Begemann 1996).

10.1 Initial Processes in the Solar System (4.57–4.54 Ga)

10.1.1 Classification of Meteorites

Meteorites are subdivided into stony, iron, and iron–stony groups from occurrence of silicate and/or metal components. The further division of meteorites is based on textural–structural features, mineral and chemical compositions (Table 10.1).

Stony meteorites are subdivided into classes of chondrites and achondrites by presence or absence of spherical inclusions (chondrules). The interest in chondrites is promoted by occurrence of micron and sub-micron size particles of diamonds, corundum, spinel, carbide of silicon, and other compounds referred to interstar dust, probably, crystallized before generation of the Solar system (Nittler 2003).

In terms of chemical composition, carbonaceous chondrites are closest to the Sun and, therefore, might represent primeval substance of the Solar system. They contain hydrous minerals and carbon, but have no native iron. The carbonaceous chondrite Allende that belongs to the subgroup CV is of special interest. It includes refractory Ca–Al inclusions, probably, crystallized from a gas–dust cloud just before or after generation of the Solar system.

Ordinary chondrites with high-, low- and ultra low-iron contents exceed 85% of all meteorites. Enstatite chondrites contain nickel iron formed in reduced conditions.

Achondrites are subdivided into Ca-rich and Ca-poor groups. Their division into subgroups is based on additional chemical characteristics. The subgroups are usually called by names of meteorites. For example, meteorites similar to the meteorite Angra

S.V. Rasskazov et al., *Radiogenic Isotopes in Geologic Processes*,
DOI 10.1007/978-90-481-2999-7_10, © Springer Science+Business Media B.V. 2010

Table 10.1 Radiogenic isotope methods applied for dating of meteorites

Type	Class	Group		Subgroup	Method
Stony	Chondrite	High-Fe (H) Low-Fe (L) Ultra low-Fe (LL)	Ordinary	EH, EL CL, CM, **CV**, CR, CO, CK, CH	U–Pb Al–Mg Hf–W
		Enstatite Carbonaceous			
	Achondrite	Ca-poor (0–3 wt. %)	Primitive	Ureilite	
			Evolved	Aubrite **Diogenite** Chassigniate	U–Pb Sm–Nd Hf–W Mn–Cr
		Ca-rich (5–25 wt. %)	Evolved	**Angrite** **Eucrite** Shergottite Howardite Nakhlite	
Iron	Octahedrite	**IA**, **IB**, IC, **IIB**, IIC, IID, IIE, **IIIA**, **IIIB**, **IIIC**, **IIID**, IIIE, IIIF, **IVA**			Re–Os Hf–W
	Hexaherdite	**IIA**			
	Ataxite	**IVB**			
Iron–stony	Pallasite Mesosiderite Siderophyre Lodranite	First, second, third			No data

The classification of meteorites is modified after Sobotovich and Semenenko (1984). Groups and subgroups discussed in the text are shown by bold font. CV – subgroup of carbonaceous chondrites with Ca–Al–Ti-rich inclusions. Diogenites – subgroup of Ca-poor achondrites with predominating hypersthene. Angrites and eucrites – respectively, augite and pigeonite-anorthite subgroups of Ca-rich achondrites. Representative chemical groups: IA – Canyon Diablo (Ni 7 wt. %, Ga 82 ppm, Ge 327 ppm), IB – Shrewsbury (Ni 8.6 wt. %, Ga 63 ppm, Ge 204 ppm), IIA – ALH A81013 (Ni 5.5 wt. %, Ga 58 ppm, Ge 192 ppm), IIB – Bilibino (Ni 6.4 wt. %, Ga 60 ppm, Ge 176 ppm), IIIAB – ALH 84165 (Ni 8.1 wt. %, Ga 20 ppm, Ge 40 ppm), IIICD – Seeläsgen (Ni 6.7 wt. %, Ga 103 ppm, Ge 493 ppm), IVA – Alvord (Ni 8.8 wt. %, Ga 2 ppm, Ge 0.13 ppm), IVB – EET 83230 (Ni 16.4 wt. %, Ga 1.3 ppm, Ge 0.075 ppm) (Choi et al. 1995; Wasson et al. 1989, 1998).

dos Reis are comprised into a subgroup of angrites (Sobotovich and Semenenko 1984). Achondrites are subdivided genetically into primitive and evolved varieties. From close values of reflection spectra, howardites, eucrites, and diogenites are hypothesized to be fragments from the large asteroid Vesta. Due to the young age and specific chemical compositions, shergottites, nakhlites, and chassigniates are examined as fragments from Mars (Shukolyukov 2003). Some evolved achondrites could originate from the Moon (not included in Table 10.1).

Iron meteorites are considered as core fragments of destroyed asteroids. Classes are determined by textural features, mineral compositions, and Ni concentrations; their chemical groups by concentrations of Ga, Ge, elements of platinum group and other elements; their genetic groups by presence or absence of silicates (respectively, magmatic and non-magmatic). Relationships between textural classes and groups are obscure. For example, the magmatic group includes meteorites of chemical groups IIA, IIB, IIIA, and IIIB, which, in turn, belong to different classes. Involvement of new analytical data inevitably leads to modifications in the classification of iron meteorites (Choi et al. 1995; Wasson et al. 1989, 1998).

Iron–stony meteorites contain approximately equal proportions of metal and silicate components showing similarities with some groups of iron or stony meteorites. Meteorites of this type might represent mantle fragments of asteroids.

10.1.2 Dating of Meteorites

At first approximation, age of meteorites was defined by measurements of Pb isotope ratios in two iron, three stony meteorites, and modern terrestrial sea sediments (Patterson 1956). In coordinates of Pb isotope ratios, data points were distributed along a straight line corresponding to the age of 4.48 Ga (recalculated relative to the currently accepted decay constants of ^{235}U and ^{238}U). Subsequent measurements of Pb isotope ratios in the same meteorites using other procedures of chemical processing and sophisticated equipment yielded a slightly older age of 4.55 Ga (Allègre et al. 1995).

Basic methods used for dating of meteorites are listed in Table 10.1. In the age determinations, problems arise because of meteorite contamination by terrestrial lead and considerable natural loss of lead by meteorite minerals (Allègre et al. 1995; Tera et al. 1997). The most reliable crystallization ages have been obtained by measurements of isotope ratios of lead in minerals and whole rocks with presentation in isochron coordinates $^{207}Pb/^{204}Pb - ^{206}Pb/^{204}Pb$, $^{207}Pb/^{206}Pb - ^{204}Pb/^{206}Pb$, and $^{206}Pb/^{207}Pb - ^{204}Pb/^{207}Pb$ (Patterson 1956; Tera et al. 1997) as well as by measurements of model $^{207}Pb/^{206}Pb$ ages for minerals (Tatsumoto et al. 1973; Göpel et al. 1985; Allègre et al. 1995).

A sequence of meteorite generation has been defined in geochronological studies with accuracy of ~1 m.y. The oldest age in the Solar system of 4566 +2/−1 Ma has been measured by the Pb–Pb method on Ca–Al-rich inclusions from the meteorite Allende. The model age is designed for samples with extremely high concentrations of radiogenic isotope ^{206}Pb (Allègre et al. 1995) (Fig. 10.1).

In these measurements, the analytical error was specified without taking into account errors in decay constants of uranium isotopes. Crystallization of the inclusions in the early Solar system or at the stage of nebula condensation just before isolation of the latter was inferred from chemical and mineral compositions of the Ca–Al-rich inclusions as well as from evidence of presence of a radioactive isotope ^{26}Al (with

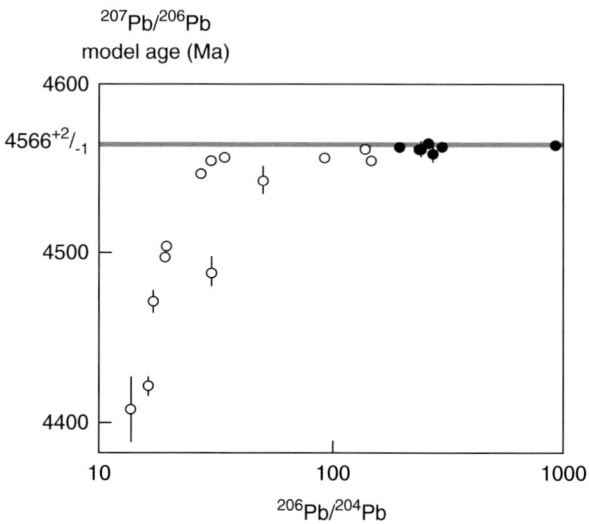

Fig. 10.1 Calculated model $^{207}Pb/^{206}Pb$ age versus $^{206}Pb/^{204}Pb$ isotope ratio in Ca–Al inclusions of the chondrite Allende (Allègre et al. 1995). With increase of the radiogenic component of lead, when $^{206}Pb/^{204}Pb > 200$, all model ages become statistically undistinguishable. Points omitted and involved in age calculations are shown by *open* and *filled circles*, respectively. A *vertical line* denotes the error, when it exceeds a size of a symbol

the period of half-decay 0.72 m.y.) at the moment of crystallization (Dickin 1997; Grossman et al. 2000; Wood 2004).

The meteorites LEW86010 and Angra dos Reis possess primary mineral compositions. The former yielded a model Pb–Pb age of 4557.84 ± 0.52 Ma and mineral Sm–Nd isochron age of 4553 ± 34 Ma, the latter a model Pb–Pb age of 4557.80 ± 0.42 Ma (Lugmair and Galer 1992) (analytical errors without uncertainty in the decay constants of ^{235}U and ^{238}U). These meteorites have similar relative Pu–Xe ages (Shukolyukov and Begemann 1996).

10.1.3 Significance of Short-Lived Radionuclides

Impressive results have been obtained on ^{182}Hf–^{182}W and ^{53}Mn–^{53}Cr systematics of stony meteorites and samples from planets of the terrestrial group with reconstruction of generation history of the proto-Earth as short as 30 m.y. (Kleine et al. 2002; Lugmair and Shukolyukov 1998; Yin et al. 2002). A brief review of these data hereafter is accompanied by discussion of a hypothesis on a longer (~100 Ma) Earth history previously inferred from calculations of Xe ages (Allègre et al. 1995; Zhang 2002).

Eight eucrites and two diogenites yielded a distinct isochron in coordinates $\in(^{53}Cr) - {}^{55}Mn/^{52}Cr$ with a slope of a line (i.e. $^{53}Mn/^{55}Mn$) $(4.7 \pm 0.5) \cdot 10^{-6}$ (Fig. 10.2). This isochron marked timing of the last differentiation in the mantle of the asteroid Vesta (Fig. 10.3). Mineral isochrons for the angrites LEW86010 and Angra dos Reis differed by the more gentle slope $(1.25 \pm 0.07) \cdot 10^{-6}$ and revealed age difference of 7.1 ± 0.8 m.y. between the angrite generation and mantle differentiation in the asteroid Vesta that corresponded to absolute age of 4564.8 ± 0.9 Ma (all errors are given without uncertainty in the decay constants of the uranium isotopes).

From all possible variations of initial values $\in(^{53}Cr)$, the age of the Solar system was estimated in a range from 4568 to 4571 Ma (Lugmair and Shukolyukov 1998).

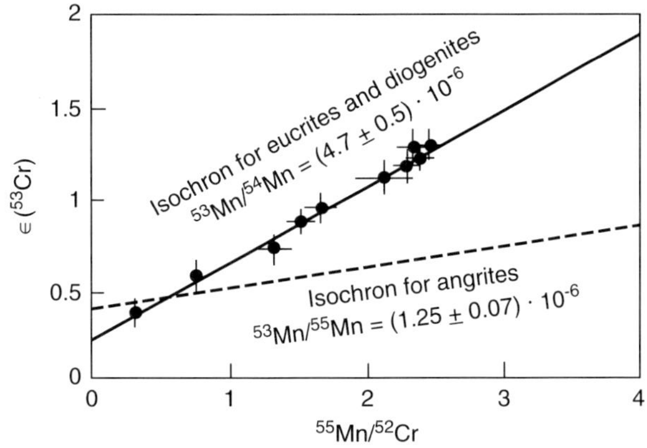

Fig. 10.2 A Cr–Mn isochron for eucrites and diogenites and its comparison with isochron for angrites. $\in(^{53}Cr) = [(^{53}Cr/^{52}Cr)_{measured}/(^{53}Cr/^{52}Cr)_{standard} - 1]\cdot 10^4$, where $(^{53}Cr/^{52}Cr)_{standard}$ – ratio for terrestrial samples 0.1134507. Modified after Lugmair and Shukolyukov (1998)

This range turned out to be a little bit older than the Pb–Pb age of 4566 +2/−1 Ma for Ca–Al inclusions of the meteorite Allende. The ^{53}Mn–^{53}Cr systematics of stony meteorites was substantiated by data on ^{26}Al–^{26}Mg systematics and the oldest ^{53}Mn–^{53}Cr age of 4571 Ma was accepted as appropriate to generation of the Solar system (Shukolyukov and Lugmair 1998).

Timing of planet fractionation has been reliably established from data on ^{182}Hf–^{182}W system, because Hf remains in a silicate mantle, whereas W concentrates in an iron core.

Firstly, Lee and Halliday (1995) defined identical values of $\in(^{182}W)$ for chondrites and terrestrial samples and suggested generation of the iron core of the Earth more than 60 Ma after generation of Ca–Al inclusions in the chondrite Allende. Quitté et al. (2000) calculated for eucrites the 11.1 m.y. younger age than for the meteorite Angra dos Reis. These results contradicted ^{53}Mn–^{53}Cr systematics of eucrites (Lugmair and Shukolyukov 1998).

More precise measurements of $\in(^{182}W)$ in chondrite meteorites showed values in all chondrites approximately 2 units lower than in terrestrial rocks, that, in turn, resulted in essential recalculation of ^{182}Hf–^{182}W ages for episodes of core segregations in planets of the terrestrial group and asteroids. From these data, core separation in

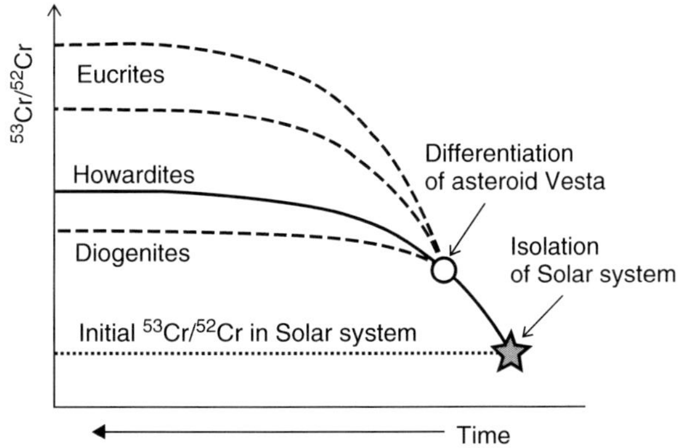

Fig. 10.3 Schematic temporal variations of the $^{53}Cr/^{52}Cr$ ratio in howardites, eucrites, and diogenites that originated from different parts of the asteroid Vesta (Lugmair and Shukolyukov 1998)

the Earth, Moon, Mars, and Vesta took place, respectively, 29–35, 24–35, 11–15, and 2.5–5.1 m.y. after generation of Ca–Al inclusions of the meteorite Allende (Kleine et al. 2002; Yin et al. 2002). These relative estimations of age can be presented in absolute values. For example, differentiation on the Moon occurred at 4537 $\pm$ 12 Ma, within error corresponding to crystallization of the oldest rocks of the Moon at 4562 $\pm$ 68 Ma inferred from Sm–Nd isochron (Alibert et al. 1994) (without error induced by decay constants of uranium and samarium), although a hypothesis on the Moon formation no earlier than ~50 Ma due to collision between proto-Earth and an impactor, named Theia, was assumed from W and Pb isotope data (Halliday 2000). Data on early differentiation of the asteroid Vesta are also consistent with variations of $\in(^{182}W)$ in iron meteorites indicating early segregation of iron cores in asteroids during the first 10 m.y. after generation of the Solar system (Kleine et al. 2002). The age sequence for planets of the terrestrial group and asteroids is shown in Fig. 10.4.

Simulation of Xe loss cessation by the terrestrial atmosphere showed dependence on assumption of a degree of the Earth degassing, iodine concentration in bulk Earth etc. (Zhang 2002). Estimates of the Xe age for the Earth yielded a narrow interval of 4.46–4.45 Ga (Allègre et al. 1995; Zhang 2002) or a little bit older age of 4.49 Ga (Ozima and Podosek 1999). The closure of Xe isotope system was connected with (1) segregation of iron core (Allègre et al. 1995) and (2) collision between the Earth and large asteroid resulted in generation of the Moon with complete reset of Xe system (Zhang 2002). Both interpretations contradicted the new ^{182}Hf–^{182}W isotope systematics (Kleine et al. 2002; Yin et al. 2002).

A xenon age was calculated from assumption that all ^{129}Xe resulted from β-decay of ^{129}I, and ^{131}Xe, ^{132}Xe, ^{134}Xe, ^{136}Xe due to spontaneous decay of ^{244}Pu (and to a lesser extent of ^{238}U). In reality, some amount of radiogenic Xe isotopes could be produced in the terrestrial crust due to neutron-induced division of ^{235}U

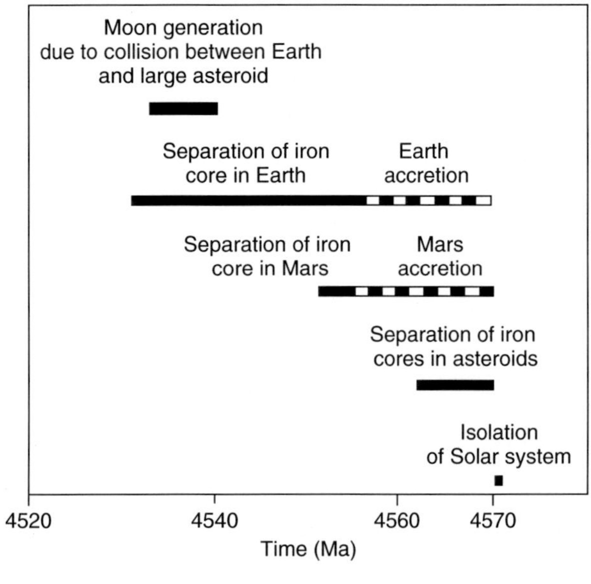

Fig. 10.4 Comparative timing of processes in the Mars, Earth, Moon, and asteroids at an early stage of the Solar system evolution

(Shukolyukov and Meshik 1995). The most favorable conditions for such process existed in an interval of 4.0–2.5 Ga, when ^{235}U contents in a natural mix of uranium isotopes was high; redistribution of uranium from the mantle to the crust resulted in high concentrations of ^{235}U, sufficient for action of natural nuclear reactors. The radiogenic admixture of Xe isotopes results in an unpredictable shift for Earth age estimates from this isotope system.

10.1.4 Evolution of Asteroids

Iron meteorites are characterized by high concentrations of Re and Os as well as wide relative variations of these elements. A closure of the Re–Os isotope system in an asteroid corresponds to an instant of an iron core crystallization, delayed from differentiation into a core and mantle. The smaller the size of an object, the less temporal shift. An asteroid with radius of 200 km cools down from 1600°C to 500°C during 70–300 m.y. and an asteroid of a twice smaller radius during 10–100 Ma (Herndon and Rowe 1973). A range of time estimates for cooling depends on the iron/silicate ratio. Relative elevation of the latter promotes increasing time of cooling. Re–Os ages of iron meteorites, presented by Shen et al. (1996) and Smoliar et al. (1996) and recalculated relative to experimentally defined decay constants of ^{187}Re (Galeazzi et al. 2001), are shown in Table 10.2.

Meteorites of the group IVA yielded two isochrons delayed from each other for 120 ± 27 m.y. The age shift was explained by reset of the Re–Os isotope system in meteorites with younger apparent age due to their secondary heating of a single asteroid. Taking into account this interpretation and assuming that the older age corresponds to the crystallization of the asteroid core, the time span between segregation and crystallization of the core is estimated as not exceeding 30 m.y. The age distinctions of different meteorites in the group IVA could be caused by a long interval of the parent asteroid cooling. In this case, Re–Os data should be examined as evidence of the asteroid cooling during 100–200 m.y. For other groups of meteorites, periods between segregation and crystallization of iron cores lasted about 65 ± 16 m.y.

Table 10.2 Re–Os ages of iron meteorites

Meteorite group	Re–Os age (Ma)	Data source	Timing of the asteroid Vesta formation
IAB–IIICD	4494 ± 35	Horan et al. (1998)	4565 ± 20 (^{53}Mn–^{53}Cr)
IIAB	4494 ± 17	Shen et al. (1996)	4562 ± 19 (^{182}Hf–^{182}W)
		Smoliar et al. (1996)	relative Pb–Pb age of the
IIIAB	4514 ± 26	Smoliar et al. (1996)	meteorite Angra dos Reis
IVA	4414 ± 39	Smoliar et al. (1996)	and Ca–Al inclusions of
IVA	4534 ± 25	Shen et al. (1996)	the meteorite Allende
IVB	4483 ± 41	Smoliar et al. (1996)	

Data of Shen et al. (1996), Smoliar et al. (1996) and Horan et al. (1998) are recalculated using experimentally obtained decay constants of ^{187}Re (Galeazzi et al. 2001). Errors are given taking into account those in constants of radioactive decay (2σ). For the Re–Os method this additional error is ± 14 Ma and for Pb–Pb method ± 17 Ma

10.2 Priscoan Crust (4.4–3.8 Ga)

Archean rocks represent approximately 15% of the total continental crust. The oldest age of 4031 ± 3 Ma was obtained for gneisses of the Acasta Complex, Slave craton, Canada. This is the only place on the Earth where Priscoan rocks occur. The gneisses are granodiorite in composition. Close ages of 4002 ± 4 Ma and 4012 ± 6 Ma were determined for gneisses of tonalite compositions. The age measurements were performed on a high-sensitivity ion microprobe with high resolution (SHRIMP) by the U–Pb method on central parts of zonal zircon crystals (Bowring and Williams 1999).

It has been proposed that these rocks reflect primary magmatic liquids of intermediate–silicic compositions (Bowring and Williams 1999). Wide variations of ϵNd from +4 to −4 in the Acasta gneisses might show their generation due to interaction of mantle liquids with the older continental crust (Bowring and Housh 1999) or due to imposed metamorphic processes (Moorbath et al. 1997).

Occurrence of the Priscoan crust was confirmed by ages in an interval from 4.0 to 4.4 Ga (U–Pb, SHRIMP) defined for detrital zircons from the Jack Hills area, Western Australia (Fig. 10.5) with the oldest age of 4404 ± 8 Ma (Wilde et al. 2001; Valley et al. 2002). Zircon ages from gneisses of the same region do not exceed

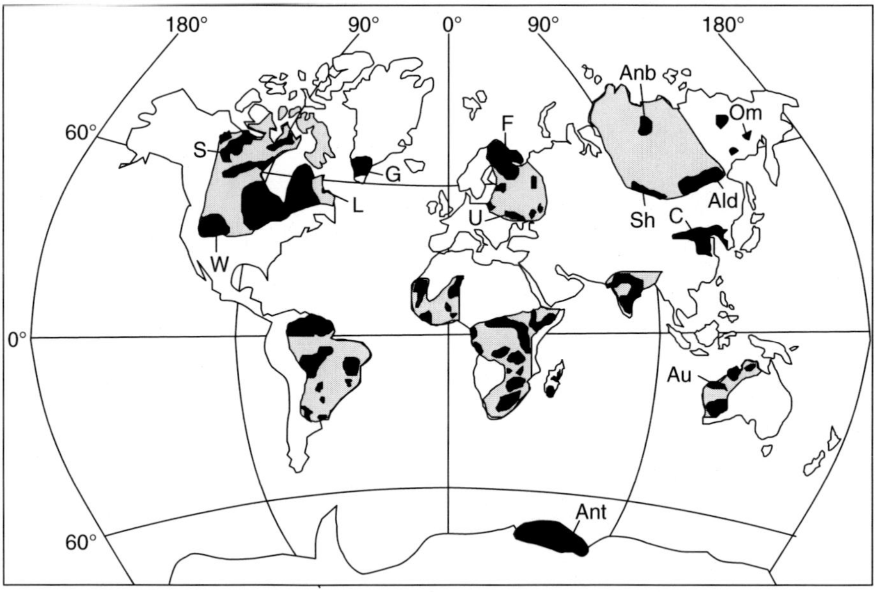

Fig. 10.5 Distribution of the early Precambrian complexes outcropped (*black fields*) and covered (*shaded fields*) (Martin 1994)

3756 ± 7 Ma (Pidgeon and Wilde 1998). High $\delta^{18}O$ in the 4.4–4.0 Ga zircons were interpreted as indicators for protoliths that were altered by interaction with liquid water at low temperatures near surface of the Earth.

It has been widely discussed that the proto-Earth melted due to heat production through transformation of kinetic and gravitational energy and differentiation of a denser metallic core from a silicate mantle as well as decay of long-lived and short-lived isotopes. A rate of heavy meteorite bombardment, estimated from craters of the Moon and Mars, indicates a high meteoritic flux at about 4.0–3.9 Ga, when primeval continental crust of intermediate-silicic composition could exist on the Earth surface (Valley et al. 2002).

10.3 Archean Crust (3.8–2.5 Ga)

Early Archean rocks are distributed in different parts of the Earth (Fig. 10.5). In North America rocks as old as 3.8 Ga of silicic composition have been found in Wyoming (Mueller et al. 1992, 1998) and of ultra basic composition in Labrador (Collerson et al. 1991). Silicic rocks of such age are known in Antarctic (Black et al. 1986), in Greenland (Nutman et al. 1996, 1997), and in Northeast China (Song et al. 1996). Sm–Nd isochron age of 3920 ± 140 Ma that was obtained for slates and amphibolites of ultrabasic and basic compositions in the Ukrainian shield (Zhuravlev et al. 1987b) was not confirmed by U–Pb dating of the host tonalite gneisses (Samsonov et al. 1996). For the latter, an age of 3196 ± 13 Ma was measured.

Global tectonic processes in the Archean have been debated for several decades. The leading roles have been attributed to: (1) tectonics of lithospheric plates operated in the past 3.0–2.5 Ga (Borukaev 1996; Kuski et al. 2001), (2) plume mechanism (Sharkov et al. 1996; Tomlinson et al. 1999), (3) cooperative effect of both plate tectonics and plumes (Kerrich et al. 1999) or (4) processes of multi-cell convection changed to plume dynamics by the late Archean and Proterozoic (e.g. Garagash and Ermakov 2004). Creation of the Archean crust is connected with komatiite magmatism (caused by anomalously hot mantle processes) in Archean greenstone belts exhibiting volcanic sequences, turbidites, and intrusive complexes. A thick lithosphere could originate beneath cratons due to upwelling of buoyant restites after removal of komatiite liquids (Whyman and Kerrich 2002).

Geodynamic hypotheses on the early Earth evolution assume similarity between Archean and Phanerozoic magmatic processes (Yogodzhinski and Kelemen 1998; Martin 1999). A complex of gabbros and lavas with U–Pb zircon age interval of 2782–2705 Ma in the North Karelian greenstone belt, Northwest Russia was interpreted as the oldest ophiolites, geochemically resembling the Phanerozoic boninite series (Schipanskii et al. 2001). Another complex, described as ophiolites in contact with gneisses of 2.5 Ga in Dongwanzi province, China (Kuski et al. 2001), probably, was emplaced in the Mesozoic (Zhai et al. 2002).

Similarity between Archean and Phanerozoic magmatic processes was inferred from occurrences of Archean adakite-like rocks of 3.0 and 2.7 Ga in the Fennoscandian shield, Northwest Russia and in the Superior province, Canada (Pollat and

Kerrich 2001; Svetov 2003). Tonalite–trondhjemite–granodiorite (TTG) complexes, geochemically resembling adakites from modern volcanic arcs, could play an important role in the Early Archean (Martin 1999). This geochemical similarity might be misleading, however; adakites are more Mg-rich and less silica saturated (Smithies 2000). In many cases, adakites were generated through melting of the lower crust (Hou et al. 2004; Guo et al. 2007 etc.).

It has been hypothesized that Archean TTG complexes could originate in three stages similar to those responsible for continental crust generation in the Phanerozoic interplate boundaries. According to the model, firstly, tholeiites appear as mantle liquids, then, tholeiitic material re-melts in H_2O-rich conditions with generation of tonalite liquids, and, at last, fractionation processes result in generation of the TTG complex (Martin 1994). Such a mechanism was supported by experiments on H_2O-rich melting of basalt in P–T conditions corresponding to a granulate phase of metamorphism (Rapp et al. 2003).

This mechanism of continental crust generation in the early Earth presumes melting of a huge amount of tholeiitic liquids during a short interval of time. Primary tholeiites are unknown, however, in the 4 Ga magmatic complexes. So, it is unlikely that early Archean tonalites have been produced by this mechanism. Alternatively, liquids of such compositions might be generated from a magmatic ocean, for example, in a mode of non-stationary convection as a buoyant fraction (foam) on top of

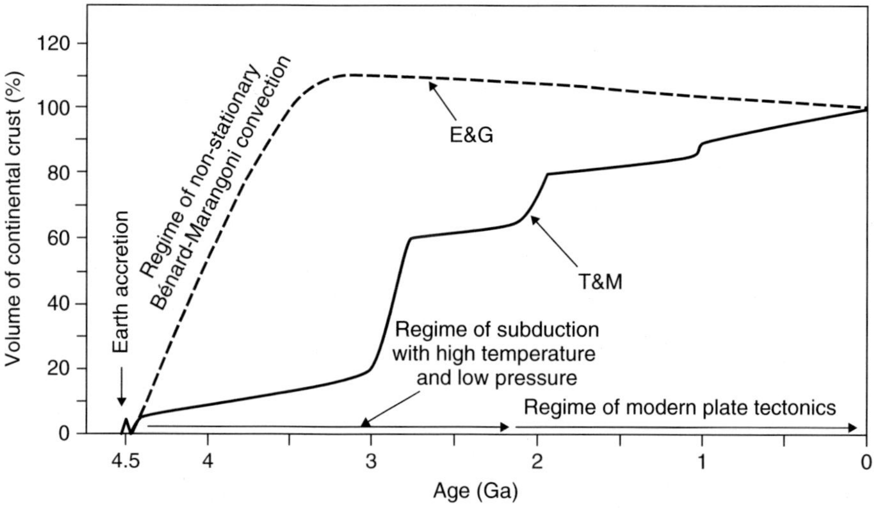

Fig. 10.6 Schematic evolution of TTG complexes. *T&M* – temporal stepwise growth of continental crust modified after Taylor and McLennan (Martin 1994). *E&G* – formation of continental crust in the early Earth with subsequent decrease of its volume (Hamilton 2003; Ermakov and Garagash 2003; Garagash and Ermakov 2004). The latter model has not been developed in terms of temporal variations of trace-element and isotope signatures (for instance, in Rb–Sr or Sm–Nd systems). Nevertheless, mechanism of the early Earth differentiation, applied in this model, seems to be plausible

a Bénard (i.e. Bénard-Marangoni type) cell system (Ermakov and Garagash 2003; Garagash and Ermakov 2004).

A model for three-stage generation of TTG complexes assumes a temporal stepwise increasing volume of continental crust, whereas that for foam-like mechanism postulates generation of maximal volume of continental crust during the first 1.0–1.5 b.y. of the Earth history with subsequent reduction of crust volume due to its thermal and mechanical erosion in convergent boundaries (Fig. 10.6).

On the one hand, incompatible element distribution in continental crust, generated over the last 3.8 Ga, has been relatively constant, so the processes of continent generation also appear to have been dominated by processes operating over this time. For example, Nb/U = 47 in the 2.7 Ga volcanic rocks from Australia, identical to that ratio in modern oceanic basalts, shows that the amount of ancient continental crust in those time was similar to the present (Hofmann 1997). On the other hand, a hypothesis on incidental growth of continental crust assumes its origin as oceanic plateau due to plume activity with subsequent accretion to a continent with essential role of terrane accumulation in transform continental margins (Abbott and Mooney 1995; Patchett and Chase 2002). It is possible that both mechanisms have contributed to continental crust growth.

10.4 Oldest Crust in Asia

Evidence of early solidification of the Earth material is provided by isotope systematics of separated leads on the conjugated Holmes–Houtermans—Concordia–Discordia model. Ore lead isotope compositions are interpreted in terms of transition from Concordia to Discordia in a time interval between 4.3–3.8 Ga. Estimates for these earliest geologic processes in Asia are obtained for ore deposits in the Gargan block and the southern margin of the Siberian craton (Chap. 14). Leads from the former ore province were separated from the 4.31 Ga protolith after initial planetary accretion and those of the latter province from the 3.82 Ga protolith generated after episode of the late heavy bombardment (Fig. 10.7).

From model TNd(DM) age estimates, notable crust-forming processes were defined in the western part of the Aldan shield between 3.8 and 3.3 Ga, in the northern part of the Sharyzhalgai block between 3.6 and 2.8 Ga, and in the Anabar shield between 3.3 and 2.3 Ga (Kotov 2003; Rosen and Turkina 2007). Isochron Pb–Pb whole-rock age of $\sim$3.2 Ma was obtained for the southern part of the Sharyzhalgai block (Rasskazov et al. 1999c). A comparable range of 3.1–3.5 Ga was estimated for Re–Os model ages of sulfide inclusions in diamonds from the Udachnaya kimberlite pipe (Taylor and Anand 2004). Archean U–Pb zircon ages were measured for metamorphic rocks of Sharyzhalgai (about 2.8 and 2.65–2.56 Ga), Aldan (3.460 $\pm$ 0.016 Ga), and Anabar (3.35–3.0 Ga) (Kotov 2003; Rosen and Turkina 2007; Sal'nikova et al. 2007).

From these data, crust-forming processes in the early Precambrian blocks of Asia appear to have been developed largely after the late heavy bombardment of the proto-Earth.

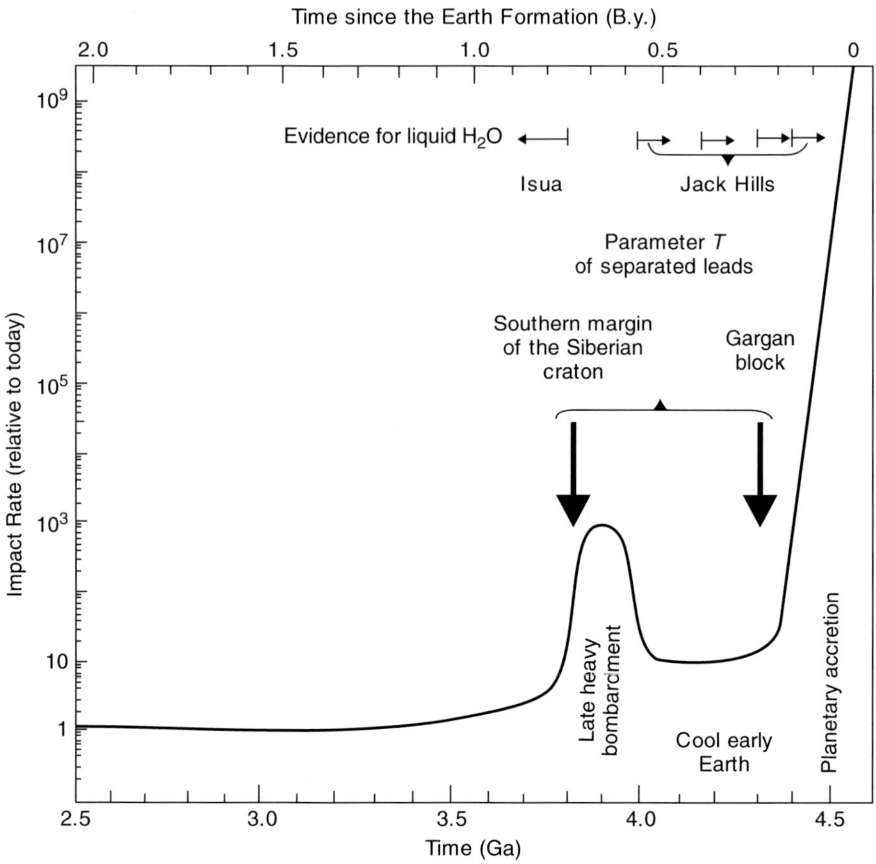

Fig. 10.7 Comparisons between major impact episodes in the early Earth and parameters T, calculated for separated leads. Estimates for temporal variations of meteorite-impact rate and geochronological data on zircons, supported hypothesis of the cool early Earth, are shown after Valley et al. (2002). Moments of transition from the Concordia to Discordia are defined at 4.31 Ga in the Gargan block and at 3.82 Ga in the southern margin of the Siberian craton using the conjugated Holmes–Houtermans – Concordia–Discordia model (Chap. 14)

10.5 Summary: Geologic Processes of the Early Earth

A problem of the Earth age—one of the major problems of natural sciences—has a long history. The solution of this problem was one of the greatest scientific discoveries in the twentieth century (Burchfield 1975; Hallem 1985; Bahcall 2000). Ideas about the early history of the Earth (e.g. Rudnick and Sobotovich 1984) underwent recently notable changes.

Here the early geologic evolution of the Earth has been highlighted by geochronological data obtained by different methods. Generation of the Solar system was determined at 4566 $+2/-1$ Ma by measurements of Pb isotopes in Ca–Al inclusions from the meteorite Allende and in a time interval of 4568–4571 Ma by study of ^{53}Mn–^{53}Cr system in stony meteorites. The values of the two independent estimates for timing of the Solar system generation mutually overlap at 4568 Ma. The initial geologic processes of ~4.57–4.54 Ga resulted in separating iron core and silicate mantle.

Due to initial planetary accretion and late heavy bombardment, the Earth underwent melting. Primeval continental crust of silicic composition could exist on its surface in a time interval of 4.4–4.0 Ga. The lower limit of crust-forming processes was specified by the oldest U–Pb age of 4404 ± 8 Ma measured on detrital zircons from Western Australia. The Priscoan continental crust as old as 4031 ± 3 Ma was defined by U–Pb zircon dating of thc Acasta gneisses in the Slave craton, northwestern Canada. The 4.31 and 3.82 Ga protoliths were identified, respectively, in the Gargan block and adjacent southern margin of the Siberian craton using the conjugated Holmes–Houtermans – Concordia–Discordia model for separated leads. The Gargan block was related to the primeval crust generated on the Earth surface after the planetary accretion, whereas the craton margin to the crust that formed after the episode of instability was caused by the late heavy bombardment.

The primeval crust could have been produced in a mode of non-stationary convection as a buoyant fraction (foam) at top of a Bénard–Marangoni multi-cell system. From TNd(DM)-model data, crust-forming processes in the early Precambrian blocks of the Earth appear to have been developed largely after the late heavy bombardment. Wide variations of Nd-isotope ratios in the Priscoan and early Archean rocks and their geochemical similarity to both rocks from the 3 Ga TTG complexes and adakites from recent subduction zones could have been caused by effective subduction-like downwelling operating in the early Earth.

Chapter 11
Important Phanerozoic Boundaries

A geologic time scale is conventionally considered as equivalent to a biological time scale that is based on evolution of life on the Earth. Using radiogenic isotopes is critical for understanding short-term geologic processes responsible for quick reorganizations of ecosystems at geologically important boundaries. In the Phanerozoic, significant events of mass extinctions occurred in the Early Cambrian, Late Ordovician, Frasnian–Famennian, Permo–Triassic, Late Triassic, Cenomanian–Turonian, and Cretaceous–Tertiary (Erwin 1998). Hereafter we focus on specifications and dating problems of three remarkable transitions: (1) Vendian–Cambrian (V–C), (2) Permo–Triassic (P–T), and (3) Cretaceous–Tertiary (K–T). The available data demonstrate that these boundaries have been caused by different geologic processes and, in fact, cannot be strictly compared to each other.

11.1 Vendian–Cambrian

11.1.1 Stratotype Sections

At the boundary between the Vendian and Cambrian (Precambrian and Cambrian), skeletal fauna appeared in mass quantities simultaneously in various parts of the Earth. The boundary was suggested, primarily, to be placed at the foot of the first biostratigraphic zone of the Tommotian with a complex of the skeletal fossils *Aj. Sunnaginicus*. The zonal complex is widely distributed in the Siberian platform and adjacent Altai-Sayany fold region. The Tommotian unit of the Siberian platform is mostly mono-facial. It contains no trilobites, but shows records of fossils with phosphate skeleton and *Archeocyatha* (Raaben 1969). The boundary is defined by $\delta^{13}C$-isotope chemostratigraphy. The Siberian stratigraphers argue that the foot of the Tommotian—the oldest biostratigraphic boundary of a high rank—should be accepted as the global stratotype for tracing the lower boundary of the Cambrian (Khomentovskii and Karlova 2005).

Versions for the global stratotype of the boundary were proposed also between the Vendian and Baltic series on the Russian platform, Yudom (Vendian) and Pestrotsvetnyi (Lower Cambrian) Formations in the Aldan and Lena river basins,

S.V. Rasskazov et al., *Radiogenic Isotopes in Geologic Processes*,
DOI 10.1007/978-90-481-2999-7_11,

and Nemakit-Daldyn and Sunnaginskii horizons (the foot unit of the Tommotian) in northern regions of the Siberian platform (Sokolov and Fedonkin 1985). As the global stratotype, the sections in Ulakhan-Sulgur (Aldan river, Siberia), Meishuchun (Yunnan, China), and Burin (Newfoundland, Canada) were considered as well (Repina and Rozanov 1992; Landing 1994; Rozanov et al. 1997). The latter section was accepted as the global stratotype by the IUGS.

11.1.2 Radiogenic Isotope Ages

Numerous radiogenic isotope ages that aimed for exact definition of the V–C boundary were published by the late 1990s (e.g. Rozanov et al. 1997). Recently, this data set was extended.

Zircons of pebbles from conglomerates, lying near the basis of the Tusera Formation in the Karaulakhskii uplift (Northern Siberia), yielded the weighted $^{207}Pb/^{206}Pb$ age of 534.6 ± 0.4 Ma (Bowring et al. 1993), which was consistent with the value of the U–Pb zircon age of 530 ± 5 Ma at the top package Byanshao in Meishuchun (China) (Sambridge and Compston 1994). Zircons from a tuff layer in a counterpart of the Chapel Formation, Saint-Johns, New Brunswick, Canada, yielded the upper intersection of Concordia at 531 $+1.8/-1.1$ Ma and average weighted $^{207}Pb/^{206}Pb$ age of 530.7 ± 0.9 Ma (Isachsen et al. 1994).

Dates of zircons (SHRIMP) from tuffs of the Mistaken Point Formation, southeastern Newfoundland with the Ediacara fossils, corresponded to three age groups: 607 ± 1 Ma, 591 ± 5 Ma, and 571 ± 4 Ma. The latter was the closest to the tuff age (Compston 1994). Zircons from Slavatyche tuffs, East Poland showed $^{207}Pb/^{206}Pb$ age of 551 ± 4 (Compston et al. 1992, 1995). Sm–Nd isochron ages of 562.8 ± 7.9 Ma, 562.1 ± 5.7 Ma, and 570.3 ± 17.1 Ma for phosphate bowls and a collophane from the top of the *Anabarites-Protoherzina* zone, China were not consistent with zircon ages of the top horizons of this zone in the Meishuchun section (Yang et al. 1996).

Carbonates of the second zone of the Tommotian lay above the Tusera Formation in the Kharaulakh uplift with a zircon age of 534.6 ± 0.4 Ma (Bowring et al. 1993). This age was considered as the best estimate of the Tommotian foot (Zhamoyda 2000). Carbonaceous rocks from the Upper-Yudom Formation, East Siberia, which underlay the Pestrotsvetnyi Formation of the Tommotian, yielded a Pb–Pb isochron of 553 ± 23 Ma interpreted as timing of the early diagenesis in sediments (Ovchinnikova et al. 2001).

In Death Valley, USA the Cambrian biota occurs in the Canyon Wood Formation, and the Vendian one in the underlying Stelling quartzites. Samples from the latter unit show positive excurse of $\delta^{13}C$ from −0.9‰ to +1.6‰ (relative to PDB). The positive values of $\delta^{13}C$ in carbonates with the Ediacara forms, found above a contact between the Stelling quartzites and Canyon Wood, are correlated with those in sediments of Namibia (Grotzinger et al. 1995; Saylor et al. 1998) and Canada (Narbonne et al. 1994). The sharp negative excurse of carbon isotopes up to −4‰ was found at the foot of the Canyon Wood Formation. Stratigraphic correlations and radiogenic isotope ages bracketed the foot of the Canyon Wood Formation and Stelling quartzites within a time interval of 548–544 Ma (Corsetti and Hagadorn 2000).

Precise dating of the V–C boundary was performed for sections of Africa and Arabia. Zircons from ash horizons lying just below and above the boundary with the ages of 543.3 ± 1 Ma and 539 Ma, respectively, were dated in Namibia, Africa (Grotzinger et al. 1995). Similarly, zircons from ash layers below and above the boundary were dated in Oman, Arabia (Amthor et al. 2003). Extinction of the Precambrian shells (*Namacalathus* and *Cloudina*) was designated by the negative excurse of $\delta^{13}C$. The results of measurements of U–Pb age for zircons from ash layers below and above the excurse yielded indiscernible ages at 542 Ma indicating rather short duration of the excurse. This value was consistent with geochronometry of ash horizons in Namibia and acceptable for definition of the V–C boundary (Bowring et al. 2003) (Fig. 11.1). In Siberia, both *Namacalathus* and *Cloudina* were found in cores of the drill hole Vostok–3 at the top of the Vendian sediments in the West Siberian lowland (Kontorovich et al. 2008).

11.1.3 Events and Hypotheses

The Vendian–Cambrian boundary is characterized by increasing global tectonothermal activity accompanied by break-up of the short-lived supercontinent Pannotia, existed near the Southern Pole at the beginning of the Vendian, i.e. at 600–580 Ma (Condie 1998). Vigorous magmatism enveloped the periphery of the West African craton, which played the role of a thermo-isolating cover in the late Precambrian (Doblas et al. 2002). During the Pan-African orogeny, the mantle and crust underwent

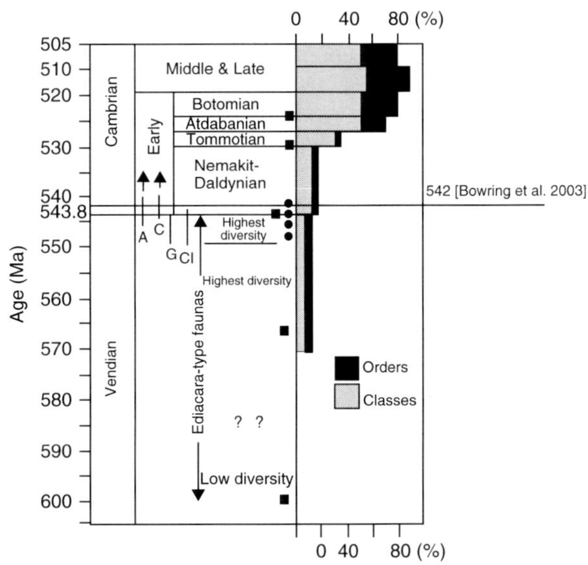

Fig. 11.1 A verified stratigraphic scale of the Vendian and Cambrian, accounted for precise U–Pb isotope dating of boundary sediments (Grotzinger et al. 1995; Bowring et al. 2003). Circles and squares denote dated samples. Intervals of the Vendian skeleton fossils: A – *Anabarites*; C – *Cambrotubulus*; Cl – *Cloudina*; G – cup-like fossils of the *Nama* group. The 542 Ma V–C boundary is shown by a heavy line at the foot of the Nemakit-Daldyn unit (Bowring et al. 2003)

differentiation expressed in inherited isotope signatures of Late Tertiary volcanic rocks. For instance, Late Miocene basalts and trachytes from the Rungwe province showed Rb–Sr errorchron of 540 ± 90 Ma (Rasskazov et al. 1999b). This value should be considered as the minimal estimate of the Pan-African differentiation process, because Rb concentrations (and Rb/Sr) in partial melts exceed values in a source region (Faure 2001).

A great amount of heat could be released from the terrestrial interior between 750 and 540 Ma due to origin of the Pacific superplume. Traps of the Volyn province, East Europe, were emplaced in the failed Middle-Baltic rift and marked separating Baltica and Amazonia at about 551 Ma. These traps are comparable in volume to those in the Deccan province, India. At the V–C boundary, polar ice melted and the level of World Ocean elevated. The increasing concentrations of CO_2 were expressed in the blooming of living organisms.

The hypothesis "of life explosion" at the Precambrian–Cambrian boundary was tested by molecular study (Bromham et al. 1998). No evolutionary chains were found for many fossilized remnants of biota, because the Vendian animals were soft-bodied and had no rigid shells favorable for fossil preservation. Fossilized skeletal fauna appeared in the early Cambrian due to change of chemical composition in ocean water in the post-glacial (post-Varanger) rising of the World Ocean level. The increasing role of water circulation and transgressions and growing concentration of CO_2 in the atmosphere resulted in Mg/Ca decreasing in ocean water and precipitating of carbonates with positive $\delta^{13}C$ values (Windley 1997).

The V–C transition had significant geodynamic consequences expressed in the Phanerozoic development of ultra-high pressure complexes that were not found in the Precambrian (Maruyama and Liou 1998). Respectively, these changes could be connected with some unclear global processes in the Earth interior, possibly marked by specific (for instance, by carbonatite) magmatism. Voluminous eruptions of the Ca-rich liquids could quickly increase Ca activity in ocean water. This assumption requires geochronological substantiation of carbonatite emplacement near the V–C boundary.

11.1.4 Alkaline Ultramafic Magmatism with Carbonatites in Cissayan, Siberia

This boundary is remarkable in respect of alkaline ultramafic magmatism with carbonatites in the Cissayan marginal uplift of the Siberian platform basement (insert in Fig. 11.2).

Data on timing of this magmatism are controversial. Forty four K–Ar dates for rocks of the Zima Complex (Konev et al. 1975; Bagdasarov et al. 1980), recalculated with constants of radioactive transformation of ^{40}K (Steiger and Jäger 1977), ranged from 700 to 500 Ma. Micas from the kimberlite-like bodies yielded $^{40}Ar/^{39}Ar$ ages of 666 ± 18 Ma and 584 ± 6 Ma. Nepheline-bearing rocks from the Bolshaya Tagna and Belaya Zima massifs showed Rb–Sr errorchron of 635 ± 18 Ma (Chernyshova et al. 1992).

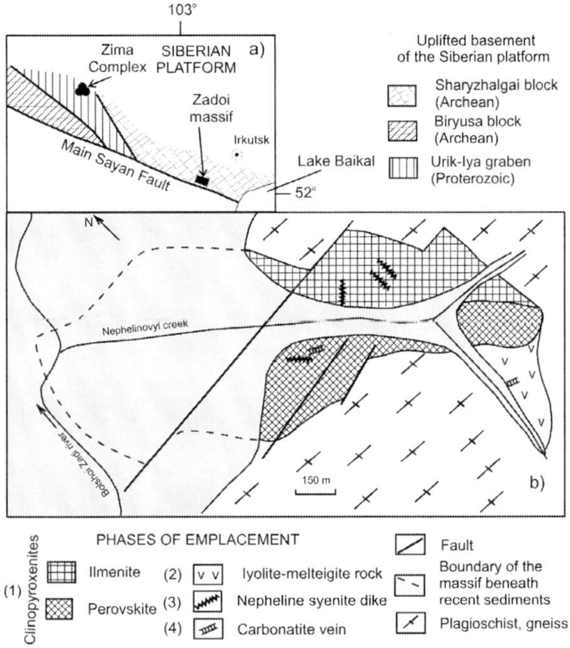

Fig. 11.2 Tectonic setting in the Cissayan marginal uplift of the Siberian platform (**a**) and a map (**b**) of the Zadoi massif. Phases of emplacement are shown from (1) to (4). Simplified after Rasskazov et al. (2007c)

As a rule, nepheline-bearing rocks and nepheline fractions involve cancrinite that possess significant excess argon (York et al. 1969), therefore K–Ar dates on whole rock samples and mineral fractions are overestimated. Amphiboles and micas from carbonatites have excess argon also. $^{40}Ar/^{36}Ar$ of the contaminating component in these minerals is as high as 1000 (Nivin et al. 2001). The rocks, used for Rb–Sr dating, have been altered; the scattering of data points does not fit the isochron model, as indicated by high MSWD. More reliable $^{40}Ar/^{39}Ar$ results by Travin et al. (2002) characterize kimberlite-like rocks having no relation to the Zima Complex composed of alkaline pyroxenites, nepheline syenites, nephelinites, and carbonatites. To define crystallization ages of these rocks, we have done additional work.

Near the Belaya Zima carbonatite massif, Vladimirov et al. (2003) found a nephelinite vein, which was not affected by secondary alterations usually accompanied by carbonatite intrusions. Separates of pyroxene, mica, and nepheline phenocrysts as well as nepheline from groundmass were used for Rb–Sr isochron measurements (Fig. 11.3).

The slope of a regression line for all minerals showed an age of 547.0 ± 8.3 Ma (±2σ) with the initial $^{87}Sr/^{86}Sr$ 0.70308 ± 0.00019 (MSWD 3.17). The points for mica and two nepheline fractions yielded an age value of 546.1 ± 2.2 Ma (±2σ) with the initial $^{87}Sr/^{86}Sr$ 0.70359 ± 0.00074 (MSWD 0.93) (decay constant of ^{87}Rb $1.42 \cdot 10^{-11}$ years^{-1}) (Steiger and Jäger 1977). Recalculations with decay constant of ^{87}Rb $1.402 \cdot 10^{-11}$ year^{-1} (Begemann et al. 2001) show ages, respectively, 553.2 and 552.3 Ma.

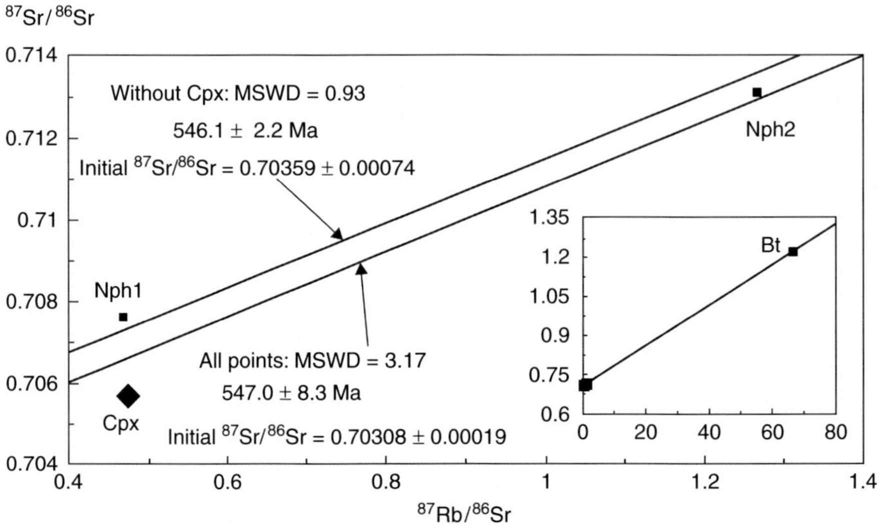

Fig. 11.3 Rb–Sr isochron diagram of minerals from a nephelinite dike, the Zima Complex. Calculations are performed by the York method. Cpx – clinopyroxene, Nph1 – nepheline phenocrysts, Nph2 – nepheline from groundmass, Bt – biotite

The Zadoi (Zhidoi) massif is composed of alkaline pyroxenites, nepheline syenites, iyolite-melteigite rocks, and carbonatites also without secondary alterations (Rasskazov et al. 2007c). From a nepheline syenite (foyaite), nepheline separates with ∼5% of K–Na feldspar admixture were used for incremental $^{40}Ar/^{39}Ar$ dating. Cancrinite was identified in both mineral phases in thin sections.

The obtained spectrum of argon release shows (a) descending steps in an interval of temperatures 670–875°C, reflecting melting cancrinite and sodalite with excess argon, (b) ascending steps in an interval of temperatures 885–1085°C, reflecting melting K–Na feldspar with partly lost radiogenic argon, (c) a plateau in an interval of temperatures 1095–1170°C, demonstrating argon release from nepheline, and (d) distortion of a spectrum in a range of temperatures 1230–1330°C due to structural transition nepheline–carnegieite. The crystallization age corresponds to a nepheline plateau of 547.8 ± 6.6 Ma (±2σ) (Fig. 11.4). The age value is given relative to the age of 128.1 Ma for the standard LP–6 (Baksi et al. 1996). Taking into account the calibration of the standards (Renne et al. 1998), the age value for the standard LP–6 should be increased up to 129.4 Ma with respective correction to the age of 98.79 Ma for the standard GA1550–biotite. From this value, the nepheline plateau yields an age of 552.7 Ma.

From the corrected decay constant of ^{87}Rb (Begemann et al. 2001) and new calibration of the standard GA1550–biotite (Renne et al. 1998), the unaltered rocks from alkaline-ultramafic complexes in the Cissayan show consistent ages in Rb–Sr and K–Ar isotope systems of 552.3–553.2 Ma and 552.7 Ma, respectively. Therefore, the dated magmatic phase took place about 10 m.y. before the event responsible for the V–C transition of 542 Ma.

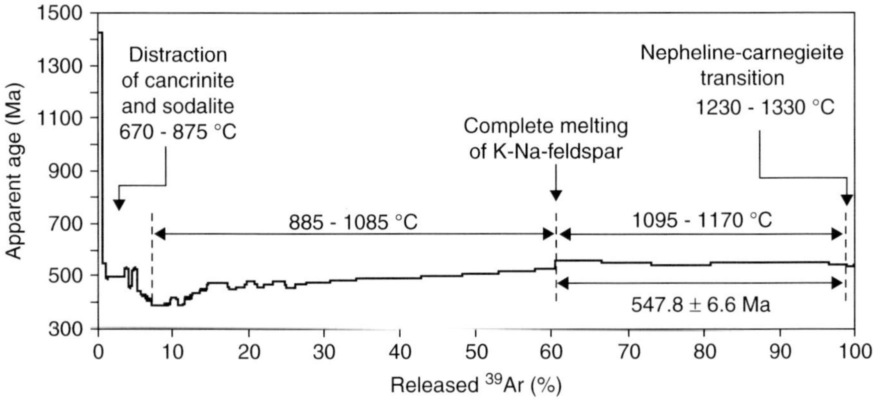

Fig. 11.4 Stepwise heating diagram of a nepheline and K–Na feldspar mixture from a nepheline syenite of the Zadoi massif. Measurements of isotopic composition were performed at the Vrije Universitat of Brussel on a mass-spectrometer MAP–216. The sample was heated in a double vacuum system with temperature control by a W/Re thermocouple and an optical pyrometer (Boven et al. 2001)

11.2 Permian–Triassic

11.2.1 Stratotype Sections

The Permian–Triassic (Paleozoic–Mesozoic) boundary is remarkable in respect of a strongly renewable animal world. Many leading forms of the Paleozoic (trilobites, goniatites, basic groups of Brachiopoda etc.) have disappeared completely, whereas new forms of life (Ammonoidea, reptiles etc.) have appeared. Firstly, Ammonoidea *Otoceras* was considered as the leading biostratigraphic marker of the Triassic sediments (Zakharov and Rybalka 1987; Tozer 1988), but then a preference was given to the conodont *Hindeodus parvus* that had wider geographical distribution (Yin et al. 2001).

Detail biostratigraphic sections of the Upper Permian and Lower Triassic have been studied in many regions (Zakharov and Rybalka 1987; Zakharov et al. 1996; Erwin 1994; Isozaki 1997; Nikishin et al. 2002 etc.) (Fig. 11.5). In the meeting of a working group on the Permo–Triassic boundary (PTBWG, Calgary 1993), only four sections were examined as its possible stratotypes: Meishan, Shansi, Selong (China), and Goryul Ravin (Kashmir). Ultimately, the stratotype of the P–T boundary was accepted by the IUGS in a section D of Meishan, Southeast China. The boundary was placed at a foot of the layer 27c, defined by appearance of the conodont *Hindeodus parvus* (Yin et al. 2001) (Fig. 11.6).

Negative excurses of $\delta^{13}C$ and $\delta^{18}O$ have been identified in all representative sections of the P–T boundary (Zakharov et al. 1996; Bowring et al. 1998; Yin et al. 2001 etc.). Multiple negative spikes of $\delta^{13}C$ recorded in terrestrial plants and marine sediments across the P–T boundary in the former Gondwana Supercontinent were explained by episodic release of methane from clathrates, possibly during disintegration

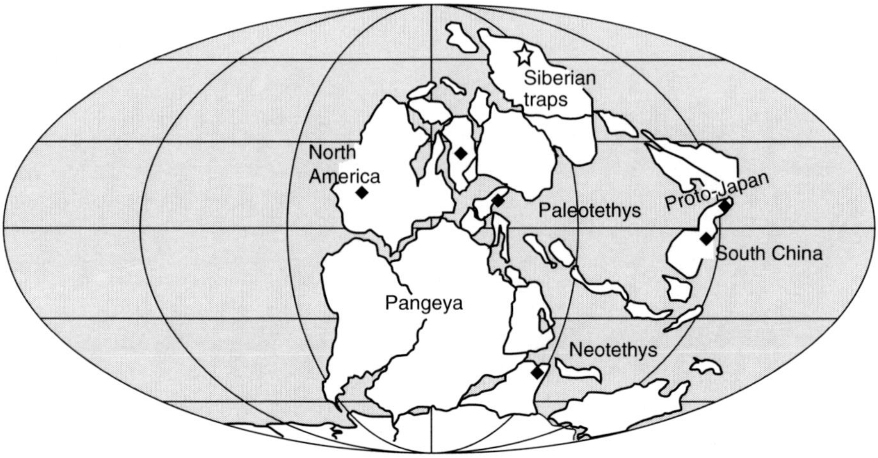

Fig. 11.5 Paleotectonic reconstruction for 250 Ma. Locations of representative sections with the P–T boundary are shown by rhombs and location of Siberian traps by a star. Modified after (Isozaki 1997)

of the continental shelf of Tethys (De Wit et al. 2002). The sharp decrease of $\delta^{18}O$ in the carbonaceous sections of the Southern Alps was interpreted as a reflection of increasing temperature in oceanic water for 5°C (Holser et al. 1989).

In the stratotype section D of Meishan, the negative excurse of $\delta^{13}C$ occurs in a layer 27a, e.g. before appearance of the conodont *Hindeodus parvus* (Fig. 11.6). In fact, the geochemically determined environmental change is a precursor for a change of biota at the P–T boundary. The temporal shift appears to be negligible, however.

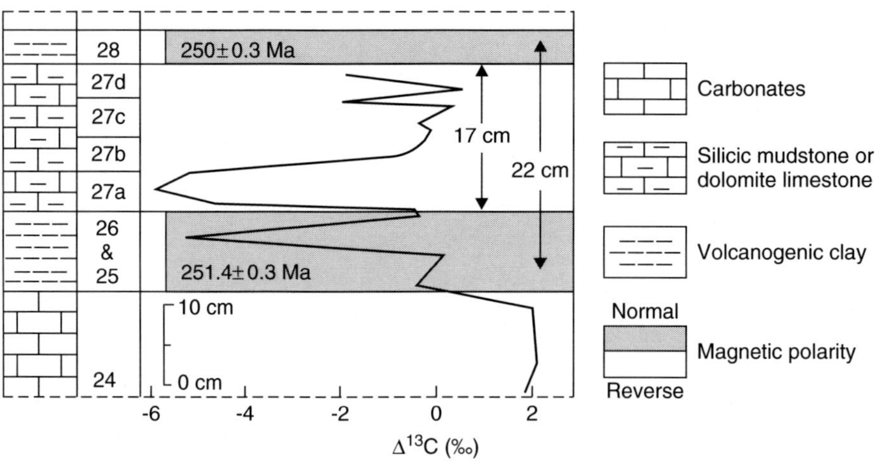

Fig. 11.6 Variations of $\delta^{13}C$ in the stratotype section D, Meishan, Southeast China. The conodont *Hindeodus parvus* appears at a foot of the layer 27c. Modified after Bowring et al. (1998) and Yin et al. (2001)

The stratotype section D in Meishan exhibits normal magnetic polarity with a short interval of reverse polarity at the layer 27 (Yin et al. 2001). The polarity data are useful for dissecting the Permian and Triassic sediments with poor biostratigraphic and chemostratigraphic records (Szurlies et al. 2003).

11.2.2 Radiogenic Isotope Ages

The exact dating of the P–T boundary was performed by the U-Pb method on zircons and the $^{40}Ar/^{39}Ar$ method on sanidine. The material for dating was selected from layers of ashes within marine sediments from Meishan. Ash beds 25–26 that predated the boundary showed a $^{40}Ar/^{39}Ar$ age of 249.9 ± 1.5 Ma (Renne et al. 1995) and U-Pb ages of 251.2 ± 3.4 Ma (Claoue-Long et al. 1991) and 251.4 ± 0.3 Ma (Bowring et al. 1998). The ash bed 28 that postdated the boundary yielded a U-Pb age of 250.3 ± 0.3 Ma (Bowring et al. 1998). From the results of dating, the duration of the negative excurse $\delta^{13}C$ was estimated in 165 +110/−160 thousand years. The biostratigraphically accepted P–T boundary was delayed relative to the excurse approximately for the same time interval.

11.2.3 Events

In the late Permian and early Triassic, fauna and flora was sufficiently changed. About 85% of marine and 70% of land species of animals disappeared in the Late Permian (Erwin 1994). The whole duration of the Paleozoic biota extinction is estimated in 5–11 m.y. and duration of the Mesozoic biodiversity appearance in 3–9 m.y. (Erwin 1998; Zhou et al. 2002). The biotic crisis of the late Paleozoic took place in the Guadalupian (Permian) and Lopingian (Permo–Triassic) (Fig. 11.7).

The late Paleozoic tectonic reorganization resulted in assembly of the supercontinent Pangaea (Wegener 2002; Dobretsov 1997a; Nikishin et al. 2002 etc.). Traps of

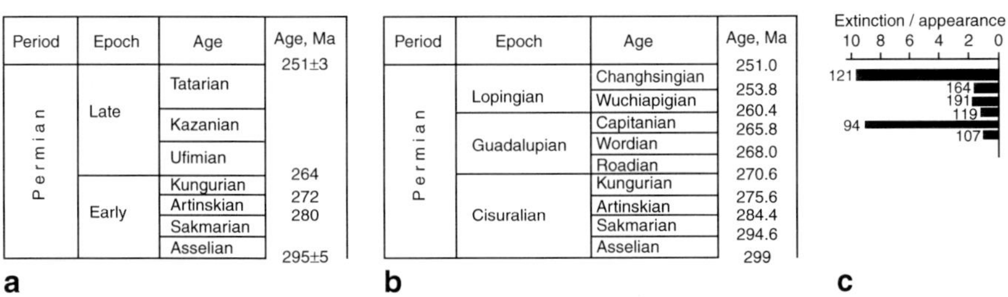

Fig. 11.7 Comparison of International Stratigraphic Charts of the Permian: **a** Oding (1994), Zhamoyda (2000), **b** Gradstein et al. (2004). The three-unit division of the Permian accepted by the International Commission on Stratigraphy (**b**) is not consistent with its two-unit division in the East European scale (**a**). The latter reflects evolution of a boreal biota (Leven 2004; Grunt 2005). Panel **c** shows the ratio of extinct species to brachiopods that appeared first in the Guadalupian and Lopingian and total numbers of species for intervals as inferred from chosen sections in China (Stanley and Yang 1994)

the Emeishan and Siberian provinces were emplaced at the Guadalupian–Lopingian and Permo–Triassic boundaries, respectively (Stanley and Yang 1994; Zhou et al. 2002; Renne et al. 1995; Reichow et al. 2002, 2009; Guo et al. 2004). Voluminous trap magmatism promoted climatic changes (Rees 2002). A short episode of global glaciation in the Late Permian was followed by warming in the Triassic (Chumakov and Zharkov 2002; Rees 2002). At the boundary, the World Ocean changed a level transgressively (Bowring et al. 1998) or regressively (Holser et al. 1989; Erwin 1994).

11.2.4 Hypotheses

Processes at the P–T boundary are explained by hypotheses on (1) fall on the Earth of a large meteorite or comet (Dao-Yi et al. 1985; Kaiho et al. 2001), (2) change of a World Ocean level resulting in destruction of the gas-hydrate layer, methane release, and formation of anoxian stratified paleoocean (Isozaki 1997), and (3) change of global ecological conditions due to voluminous trap magmatism (Canaghan et al. 1994; Renne et al. 1995; Dobretsov 1997a; Reichow et al. 2002).

The first hypothesis was substantiated by the catastrophic character of the event: drastic changes of isotope compositions of sulfur and strontium, elevated concentrations of platinum group elements in sediments of the P–T boundary (Dao-Yi et al. 1985; Kaiho et al. 2001). The data on platinum group elements were not confirmed, however, and those on isotopes of sulfur and strontium were interpreted ambiguously (Holser et al. 1989; Koeberl et al. 2002; Sect. 11.2.5.3; Chap. 13).

The second hypothesis assumed that a gas-hydrate layer was destroyed due to global uplifting of a shelf zone at assembly of continents into the supercontinent, which was accompanied by decreasing sea level in the epoch of glaciation or increasing temperature of the ocean at global warming after a period of glaciation (Dickens 2003).

The third hypothesis considered the Guadalupian mass extinction as caused by the trap magmatism in the Emeishan province and the subsequent Lopingian one by trap magmatism in the Siberian province (Renne et al. 1995; Bowring et al. 1998; Reichow et al. 2002; Zhou et al. 2002). A huge amount of erupted magmatic liquids could result in falling temperature due to infilling of the stratosphere by volcanic dust (at low sea level) and in subsequent long warming caused by the greenhouse effect in expense of emissions of CO_2 and SO_4. These effects, in turn, broke the balance of carbon, oxygen, and sulfur in the atmosphere and caused extinction of biota (Canaghan et al. 1994; Renne et al. 1995; Dobretsov 1997a). Additional factors are required also to explain an observable shift of the carbon isotope composition (Bowring et al. 1998).

11.2.5 Siberian Traps

11.2.5.1 Northern and Western Parts of the Province

Before the 1990s, the Siberian trap magmatism was considered as a long-term process that began in the Permian and proceeded all through the Triassic as was inferred from results of conventional K–Ar dating (Zolotukhin et al. 1986; Feoktistov 1979). High precision geochronological information on traps is concentrated mostly on the northern part of the province, in the Noril'sk area (Fig. 11.8).

The first results of $^{40}Ar/^{39}Ar$ dating by stepwise heating of plagioclase separates from two samples showed a short time interval (~1 Ma) between accumulation of the lower part of a lava stratum in the Noril'sk area and the upper part in the Plateau Putorana (Renne and Basu 1991). These results were interpreted as indication of short-term trap magmatism in the whole Siberian province. The major purpose of

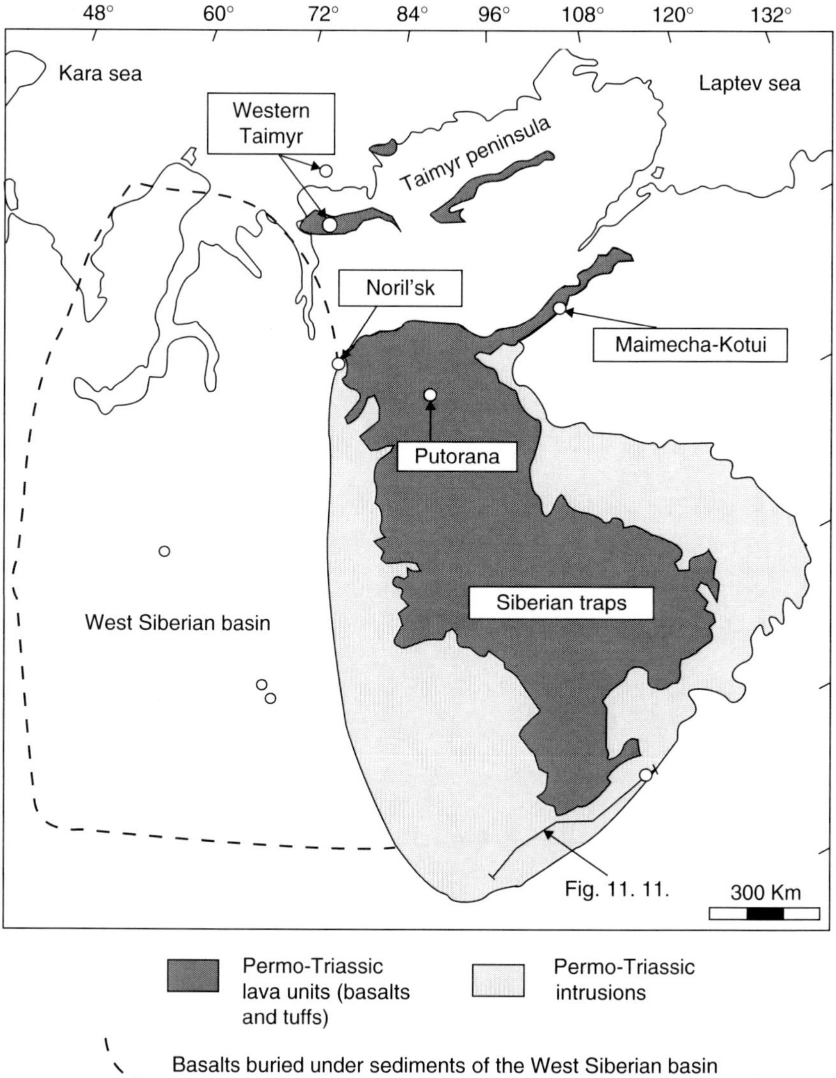

Fig. 11.8 The Siberian trap province. Locations of dated samples are shown by open circles

subsequent studies was to confirm this time interval. Additional $^{40}Ar/^{39}Ar$ and U–Pb dating showed, however, that the trap magmatism could last longer than it was firstly proposed (Dalrymple et al. 1995; Kamo et al. 2000; Ivanov et al. 2005).

In this work, all $^{40}Ar/^{39}Ar$ data are considered relative to the age of 98.79 Ma for the standard GA1550–biotite (Renne et al. 1998). The error connected to the J-factor is excluded. At such an approach, the age of the P–T boundary, determined by the $^{40}Ar/^{39}Ar$ method, is equal to 249.75 ± 0.15 Ma, instead of value 249.9 ± 1.5 Ma (Renne et al. 1995). This value differs from the U/Pb age of 251.4 ± 0.3 Ma measured for the same stratigraphic horizon (Bowring et al. 1998).

Figure 11.9 illustrates $^{40}Ar/^{39}Ar$ ages of traps from different areas of the province, including burred strata of the West Siberian lowland. All the used dates show distinct plateaus conterminous within the errors with isochron ages. Using only this criterion, however, is not enough for distinction of distorted argon spectra. Applying A.I. (alteration index) method, Baksi (2007) pointed out that some altered trap samples were involved in age interpretations by Dalrymple et al. (1995), Venkatesan et al. (1997), and Walderhaug et al. (2005).

Figure 11.9 does not present dates on biotite from the Nizhniy Talnakh intrusion with a plateau 253.7 ± 1.4 Ma and veins of the Noril'sk–1 intrusion with plateaus

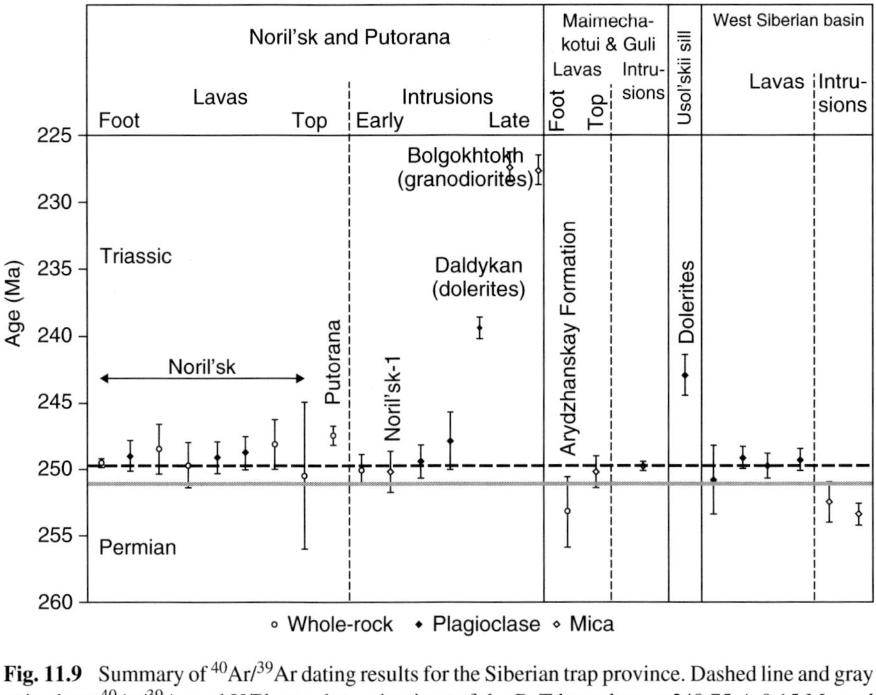

Fig. 11.9 Summary of $^{40}Ar/^{39}Ar$ dating results for the Siberian trap province. Dashed line and gray strip show $^{40}Ar/^{39}Ar$ and U/Pb age determinations of the P–T boundary at 249.75 ± 0.15 Ma and 251.4 ± 0.3 Ma, respectively (Renne et al. 1995; Bowring et al. 1998). (Data sources: Renne and Basu 1991; Basu et al. 1995; Renne 1995; Dalrymple et al. 1995; Venkatesan et al. 1997; Reichow et al. 2002; Ivanov et al. 2005). Locations of dated samples are shown on Fig. 11.8

253.5 ± 1.2 and 254.0 ± 1.2 Ma, although these values are consistent with respective isochrons and no excess argon is detected in the samples on the diagrams of inverse isochron coordinates. In fact, these dates contradicted the younger $^{40}Ar/^{39}Ar$ ages on plagioclases from lavas, which were cut by these bodies (Dalrymple et al. 1995). To find out the reason for this discrepancy, Renne (1995) analyzed several biotite aliquots from the Noril'sk–1 intrusion using a laser for stepwise heating. The obtained satisfactory argon spectra span an age interval of 254–251 Ma. In inverse isochron coordinates, however, 115 high-temperature steps of plateaus of all biotite aliquots showed presence of excess argon with $^{40}Ar/^{36}Ar = 565 \pm 39$ and yielded the common isochron with the age of 250.2 ± 1.5 Ma. This value was accepted as the true age for Noril'sk–1 intrusion.

Figure 11.9 demonstrates a short time interval for lava stratum in the Noril'sk area at the P–T boundary. Duration of the magmatic events cannot be estimated satisfactorily so far because of the large analytical uncertainty. Dates for traps from the Maimecha-Kotui, Noril'sk, and Putorana areas (Renne and Basu 1991; Basu et al. 1995; Renne et al. 1995) were obtained in one laboratory and can be compared to each other. These data show that the lower part of the trap section (Arydzhan Formation) in Maimecha-Kotui was generated in the late Permian, whereas the upper part of the section in Putorana Plateau in the early Triassic. The trap stratum could accumulate as long as 5.7 ± 3.3 Ma. According to U–Pb dating (Kamo et al. 2000), lavas from the Arydzhan Formation should be placed above the P–T boundary (Fig. 11.10).

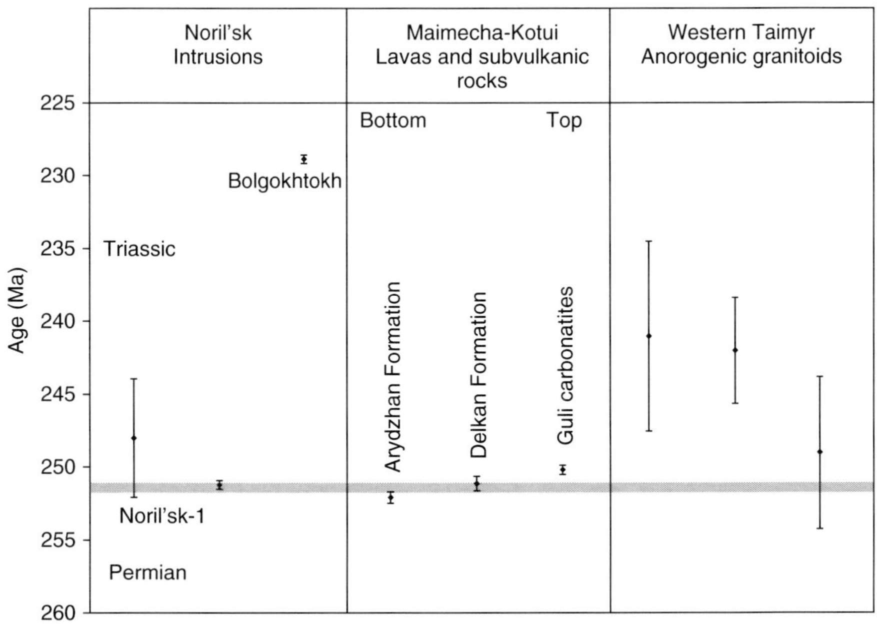

Fig. 11.10 Summary of U–Pb dating results for the Siberian traps. A gray strip denotes the age of the P–T boundary after Bowring et al. (1998). (Data sources: Campbell et al. 1992; Kamo et al. 1996, 2000; Vernikovsky et al. 2003)

Dalrymple et al. (1995) dated practically all major effusion and intrusion unites in the Noril'sk area and showed that dolerite sills and dikes of the Daldykan type were considerably delayed relative to the main lava stratum. The determined time span between the Syvermin Formation, which lay above the oldest Ivakin Formation, and Daldykan intrusions was 9.6 ± 2 m.y. The Bolgokhtokh granodiorite intrusions, connected with trap magmatism (Zolotukhin et al. 1986; Ryabov et al. 2001), emplaced approximately 10 m.y. after the Daldykan intrusions. $^{40}Ar/^{39}Ar$ ages of the Bolgokhtokh intrusions are consistent with results of U–Pb dating (Fig. 11.10).

Drilling in territory of the West Siberian lowland revealed thick effusion strata and intrusions overlain by the Mesozoic sediments. According to $^{40}Ar/^{39}Ar$ dating of lavas on plagioclase separates (Reichow et al. 2002), timing of magmatic events here is statistically undistinguishable from the age of the P–T boundary. Two biotite dates for intrusive rocks fall into the Late Permian (Fig. 11.9).

11.2.5.2 Southern Part of the Province

Several large sills, such as the Chuno-Birusa, Padun, Tolstyi Mys, Tulun, Zayar, and Usol'e, yielding a total volume of 67 km^3, occupied the Kansk-Taseevka basin (Feoktistov 1979; Vasil'ev et al. 2000). The Chuno-Birusa sill outcrops in the western part of the basin and penetrates into the Silurian and Ordovician units in its eastern part. The Padun and Tolstyi Mys sills have connections with the Chuno-Birusa one. They cut through volcanogenic sediments of the Tutonchan and Korvuchan Formations of the early Triassic (Domyshev 1974). The Tulun sill occurs on the top of the upper Cambrian unit, penetrates partly into the lower Ordovician one, and outcrops in large river valleys. The Zayar and Usol'e sills occur within the lower Cambrian unit (Fig. 11.11). The former locates above the latter exhibiting its fragment.

Geologic position of the Usol'e sill indicates its youngest age confirmed by $^{40}Ar/^{39}Ar$ dating performed on plagioclase separates of two dolerite samples from the Severo-Markovskii drill hole N 2. A sample 2840 yielded the plateau age of 243.0 ± 1.5 Ma on nine steps with 98.3% of ^{39}Ar released. The age value is consistent with an

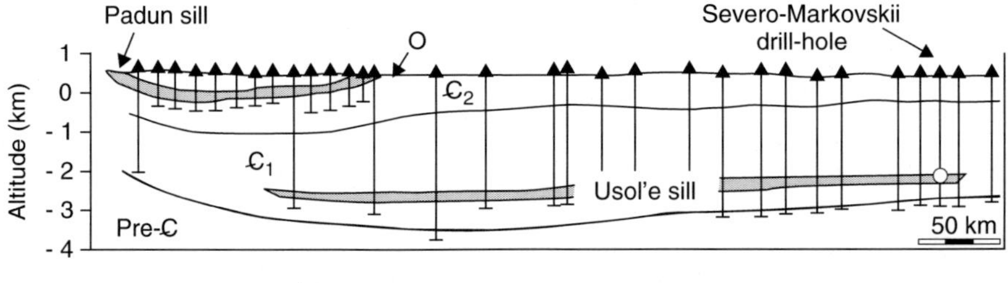

Fig. 11.11 Cross-section of the Kansk-Taseevka basin adopted from Feoktistov (1979). Location of dated samples of the Usol'e sill is shown by an open circle

isochron age of 244.3 ± 1.5 Ma with relatively low initial $^{40}Ar/^{36}Ar$ ratio of 283.3 ± 2.4 (MSWD 1.75). Another sample (2848) showed argon spectrum with lower age values at initial steps of heating. The last seven steps exhibited a 244.6 ± 1.3 Ma plateau by 65.5% ^{39}Ar released. An isochron age of 240.6 ± 1.5 Ma was a little bit younger than plateau. Its initial $^{40}Ar/^{36}Ar$ was also low (286.3 ± 1.6, MSWD 2.25) (Fig. 11.12).

Ages of plateaus in the two samples are comparable within error. The mean weighted age of the plateaus is 243.9 ± 1.0 Ma, e.g. 7 m.y. younger than the P–T boundary.

Some doubts arise about possible distortion of both spectra. Its form correlates neither with the Ca/K ratio, nor with the amount of radiogenic argon. Mineral separates that underwent non-uniform losses of radiogenic argon can show fictitious plateaus. The true $^{40}Ar/^{39}Ar$ age for such samples is characterized by the higher contents of radiogenic argon as compared with apparent underestimated $^{40}Ar/^{39}Ar$ ages (Min et al. 2000). Plagioclase samples 2840 and 2848 show the oldest age values, respectively, at temperature steps 1225–1270°C (249.3–247.6 Ma) and 1330°C (252.0 Ma) and the younger ones at lower- and higher-temperature steps (Fig. 11.12). Possible underestimation of both ages is indicated also by anomalously low initial $^{40}Ar/^{36}Ar$ rations, which might result from specific activation energy of nucleogenic argon (lower than radiogenic and higher than contaminating) or loss of radiogenic argon in the geologic past (Chap. 5).

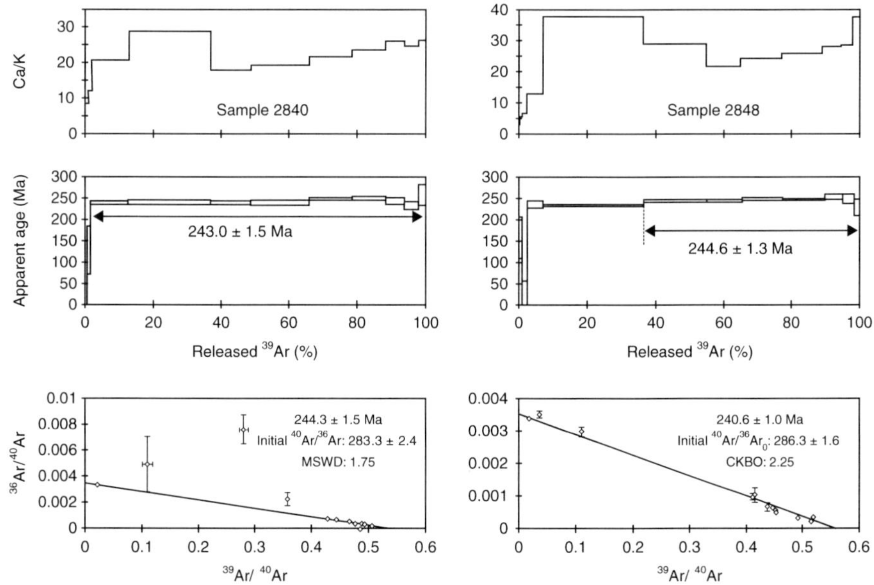

Fig. 11.12 Diagrams of ^{39}Ar release versus Ca/K and apparent age with respective diagrams in inverse isochron coordinates for two plagioclase separates from dolerites of the Usol'e sill (Ivanov et al. 2005)

11.2.5.3 Correlations

The presented overview of data on timing of magmatic events in the Siberian province demonstrates long-lasting processes from the P–T boundary to the late Triassic. The magmatic events, coeval to the trap magmatism in the Siberian province, were widespread in the adjacent terranes to the south of the Siberian platform. Basalts and gabbros from the Kuznetsk basin showed $^{40}Ar/^{39}Ar$ time span of isochrons 251.3–249.8 Ma relative to sanidine feldspar standard FC–1 at 28.02 Ma (Reichow et al. 2009). $^{40}Ar/^{39}Ar$ ages in a range of the late Permian and early Triassic were obtained for monzonite, dolerite, and lamprophyre (kersantite) dikes of the Tomsk area in the northern part of the Kolyvan-Tomsk fold zone (Fedoseev and Sotnikov 2003). Ages about 244 Ma and younger were reported for lamprophyre dikes from Gornyi Altai (Vrublevskii et al. 2004). A relatively young age of 243 Ma was obtained for basalts from the Chelyabinsk area (Reichow et al. 2009).

The P–T boundary was predated by emplacement of lithium-fluorine, alkaline, and subalkaline granitoids of the Samsal and Khonchin Complexes in Eastern Sayan, near the southwestern edge of the Siberian platform. Timing of this magmatism was constrained between 307 and 261 Ma by K–Ar and Rb–Sr dating (Rasskazov et al. 2003d). Coeval motions in the Irtysh shear zone of 283–276 Ma and 272–265 Ma were recorded by $^{40}Ar/^{39}Ar$ study of tectonites (Travin et al. 2001). A significant rotation of the Siberian craton between 360 and 250 Ma was inferred from paleomagnetic data (Kravchinsky et al. 2002).

In Southeast Asia, in the Emeishan province, trap magmatism was initiated at ca. 262 Ma and peaked at 253–251 Ma (Lo et al. 2002; Guo et al. 2004). This magmatism preceded and accompanied the main magmatic pulse in the Siberian province. In marine isotope stratigraphy, the initial Emeishan trap event of 262–258 Ma was contemporaneous to a minimum of $^{87}Sr/^{86}Sr$ at 0.7070 that occurred between maximums of 320–290 Ma ($^{87}Sr/^{86}Sr$ 0.7080–0.7084) and 250–210 Ma ($^{87}Sr/^{86}Sr$ $\sim$0.7078). Transition from the Emeishan to Siberian trap magmatism accompanied persistently increasing $^{87}Sr/^{86}Sr$ which lasted from $\sim$258 Ma to $\sim$239 Ma (McArthur et al. 2001; Huang et al. 2008). These data demonstrate no influence of trap emplacement on the sea water composition.

Tectonic motions, related to collision between the Sino-Korean (North China) and Yangtze cratons, lasted in the East Asian continental margin from 320 to 210 Ma (Ernst et al. 2007). Biotite separates from stress-granites in south Primorye were dated by K–Ar and Rb–Sr methods at the P–T boundary and phengite separates from pelitic schists within the HP terrane from this area within an age interval of 250–230 Ma. Similarly, strong deformations and cooling of high- and ultra high pressure rocks with fast exhumation were constrained in different isotope systems in the Quinlin-Dabi orogen, East China after the P–T boundary, at 245–230 Ma (Webb et al. 1999).

The U–Pb zircon age of 248 $\pm$ 6 Ma was reported for stratified mafic-ultramafic massifs of the Lukinda Complex generated in the central part of the Dzhugdzhur-Stanovoi fold region (Larin et al. 2003). The close U–Pb zircon ages of 251 $\pm$ 3 Ma and 254 $\pm$ 4 Ma were obtained for granitoids of the Tokur terrane from the accretional complex in the eastern part of the Mongolia-Okhotsk belt (Sorokin et al. 2003).

In the Western Taimyr, anorogenic intrusions of syenite and granosyenite compositions yielded U–Pb zircon ages older and younger than the P–T boundary: 259.0 ± 5.2 Ma, 241.0 ± 6.5 Ma, and 242.0 ± 3.6 Ma (Vernikovsky et al. 2003).

From the presented short overview, the P–T boundary events, marked by magmatism in the Emeishan and Siberian large igneous provinces, appear to have exhibited specific conditions in a long-term sequence of geologic processes occurring in Asia between 320 and 210 Ma.

11.3 Cretaceous–Tertiary

11.3.1 Definition of the Boundary

The Cretaceous–Tertiary (Mesozoic–Cenozoic) boundary is characterized by disappearance of dinosaurs and tropical-subtropical plankton forms of foraminifers in marine carbonaceous sediments (Keller et al. 1995). The boundary is traced by a clay layer, drastic decreasing carbonate content, increasing concentration of total organic carbon, and negative excurse of δ^{13}C (Fig. 11.13).

The remarkable geochemical feature of the boundary clay layer is the, so-called, iridium anomaly. The crust contains about $2 \cdot 10^{-10}$% of iridium and meteorites (chondritic and iron) in an interval of $(4–8) \cdot 10^{-7}$%. Therefore, the iridium anomaly could

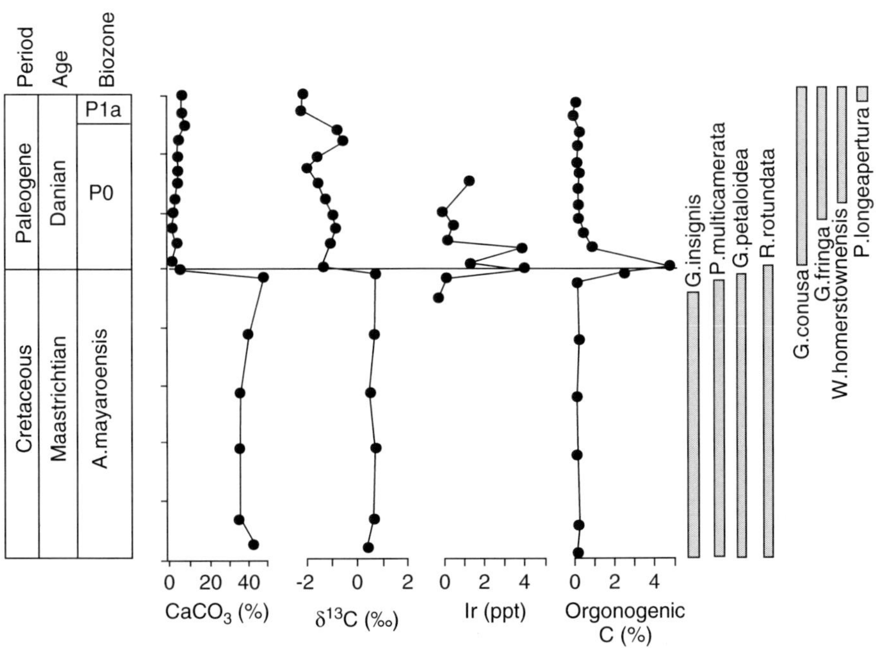

Fig. 11.13 Paleontological and chemical markers of the K–T boundary in the stratotype El Kef, Tunisia. Simplified after Keller et al. (1995)

originate due to contamination of the clay layer by a meteoritic dust (Alvarez et al. 1980). The main accent on this chemical element was made, because it was reliably analyzed by neutron activation in laboratories of the 1980s. The subsequent ICP-MS trace-element study of the clay layer showed elevated concentrations of all elements of the Pt-group (Evans et al. 1993).

Isotope compositions of osmium and chromium were found to be shifted to meteoritic values (Koeberl and Shirey 1997; Shukolyukov and Lugmair 1998). The clay layer of impact origin was traced in Brazil, Turkey, and other parts of the world (Arakawa et al. 2003; Albertão et al. 2004 etc.). As a stratotype of the boundary, a section El Kef in Tunisia was accepted in the 28th International Geologic Congress, Washington, 1989. Representative sections were studied also in North America (Texas and Mexico), Europe (Spain, Italy, and Denmark), and Asia (Russian Far East).

11.3.2 Radiogenic Isotope Ages

The K–T boundary have been dated by $^{40}Ar/^{39}Ar$ technique on impact glasses from the Chicxulub Crater and on tektites from sedimentary sections in Mexico and Haiti (Dalrymple et al. 1983; Swisher et al. 1992; Sharpton et al. 1992). The obtained four ages are comparable within error. Their weighed mean value 65.00 ± 0.04 Ma was defined relative to the standard MMhb–1 with the reference age of 520.4 Ma (Samson and Alexander 1987). The standard was found to be heterogeneous at a level of 15 mg and less (Baksi et al. 1996). For this standard, the value 523.1 Ma relative to the primary standard GA–1550–biotite with the age of 98.79 Ma was reported (Renne et al. 1998), so with the updated standard, the age of the K–T boundary increases up to 65.4 Ma. From underestimated Ar–Ar ages relative to U–Pb ones (Fig. 11.9), due to inaccurate constant of ^{40}K radioactive transformation, this age should be slightly older. The boundary is placed at 65.5 Ma (Gradstein et al. 2004).

11.3.3 Events and Hypotheses

The mass extinction of biota at the K–T boundary was explained by a meteorite fall (Alvarez et al. 1980; Clube and Napier 1982; Erwin 1998) or trap magmatism of the Deccan province (Courtillot et al. 1988; Duncan and Pyle 1988).

The former hypothesis is substantiated by distinct cause-consequence relation between an impact event (or several temporally close events) and the clay layer that marked the K–T boundary (Evans et al. 1993; Koeberl and Shirey 1997; Shukolyukov and Lugmair 1998). The instant and long-term extinction mechanisms, caused by an impact event, are easily recognized (Toon et al. 1997). The former exhibit shock waves, earthquakes, tsunami, and fires, the latter chemical intoxication and global glaciation with subsequent global warming.

After publication of the impact hypothesis, the Chicxulub Crater in Yucatan peninsula, Mexico was described as a structure with diameter of 200 km buried under a 1 km thick layer of sediments (Penfield and Camargo 1981). The hypothesis on origin of this crater at the K–T boundary was supported by data obtained from a drill hole of the Mexican petroleum company Pemex (Hildebrand et al. 1991). Similarly, it was hypothesized that the impact event of the K–T boundary was exhibited by the crater Manson in Iowa, USA, but subsequent radioisotope dating showed that this crater formed 10 m.y., later the K–T boundary event (Izett et al. 1993, 1998).

Biostratigraphic data do not promote unequivocal acceptance of one or another hypothesis. On the one hand, benthos foraminifers are indicative for instant destruction of bottom and littoral ecological systems (Coccioni and Galeotti 1994). On the other hand, consecutive and selective extinction of plankton foraminifers (Fig. 11.13) is not consistent with hypothesis of global catastrophe (Keller et al. 1995). From analysis of Mesozoic echinoderms and fishes, only 25% of the available data specify real mass extinctions (complete disappearance of monophyletic groups). 75% of data represent specific noise (Patterson and Smith 1987). The Cretaceous–Tertiary extinction, lasted for tens of millions of years, assumes various scenarios (Sepkoski 1987). A fall of a large meteorite appears to occasionally coincide with the extinction at the K–T boundary (Courtillot 1994).

11.3.4 Deccan Traps

Occurrence of a sedimentary horizon with an iridium anomaly between lavas of the Deccan province unequivocally specified a beginning of trap magmatism before the impact event of the K–T boundary (e.g. Courtillot et al. 2000).

According to $^{40}Ar/^{39}Ar$ geochronometry, thick lava strata of the province were formed within a short time interval. For instance, 33 dates obtained in different laboratories for a 2 km thick unit of Western Ghats fall in a range of 70–62 Ma with a peak at 65.5 Ma relative to the 520.4 Ma age for the standard MMhb–1 (Hofmann et al. 2000). Dates of lavas from other areas are within the limits of this interval (Courtillot et al. 1988, 2000; Knight et al. 2003 etc.).

Duration of trap magmatism in the province was argued not to exceed 2 m.y., whereas scattering of dates from 70 to 62 Ma was explained by losses of radiogenic argon and deviations of $^{40}Ar/^{39}Ar$ ratios due to recoil effect (Hofmann et al. 2000; Knight et al. 2003). In limits of the trap province, the younger dolerite sills, effusive and intrusive evolved rocks were recorded, however. The Bombay rhyolites yielded an Rb–Sr isochron age of 61.5 ± 1.9 Ma (Lightfoot et al. 1987) and trachytes from the same area $^{40}Ar/^{39}Ar$ ages of 60.4 ± 0.6 Ma and 61.8 ± 0.6 Ma. A dolerite from a sill in the Bombay area showed a $^{40}Ar/^{39}Ar$ age at 60.5 ± 1.2 Ma relative to the 520.4 Ma age for the standard MMhb–1 (Sheth et al. 2001a,b). Thus, trap and related evolved magmatism lasted in the Deccan province more than 6 m.y. With subsequent evolution of the Reunion hotspot (Mahoney et al. 2002), the whole duration of magmatic processes, related to the province, exceeded 65 m.y.

11.4 Summary: Specifications of the Important Phanerozoic Boundaries

The Vendian–Cambrian boundary—one of the most important in the Earth's history—is characterized by simultaneous enormous appearance of skeletal fauna in various regions. The results of the molecular analysis, however, do not support a hypothesis on the Cambrian life explosion and show probable existence of many evolutionary chains of skeletal fauna at the beginning of the Cambrian. Its appearance was likely due to change of water chemical composition in World Ocean. The 542 Ma age of the boundary was measured on zircons from ash layers in Oman. The Precambrian–Cambrian transition was preceded by break-up of the supercontinent Pannotia and Pan-African orogeny. Approximately 10 m.y. before the major boundary event, alkaline ultramafic magmas with carbonatites were emplaced in the Siberian platform margin. Additional work is required to connect the change of water composition in World Ocean at 542 Ma with possible voluminous release from the Earth interior of Ca-rich carbonatite and/or other magmatic liquids.

Both the Permo–Triassic and Cretaceous–Tertiary boundaries were exhibited by remarkable reorganizations of ecosystems. The biotic responses were coeval to voluminous trap magmatism in the Emeishan and Siberian provinces (262–250 Ma and ~251 Ma, respectively) and in the Deccan province (~66–65 Ma). Both boundaries were predated and postdated by magmatic events. Evolution of biota was not influenced by temporally dispersed magmatic sequences, but was affected by impacts of short-term voluminous magmatic phases. The P–T and K–T boundaries were expressed, respectively, by the longest and shortest reorganizations of biota.

The P–T boundary appears to have resulted in magmatism of the Emeishan and Siberian large igneous provinces due to critical conditions in evolution of geologic processes in Asia lasting between 320 and 210 Ma and expressed by collision between the Sino-Korean and Yangtze cratons. Similarly, the K–T boundary, marked by magmatism in the Deccan large igneous province, seems to have related to geologic processes that accompanied collision between India and Asia (Chaps. 12 and 13).

The K–T boundary has been traced in marine carbonaceous sediments by the clay layer resulted from the large impact event. No evidence of a similar event has been found at the P–T boundary.

The 251.4 ± 0.3 Ma age of the P–T boundary was specified and precisely dated by the U–Pb method on zircons from ash layer in the stratotype section D in Meishan. The K–T boundary was dated by $^{40}Ar/^{39}Ar$ method on impact glasses from the Chicxulub Crater and on tektites from sedimentary sections in Mexico and Haiti. The 65.00 ± 0.04 Ma age was measured relative to the standard MMhb–1 value of 520.4 Ma and was elevated up to 65.4 Ma due to change of the latter to the value of 523.1 Ma and up to 65.5 Ma because of inaccurate constant of ^{40}K radioactive transformation.

Chapter 12
Late Phanerozoic Magmatic Evolution of Asia

A comprehensive approach to study of the late Phanerozoic, as a time interval with common geologic evolution, is substantiated by marine Sr-isotope stratigraphy demonstrated persistently increasing $^{87}Sr/^{86}Sr$ ratios in sea water from 0.7068 to 0.7092 in the past 160 Ma. The preceding time interval of 320–210 Ma was marked by $^{87}Sr/^{86}Sr$ ratios as high as 0.7078–0.7084 with a single minimum of 0.7070 centered at the early events of 262–258 Ma in the Emeishan large igneous province (McArthur et al. 2001; Chaps. 11 and 13).

Directions and velocities of lithospheric plate motions, at the first approximation, were connected with driving forces created by mantle density and viscosity heterogeneities. From retrospective analysis of plate kinematics in coordinates of fixed hotspots, high kinetic energy of plates was inferred to exist in a time interval of 100–84 Ma. The subsequent decreasing kinetic energy occurred mainly at the expense of reduction of poloidal (spreading/subduction) component of plate motions at a relatively constant component of their toroidal (transform/spin) motions with transition to a minimal kinetic energy at $\sim$43 Ma. This minimal energy was considered to be maintained until the present (Lithgow-Bertelloni and Richards 1998; Lithgow-Bertelloni and Guynn 2004).

Globally, kimberlite magmatism has been recorded in time intervals of: (1) 240–215 Ma, (2) 160–140 Ma, (3) 105–95 Ma, and (4) 60–45 Ma. The last three intervals are characteristic for North America and Yakutia. Kimberlites are unknown in time ranges of 324–240 Ma and <45 Ma. Unlike North America and Yakutia, in South Africa, kimberlites show continuous sequences of 120–105 Ma and 95–70 Ma and less representative events of $\sim$235 Ma, 190–150 Ma, and 53–50 Ma (Brakhfogel 1995; Brakhfogel et al. 1997; Heaman et al. 2003).

Timing of magmatic processes was an object of global and regional studies. Kennett et al. (1977), Johnson and Rich (1986), and Dobretsov (1997b) emphasized globally significant geologic processes with periodicity of $\sim$30 m.y. Segev (2000) systematized anorogenic magmatic events, accompanied disintegration of Gondwana during the time interval of the past 205 Ma, and came to a conclusion on the monotonic character of magmatic evolution with quasi-periodicity of 13 m.y. Rasskazov et al. (2000b) and Rasskazov (2002) reported geochronological data on late Tertiary

S.V. Rasskazov et al., *Radiogenic Isotopes in Geologic Processes*,
DOI 10.1007/978-90-481-2999-7_12,

and Quaternary volcanic rocks from southern Siberia and adjacent southeastern Asia and recognized intervals of more frequent volcanic quasi-periods from 2.6 to 0.1 Ma.

Causes of late Mesozoic through Cenozoic tectonic and magmatic processes in Asia have been debated for several decades. Leading driving mechanisms have been attributed to mantle plumes, plate subduction, or collision. The first hypothesis assumes the origin of magmatic liquids to be upwelling of hot material from the core–mantle boundary, the second emphasizes the role of convergence between plates of the Pacific or Neo-Tethys and Eurasia, and the third gives priority to collision and post-collision interaction between Eurasia and other tectonic units (Molnar and Tapponnier 1975; Tapponnier et al. 1982; Windley and Allen 1993; Yarmolyuk et al. 1994; Rasskazov 1991, 1993, 1994a, b; Enkhtuvshin 1995; Sengör and Natal'in 1996; Rasskazov et al. 1998, 2000b, 2003b, c, 2004a; Rasskazov and Taniguchi 2006).

The relative roles of these geodynamic factors can be estimated by comparative study of magmatic evolution in such key regions as Tibet, Sea of Japan, and Central Mongolia (Fig. 12.1). On the one hand, Cenozoic magmatism in the latter region distinctly shifts northwards (Devyatkin 1981), assuming a southward motion of the lithosphere above a deep mantle plume (Yarmolyuk et al. 1994, 2007). On the other hand, Central Mongolia is situated in front of Indo-Asian collision zone and volcanic rocks reveal here geochemical signatures similar to those of volcanic rocks from Tibet (Rasskazov et al. 2008b). At last, in spite of its large distance from the Sea of Japan (ca. 2500 km), a possible explanation of Cenozoic deformations and volcanic eruptions in the whole territory of Inner Asia assumes also a mechanism of block extrusion towards this part of a "free continental margin" (Tapponnier et al. 1982; Jolivet et al. 1994).

In fact, seismic images of the mantle and geochemical data on volcanic rocks from Central Mongolia (Barry et al. 2003, 2007; Yarmolyuk et al. 2007; Chuvashova et al. 2007; Mordvinova et al. 2007; Barruol et al. 2008; Rasskazov et al. 2008b) are not plausible for the plume hypothesis. High-resolution seismic tomography shows a low velocity domain, which extends not deeper than 200 km in the upper mantle with significant seismic anisotropy. The measured $^{3}He/^{4}He$ ratios in basalts range within values of the shallow mantle sources. Paleomagnetic data on the Early Cretaceous and Early Miocene volcanic rocks from the area (Kovalenko et al. 1997) yield no reliable solution on the lithosphere displacement above the "fixed plume" because of negligible change of paleolatitudes (within error of measurements).

On the contrary, data on influence of the Indo-Asian convergence on recent tectonics and seismicity in Central Asia have been presented in numerous papers. It is obvious that tectonic strains propagate from the Indo-Asian interplate boundary northwards, eastwards, and southeastwards. Uniform crustal deformations are traced in Western and Central China through earthquake mechanism solutions and GPS measurements (Zoback 1992; Qin et al. 2002; Calais et al. 2003; Chen 2007). From the Cretaceous to the present, essential crustal shortening between India and Mongolia with northern limits in Mongolian Altay and Gobi Altay have been recorded by paleomagnetic data (Halim et al. 1998). The maximal integrated contraction effect of the Indian indenter in Tibet and its negligible role in Central Mongolia with a transition

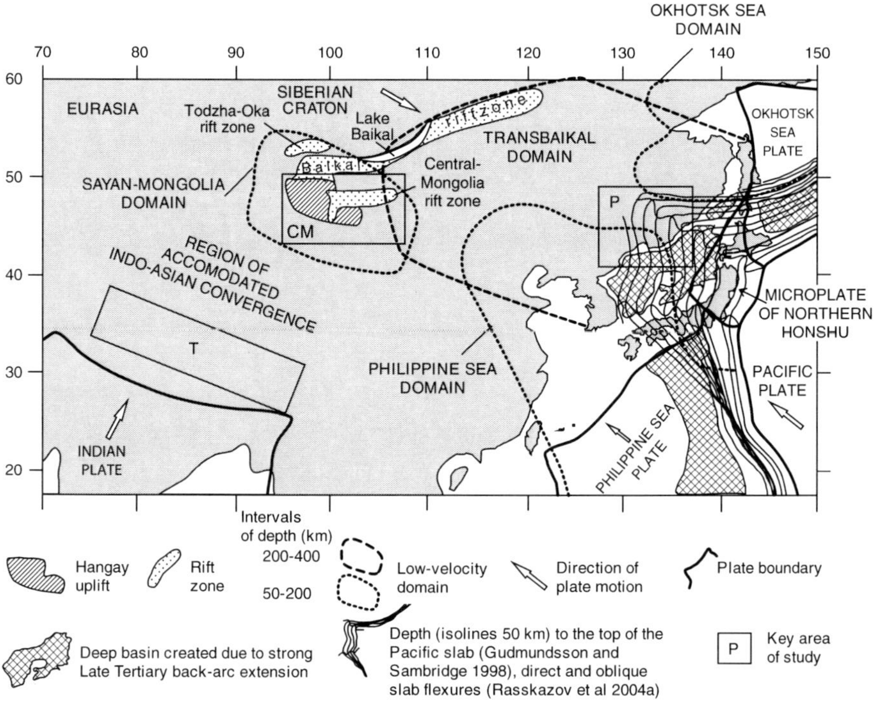

Fig. 12.1 Location of the key areas for study of magmatic geodynamics in Asia. Primorye (*P*) is located at an extended back-arc region of the East Asian continental margin above directly subducted flexure of the Pacific slab, Tibet (*T*) next to the Indian plate that collided with Eurasia, and Central Mongolia (*CM*) in a region which is affected by both convergence due to Indo-Asian interaction and extension in the East Asian continental margin. The scheme is modified after Rasskazov et al. (2003b, 2004a) and Chuvashova et al. (2007)

to predominating extension in the Baikal rift zone is emphasized by results of fractal analysis of Quaternary slip rates on faults (England and Molnar 1993).

Mo et al. (2007) considered Indo-Asian interaction as an irregular process and suggested that a "soft" phase of a collision at ca. 70–65 Ma was followed by a "hard" phase of a post-collision convergence that commenced between 45 and 40 Ma. Molnar (2005) discussed evidence on essential structural reorganization of convergence processes in Asia at ca. 10 Ma.

In this chapter, we present data on magmatism of Asia as a basis for consideration of the geodynamic conditions responsible for the globally common trend of the late Phanerozoic evolution. From compilation of geochronological data on Cenozoic volcanic rocks from Central Mongolia, we show that temporal evolution of volcanism in this region and Tibet was contemporaneous. To explain spatial–temporal particularities of volcanic events in Central Mongolia, we discuss also the possible role of rifting, which was initiated due to extension in the East Asian continental margin. Magmatic response to the late Phanerozoic plate subduction beneath

East Asia, which can only be considered briefly here, was examined by Rasskazov et al. (2004a) and Rasskazov and Taniguchi (2006). We emphasize temporal transition from Mesozoic to Cenozoic magmatic processes in Asia; the former processes reflect its convergence with plates of Pacific and Neo-Tethys, the latter demonstrate a role of Indo-Asian collision and post-collision convergence. An outline of geochronological data on volcanic and volcano-sedimentary units focuses on geologically important processes of Central and East Asia in the Middle–Late Cretaceous, early Tertiary, and middle–late Tertiary through Quaternary.

12.1 Hierarchy and Periodicity of Magmatic Events

From geochronological data, we arrange Mesozoic and Cenozoic magmatic events into a hierarchical sequence of five levels with the readout from the present and characterize each level by representative series of events in a selected territory (Fig. 12.2).

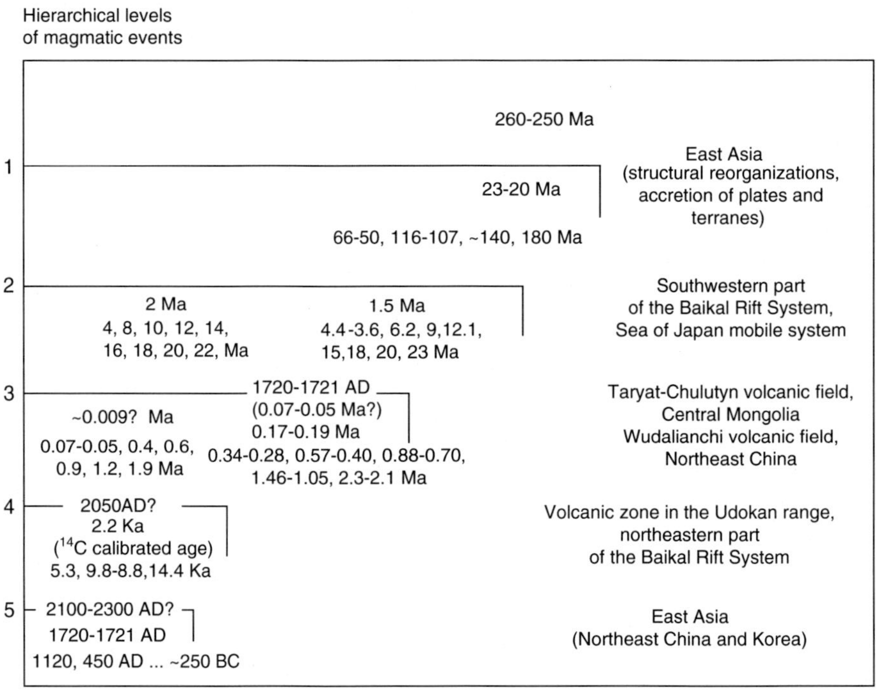

Fig. 12.2 Hierarchy of late Phanerozoic magmatic events in Asia. The last third-order events are exhibited in Taryat-Chulutyn and Wudalianchi by recent eruptions of ~9 Ka and 1720–1721 A.D. The future fourth-order event is predicted at ~2050 A.D. from the recorded volcanic sequence of Udokan. The future fifth-order events are shown by assumed interval of increasing activity between 2100 and 2300 A.D. from records in East Asia

An interval of a higher-order hierarchical level is limited by an event of a lower-order level. Respectively, the fifth-order events are limited by the past ~2.2 Ka (~250 B.C.), fourth-order—14.4 Ka, third-order—2.5–2.1 Ma, second-order—23–20 Ma, and the first-order—260–250 Ma.

The first-order hierarchical level of magmatic evolution is connected with the most important structural reorganizations of about 116–107, 66–50, and 21–15 Ma. Initial events of this level are exhibited also by: (1) emplacement of huge amounts of traps in the Siberian and Emeishan large igneous provinces about 260–250 Ma, (2) origin the Pacific plate at a triple junction of plates Kula, Phoenix, and Farallon about 180 Ma with the beginning of accompanying processes on perimeter of the Pacific Ocean, and (3) the closing Mongolia-Okhotsk Ocean about 140 Ma (e.g. Rasskazov and Taniguchi 2006).

The fifth-order hierarchical level that has a relation to estimation of a present-day magmatic activity is characterized by recent volcanic eruptions in Northeast China and Korea. According to ^{14}C dating, frequent volcanic eruptions began in this territory at about 2.5–2.2 Ka (Liu and Taniguchi 2001; Chuvashova et al. 2007). From historical data, volcanic cycles are exhibited by periods of 700, 670, and 600 years (Fig. 12.3). The modern events are considered in terms of similarity with preceding time intervals of 450–1120 and 1120–1720 A.D., that demonstrate relatively low volcanic activity during 400–450 years and its increase during 200–250 years of each interval. The readout of the modern events commenced from volcanic eruptions of 1720–1721 A.D. in Wudalianchi. The current 2009th year corresponds to the 289th year of the relatively quiescent period of the cycle with probable transition to a state of increasing activity approximately between 2100 and 2300 A.D.

Volcanic episodes in the Udokan range, Northern Transbaikal are considered as the most indicative for the Holocene events (Rasskazov 1999; Chuvashova et al. 2007), i.e. of the fourth-order hierarchical level. On 95%-probability of a calibrated ^{14}C age, referred to mean value of a statistically constrained interval, volcanic activity began here at ~14.4 Ka by eruptions of plagioclase trachytes in volcano Trachytovyi and was followed by eruptions of more evolved biotite trachytes in volcano Dolinnyi at ~8.8 Ka, in volcano Aku at ~5.3 Ka, and in volcano Chepe at ~2.2 Ka. Alkaline basaltic lavas erupted in volcano Syni at about ~9.8 Ka (average of four calibrated dates with probability more than 94%).

The calibrated dates of eruptions in basaltic volcano Syni and trachytic volcano Dolinnyi, overlapped by confidential intervals, are comprised into a single volcanic episode of 9.8–8.8 Ka. Eruptions of volcano Dolinnyi are characterized by high explosive activity with release of dispersed pumice, which includes rounded partially crystallized basic blobs. Such specific eruptions with drastic release of volatile components were likely triggered by emplacement of alkaline basaltic liquid into a shallow chamber with evolved trachytic material. The 9.8–8.8 Ka edifices are considered, therefore, in a frame of a single episode of contemporaneous basaltic and trachytic eruptions.

Four pulses of volcanic activity in Udokan occurred in the past 14.4 Ka with relative reduction of quiescence periods from 5.6 thousand years between eruptions of 14.4 and 8.8 Ka to 3.1 thousand years between those of 5.3 and 2.2 Ka.

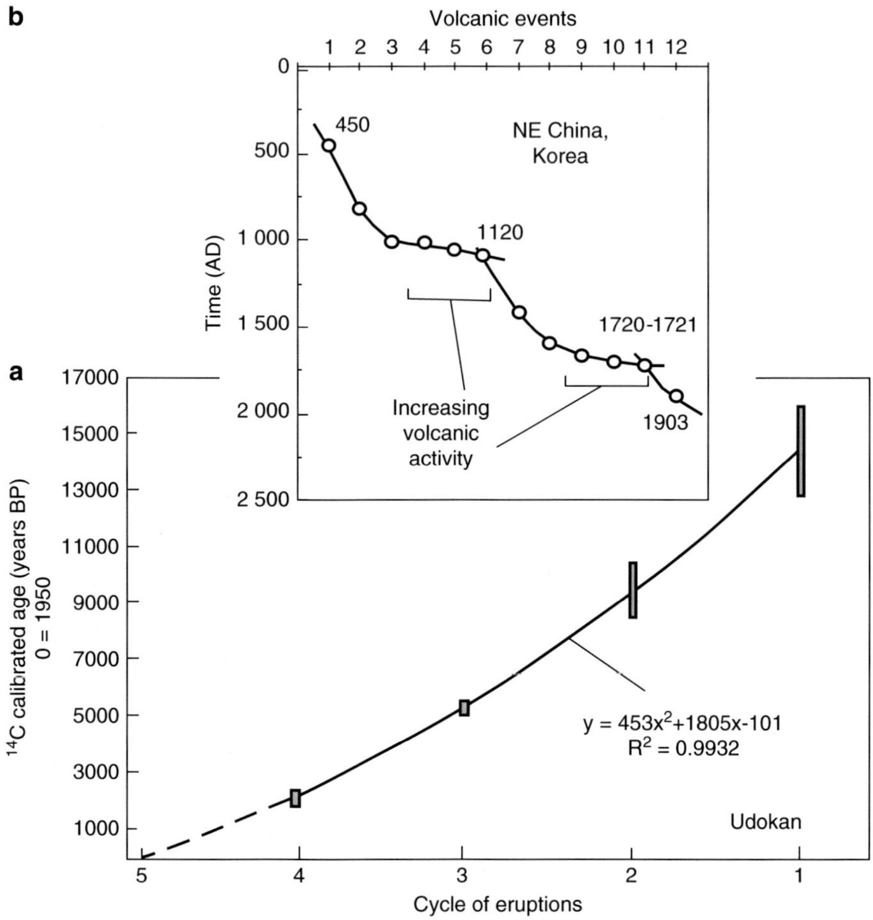

Fig. 12.3 Intervals of calibrated radiocarbon ages of eruptive cycles in Udokan in the past 14.4 Ka (**a**), and succession of historic eruptions in Northeast China and Korea in the past 2 Ka (**b**). A curve of polynomial regression on the panel **a**, shows that the next (fifth) cycle of eruptions may occur in ~2050 (100 years after the reference point of the ^{14}C calendar scale at 1950). Years of volcanic events that are the most important for analysis of their periodicity are shown on the panel **b**. (Data sources: Devirts et al. 1981; Simkin and Siebert 1994; Rasskazov 1999; Liu and Taniguchi 2000; Rasskazov et al. 2000b)

Anomalous fading of S-waves between small epicenter distances shows probable present-day occurrence of a magmatic chamber under these volcanoes at depth of 7 km (Solonenko 1988), which makes real future volcanic eruptions inevitable. For timing estimation, we have applied an equation of polynomial regression obtained for intervals of calibrated ages with a part of more than 0.9 from the common 95% interval of probability. The next volcanic pulse might occur in the middle of the twenty-first century. The regression is obtained for a case of Gauss (normal) distribution of probabilities. Real distribution of probabilities differs from normal (Fig. 3.7), therefore, the result of calculations is considered only as a preliminary approximation. By more correct statistical processing of the data set, a year of the future volcanic eruptions in Udokan can be specified more precisely.

Frequency of eruptions increased in Northeast China about 2.5–2.2 Ka, i.e. at the time of the latest eruption in Udokan. A possible future volcanic event in the middle of the twenty-first century in Udokan (the fourth-order hierarchical level, quasi-periodicity 2–6 thousand years) might be contemporaneous with increasing frequency of eruptions in Northeast China and Korea (the fifth-order hierarchical level, quasi-periodicity 600–700 years).

One more point should be taken into account in respect of the fourth-order hierarchical level. Initial, mostly effusive, volcanism in Udokan showed extreme activity by increasing volume of both lava and pyroclastic material at 9.8–8.8 Ka (calibrated age) and its subsequent decrease with predominating explosive activity. From analysis of the latest Pleistocene through Holocene magmatic events in Central Mongolia, Northern Transbaikal, and Northeast China, we suggested that their regular temporal relations were controlled by a common key trigger, which might be a strong tectonic pulse propagated from the Indo-Asian collision zone. The maximal tectonic stress was connected with the structural reorganization of the volcanic zone in Udokan at 8.8 $\pm$ 0.3 Ka (calibrated age). In Central Mongolia, the volcanoes Odnobokii, Listvennichnyi, and Sosnovyi of the east–west volcanic line were active at ca. 70–50 Ka, i.e. before the reorganization, while the eruptions at the north-northeast line of the volcano Khorgo and related lava flows, which flooded the area, occurred either at the time of the structural reorganization or just after (not later than 7.71 $\pm$ 0.81 Ka calibrated age). In Northeast China, the Holocene volcanism was initiated at 8.74 $\pm$ 0.40 Ka (calibrated age), i.e. simultaneously with the structural reorganization in Udokan (Chuvashova et al. 2007). Similar activity was recorded by radiocarbon dating of eruptions in volcano Malyi Semyachic (Kamchatka) that produced basalts, basaltic andesites, and andesites in the past 14.4 Ka (calibrated age) with a change of the eruption character about 9 Ka (Laverov 2005).

Study of a transition between the fourth- and third-order hierarchical levels of magmatic events is complicated by geochronological uncertainty in a time interval between 0.30 and 0.06 Ma. ^{14}C ages are not obtained for events older than 60 Ka, whereas K–Ar age estimates assume interpretations with wide deviations of the results due to analytical errors and occurrence of excess argon. The doubtful K–Ar dates have been revealed by parallel dating of basalts using the ^{40}Ar/^{39}Ar and U–Th disequilibrium methods.

A lack of retrospective continuation for 2.2–14.4 Ka sequence indicates that the third-order hierarchical level should have repeatability of events not less than 14 thousand years. So far, we have not recorded any series of volcanic eruptions with such quasi-periodicity, but have defined events of 70–50 Ka by the U–Th disequilibrium dating of basalts from Central Mongolia and Tuva (Rasskazov et al. 2008a).

Quasi-periodicity about 10–19 Ka is characteristic of tectonic pulses in the Muya-Kuanda rift basin, where sedimentation processes and soils formation renewed in time intervals of 58.0–56.5, 39.0–36.5, 24.0–22.3, and 13.5–12.9 Ka (^{14}C ages) due to accelerated uplift of the northern shoulder of the basin. The volcanic episodes of 70–50 and <14.4 Ka are separated from each other by the three tectonic quasi-periods manifested in the northeastern part of the Baikal Rift System (Fig. 12.4b).

Further, the 110 Ka episodes are characteristic for East Tuva and those of 200–180 Ka for Udokan and Wudalianchi. These events belong to a retrospective sequence of volcanic events in Asia (Fig. 12.4a).

Volcanic eruptions in Wudalianchi show exponentially increasing frequency in the past 2.3–2.0 Ma with decreasing intervals of volcanic quiescence from ~800 Ka to ~50 Ka. Volcanic episodes in Keluo preceded those in Wudalianchi in a time interval of 2.5–0.7 Ma and afterwards volcanic timing in these areas was undistinguishable. The initial volcanic episode of 1.9 Ma in Taryat postdated the initial episode of 2.3–2.0 Ma in Wudalianchi, the next Taryat episode of 1.2 Ma was within a 1.42–1.05 Ma interval of volcanism in Wudalianchi, and the subsequent Taryat episodes of 0.9 Ma and younger predated those of Wudalianchi up to the Holocene.

The transition from asynchronous magmatism of the Late Pliocene–Early Quaternary to synchronous magmatism of the Late Quaternary, was illustrated in

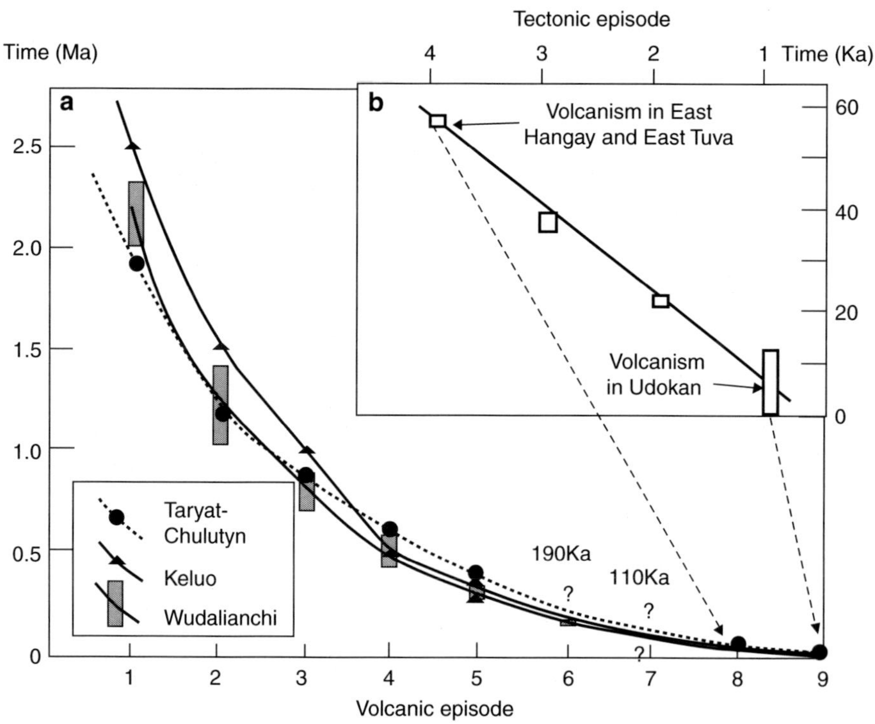

Fig. 12.4 Sequences of the third-order hierarchical level (**a**), and succession of tectonic events in the Muya-Kuanda basin of the northeastern Baikal Rift System (**b**). (Data sources: Wang et al. 1996; Wang and Chen 2005; Zhang et al. 1995; Rasskazov et al. 2000b and references therein)

comparative geochronological study of the Quaternary eruptions in volcanic fields of Inner Asia and those located near its southern and eastern interplate boundaries (Rasskazov et al. 2000b). The volcanic events were asynchronous between 2.1 and 0.7 Ma, synchronized for a short episode of 1.2–1.0 Ma, and became strictly coeval in the past 0.6 Ma with peaks ~0.5 Ma, ~0.2 Ma and <15 Ka (Fig. 12.5).

The temporal change of magmatism was explained by transition from independent withinplate processes, which did not show any specific relation to events that occurred at interplate boundaries, to those sufficiently dominated by the plate interaction. An effect of the Pliocene–Quaternary structural reorganization is well expressed in

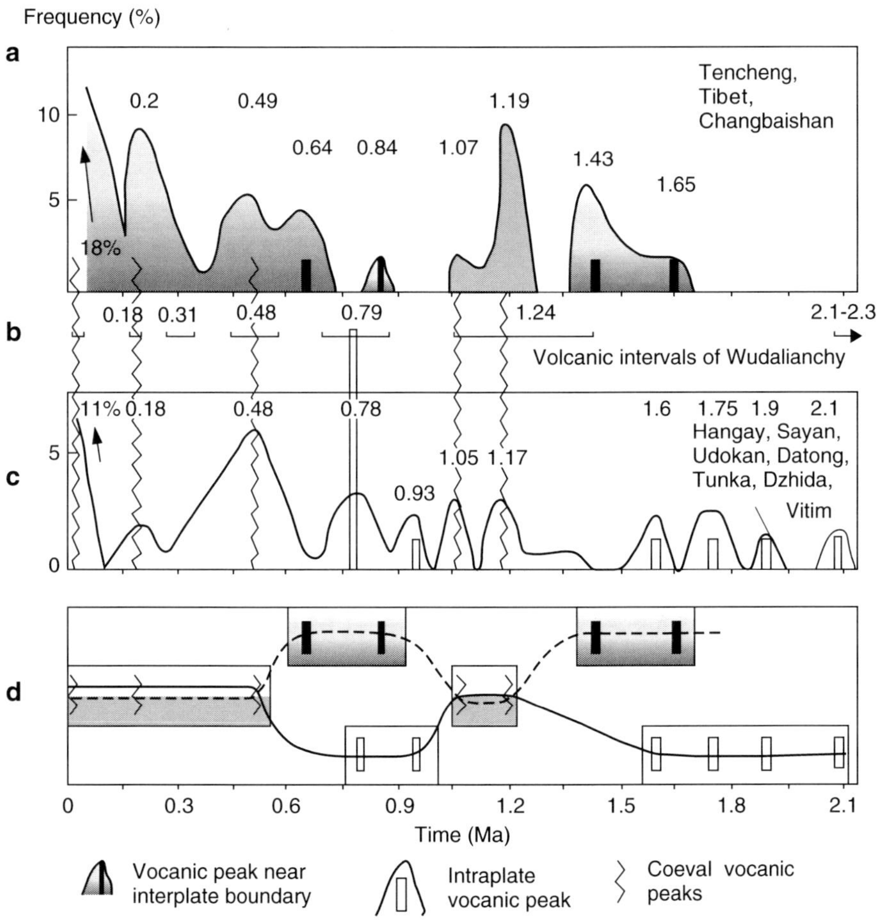

Fig. 12.5 Comparisons of geochronological data on volcanic peaks near plate boundaries (**a**) and within Asia (**c**), (Rasskazov et al. 2000b and references therein). Data set for **a** includes 104 ages, **b** 60 ages, **c** 79 ages. The width of sliding window is 0.06 m.y. The panel **d** emphasizes independent development of interplate and withinplate volcanism during the first half of the Quaternary and resonant activity during its second half—coincidence of volcanic peaks in time intervals of 1.2–1.0 Ma and of the past 0.6 Ma

magmatic evolution of Central Mongolia, where eruptions of 1.9 Ma lavas from the unique magmatic source were followed in Taryat by magmatic lull with subsequent rejuvenation of eruptions since 1.2 Ma from a homogeneous magmatic source with transition from high-K compositions to those with wide variations of potassium at 0.6–0.5 Ma (Sect. 12.4).

One more key area for definition of the Pliocene–Quaternary magmatic change occurs in Central Japan to the west of the triple junction between the Pacific, Philippine Sea plates and Eurasia, where magmatic episodisity in the Ryhaku-Hida Mountains demonstrates a discrete character of subduction-related magmatic evolution by four time intervals: (1) 5.0–2.7 Ma, (2) 2.7–1.6 Ma, (3) 1.5–0.8 Ma, and (4) <0.9 Ma. The first and third episodes exhibited activity of stratovolcanoes in W–E trending volcanic chains related to subduction of the Philippine Sea plate beneath the southwestern Honshu. The 2.7–1.6 Ma episode reflected a short-term change of compression to extension accompanied by eruptions of monogenic volcanoes. The volcanic interval of the past 0.9 Ma marked concentration of volcanic activity along N–S striking chains of stratovolcanoes that is characteristic of magmatism related to subduction of the Pacific plate subduction processes (Shimizu and Itaya 1993). The final magmatic interval began simultaneously with minor eruptions of high-K basalts in Central Mongolia between 1.2 and 0.6 Ma.

The whole data set for volcanic areas of Central Asia yields general quasi-periodicity of ~300 thousand years all through the Quaternary. According to this periodicity, lava eruptions finalized in Tokinskii Stanovik—~300 Ka, Vitim and Dzhida in the central part of the Baikal Rift System—~600 Ka. Lavas from Taryat-Chulutyn of Central Mongolia show peaks of high-K compositions at ~1900, ~1200, ~900, 400, 70–50, and ~9 Ka. Events of the third-order hierarchical level began between 2.6 and 1.9 Ma. At that time, volcanic eruptions redistributed in volcanic fields of Udokan, Eastern Sayan, Northeast China, and Central Mongolia (Rasskazov et al. 1996, 2000a, b).

Events of the second-order hierarchical level with quasi-periodicity of about 2–3 m.y. characterize late Tertiary volcanic evolution in the southwestern part of the Baikal Rift System and in the Northeast Japan Arc in the past 23–20 Ma. Events of this hierarchical level were subjected to detail examination earlier (Rasskazov et al. 2000b, 2004b etc.) and will be considered hereafter in terms of potassium variations in lavas from Central Mongolia and Primorye (Sect. 12.4). This kind of quasi-periodicity was maintained, for instance, in the Hawaiian Chain during the past 47–43 Ma due to oscillatory rotating stress field (Jackson et al. 1975; Lynch 1999; Rasskazov and Taniguchi 2006 and references therein).

12.2 Middle–Late Cretaceous: First-Order Processes

Stratigraphic schemes of Cretaceous sediments in Southern Siberia and Far East of Russia are based mainly on paleontological records. The Aptian sediments are defined from radiolarian classification. This approach is not unequivocal, however. For example, Vishnevskaya et al. (2003) argued that the proper Aptian radiolarians are unknown in Russia. Age interpretations of sections are often ambiguous especially in accretional complexes, where normal sequences of sediments are rare and tectonic fragments of different age are spatially overlapped (Kirillova 2000, 2002). High precision radiogenic isotope dating of rocks in this case becomes especially useful. Controversies in age interpretations of sedimentary-volcanogenic complexes based on biostratigraphic approaches and radioisotope dating should focus on probable distortion of normal stratigraphic sequences and promote additional studies.

12.2.1 Dating

In this section, we present some results on $^{40}Ar/^{39}Ar$ and K–Ar dating of volcanic rocks from the Amur and Transbaikal areas.

In the Selemdzha river basin, ages of volcanic rocks have been measured for the Burunda stratum in the Sergilen volcanic field (coordinates: 52°N 132°E), where high-Mg andesites (MgO 2.6–4.7 wt%) predominate. The Lower Burunda stratum as thick as 400 m was dated back to the Hauterivian–Barremian and lower part of the Aptian based on records of fishes *Licoptera middendorfii* Muller and flora *Sphenobaiera* sp., *Cycadophyta*. The Upper Burunda stratum as thick as 660 m was considered as Aptian–Albian due to occurrence of flora *Coniopteris burejensis* (Zalesskay) Sewakos, *Cladophlebis* sp., *Pityophyllum* sp., *Elatocladus brevifolius* (Fontaine) Bell, *Araliaephyllum* sp., *Hausmannia* sp.

Plagioclase separates from an andesite of the Lower Burunda stratum (sample 8188) showed $^{40}Ar/^{39}Ar$ plateau age of 105.0 ± 6.5 Ma with 71.2% of released ^{39}Ar). Another plagioclase separate (sample 8211) from a vein, which cut the stratum, yielded a plateau age of 108.2 ± 6.6 Ma with released ^{39}Ar 96.5%. A whole rock age of a sample 1131–3, represented the top of the Upper Burunda stratum, showed a plateau age of 107.0 ± 1.7 Ma. The ages of all three samples were comparable within error and constrained volcanic activity in the early Albian, at 108–105 Ma. The results obtained are consistent with age estimates from biota recorded in the Upper Burunda stratum.

In the Middle Amur basin, the Aptian–Albian rocks are andesitic basalts (Simanenko 1986). We have dated a core selected in a deep hole 1E (depth 1079.8 m) near Pereyaslavka (coordinates: 47°N 135°E). The sample yields a whole rock plateau of released ^{39}Ar 64.6% with an age of 106.8 ± 1.3 Ma, i.e. in the age interval of the Burunda stratum.

In Transbaikal, a basaltic stratum occurs in the Khushinda paleovalley (coordinates: 52.5°N 113°E) located at the main watershed between Central and East Asia (i.e. between basins of the Vitim and Amur rivers). A thickness of the dated volcanic stratum exceeds 150 m. Olivine tholeiitic lavas of a sequences are monotonous in composition. Core samples selected from the drill hole 31 at depths 140, 113, and 104 m show whole rock K–Ar ages of 99 ± 4 Ma, 96 ± 4 Ma, and 103 ± 4 Ma, respectively. The results of measurements are comparable within error and indicate accumulation of the lava stratum at the end of the Albian—beginning of the Cenomanian.

The obtained ages fall within the time interval of the Albian constrained at 112.0–99.6 Ma (Gradstein et al. 2004). This time interval corresponds to the middle part of a quiet magnetic period with normal polarity between 119 and 83 Ma (Zhamoyda 2000) or 125.0 and 83.5 Ma (Gradstein et al. 2004).

12.2.2 Geologic Background

Cretaceous magmatic events in Asia of 120–65 Ma are often connected with its interaction with plates of the Pacific or Neo-Tethys. For example, the southern margin of Asia was affected by possible collision with the ensimatic Kohistan island arc between 102 and 85 Ma (Petterson and Windley 1991). The margin of East Asia was influenced by processes related to plate motion in Pacific that resulted in the most pronounced Albian structural reorganization in the Far East of Russia (Maruyama et al. 1997; Natal'in 1993; Kirillova 2000; Parfenov and Kuzmin 2001; Parfenov et al. 2003; Kemkin 2006; Khanchuk 2006; Rasskazov and Taniguchi 2006). From geochronological data, presented in this section for the East Asian continental margin and Transbaikal, these events are bracketed within a time interval of 116–107 Ma.

Paleozoic and Mesozoic accreted terranes of the southern Sikhote Alin are well correlated lithologically, biostratigraphically, and geochronologically with those of the Inner Zone of Southwest Japan (Khanchuk 2000; Kojima et al. 2000; Ishiwatari and Tsujimory 2003; Kemkin 2006). This link allows searching coeval late Mesozoic tectonic processes in frontal accreted complexes and within the continent presuming the common geodynamic context. K–Ar and Rb–Sr ages of gabbro and diorite intrusions and Mg-andesites from the Inner Zone (San-in and San-yo belts) range from 110 to 101 Ma (Iizumi et al. 2000). These rocks are coeval to lavas that erupted within the Early Cretaceous Khingan-Okhotsk continental margin. The sin-collisional structural reorganization and magmatic pulse at 110–107 Ma in East Asia gets an interesting interpretation, when compared with metamorphic evolution of the Sanbagawa Belt in Southwest Japan.

This metamorphic complex exhibits an accreted wedge of the Kula–Izanagi plate. Fossil records and a maximum zircon age show deposition of sediments on oceanic plate between 148 and 134 Ma (Okamoto et al. 2004). Metamorphic rims of zircons from quartz eclogites and associated metasandstones indicate downgoing wedge material in a time interval of 132–112 Ma with eclogite faces metamorphism peaked between 120–110 Ma. This is consistent with Rb–Sr whole-rock isochron ages of high-grade pelitic schists at 116 ± 10 Ma. Subsequent exhumation by wedge extrusion with re-equilibration in conditions of epidote-amphibolite faces resulted in a 100–90 Ma $^{40}Ar/^{39}Ar$ age of amphibole. Further development was expressed by domal uplift between 90 and 80 Ma (K–Ar and $^{40}Ar/^{39}Ar$ ages of white mica) and ultimate emplacement to surface in the latest Cretaceous–Eocene.

There is a distinct correlation between late Mesozoic magmatic and tectonic events of the mainland of East Asia and change in kinematics from descending subduction to extrusive exhumation of the Sanbagawa accretional wedge. The subduction processes developed simultaneously with magmatism of the Aptian–Albian Moneron-Samarga arc system and accretion of the Kema terrane to continental margin, whereas the subsequent exhumation accompanied activity of the East-Sikhote-Alin volcano-plutonic belt (e.g. Rasskazov et al. 2004a).

In the Amur region, the Aptian–Albian volcanic rocks were widespread in the south flank of the Mongolia-Okhotsk suture (Sukhov 1975). Structural study and $^{40}Ar/^{39}Ar$ dating of mica from metamorphic terranes and granites in Central Sikhote Alin showed ophiolites obducted onto the Anui metamorphic dome at about 110–108 Ma. The Gobillin pluton of syncollisional high-alumina granites of the Khungarii Complex postdated the ophiolites at about 107 Ma (Natal'in et al. 1994). The Burunda stratum and coeval Early Cretaceous unites dated in a time interval of 108–105 Ma reflected activity on the Khingan-Okhotsk continental margin (Natal'in 1993). The same time interval was defined by the $^{40}Ar/^{39}Ar$ method for the Stanolirskii volcanic complex composed of calc-alkaline and transitional (to moderate alkaline) basalts, andesitic basalts, andesites, and dacites from Malyi Khingan. The stratigraphically upper rhyolites of the Solnechnyi, Yaurinskii, and Obmaniyskii volcanic complexes were dated in a range of 101–99 Ma (Sorokin et al. 2003). The age of 101 $\pm$ 4 Ma was measured by U–Pb method on zircons of porphyritic granosyenites, which form small sub-volcanic bodies of the Nevachan complex in the central part of the Dzhugdzhur-Stanovoi fold region (Larin et al. 2003).

Granite magmatism of the Pribrezhno-Taigonos and Vostochno-Taigonos belts (southern part of the Taigonos peninsula) occurred in a time interval of 106.5–97.0 Ma as inferred from $^{40}Ar/^{39}Ar$ and U–Pb dating (Luchitskaya et al. 2003). Magmatism of the Okhotsk-Chukotka volcano-plutonic belt could begin at the Aptian–Albian boundary (Kotlyar et al. 2001), but the main peak of activity here took place in the time interval between 86 and 83 Ma (Akinin and Hourigan 2002).

Volcanism of East Transbaikal dated in an interval of 122–113 Ma (Gordienko et al. 1999; Rasskazov and Taniguchi 2006) was finalized simultaneously with tectonic motions of 114–112 Ma. A K–Ar age of 114 Ma was obtained for tectonized trachybasalts from the Chukchudu graben, Northern Transbaikal (Rasskazov et al. 2000b). A close age of 112 Ma was determined by both K–Ar and $^{40}Ar/^{39}Ar$ techniques on a biotite separate from middle-grained slates located among melonitized gneisses in the southern part of the Zagan block (Sklyarov et al. 1997). The Albian events were postdated by the 103–96 Ma basalts from the Khushinda paleovalley.

Lavas of a high-K shoshonite–latite series were widespread in Central Mongolia and adjacent regions of Asia at the beginning of the Cretaceous. Sparse lavas of high-K group erupted also in the Late Cretaceous. In the Arts-Bogd ridge, the youngest latite with potassium content of 3.35 wt% was dated by the conventional K–Ar method at 94.3 $\pm$ 2.2 Ma. Trachybasalt–benmoreite series and basanites with moderate concentrations of potassium from 1.5 to 2.0 wt% span the younger time intervals of 91–82 and 76–71 Ma (Shuvalov and Nikolayeva 1985; Enkhtuvshin 1995; Yarmolyuk et al. 1994, 2007).

12.3 Early Tertiary: First-Order Processes

Tectonic and magmatic events of a time interval from the Late Cretaceous (Maastrichtian) to the Early Eocene (Ypresian) are comprised of the Laramide epoch (Milanovskii 1995). In overview of this epoch in Eurasia, Tsekhovskii and Akhmetev (2003) have suggested that volcanism occurred from the Maastrichtian to Lutetian continuously. This assumption was based mainly on paleontological records in sediments. Results of radiogenic isotope dating demonstrate, however, that widespread volcanism peaked within time intervals of 59–56 and 47–43 Ma, i.e. in the Thanetien and Lutetian, respectively. The Thanetian, constrained between 59 and 55 Ma (Zhamoyda 2000) and between 58.7 and 55.8 Ma (Gradstein et al. 2004), corresponds to the Laramide epoch. The Lutetian, referred to 48–40 Ma (Zhamoyda 2000) and 48.6–40.4 Ma (Gradstein et al. 2004), could postdate this epoch or belong to its late phase (Chap. 13).

12.3.1 Thanetian Magmatic Events

No reliable geochronological substantiation is available in the Far East of Russia for volcanic events within a Pre-Cenozoic time interval of 75–65 Ma, but subsequent events have been dated practically in all areas of the continental margin and islands of the adjacent western Pacific (Rasskazov and Taniguchi 2006).

In the Okhotsk Sea region, K–Ar dates were reported for intrusives, dikes, and volcanic rocks from the Shikotan Island in a range from 62 to 55 Ma (Govorov 2002). K–Ar ages were defined for potassic alkaline and subalkaline intrusions of Western Kamchatka within a time interval of 56–52 Ma (Shantser and Gladenkov 1997). This magmatism was coeval with accretion of the Ozerny-Valanginsky arc system to the continental margin in the latest Paleocene–Early Eocene (Konstantinovskaya 2002; Perepelov et al. 2003). K–Ar ages within a similar time interval were mentioned for the latest magmatic rocks in the Maastrichtian–Paleocene Penzhin–Anadyr belt, which overlapped the Penzhin forearc basin and adjacent accretional terranes (Palandzhyan 2002).

Ages at 58 ± 2 and 56 ± 2 Ma were measured, respectively, for dacite and rhyolite samples collected in the southern part of the Sakhalin Island (Starodinka River) (data of authors). A time interval of 60–53 Ma was obtained for tuffs and ignimbrites of the Bogopolye Formation dated by Rb–Sr isochron method on minerals (Popov and Grebennikov 2001). K–Ar ages of 58.5 ± 3.2 Ma and 56.8 ± 3.4 Ma, comparable with those of the Bogopolye Formation, were measured for biotite separates from the Kanayama rhyolite welded tuffs, Central Japan (Takahashi et al. 1991). Similar K–Ar ages of 57 ± 3 Ma and 54.8 ± 1.8 Ma were determined for rhyolite and basalt samples in the southern Sikhote-Alin (Arakelyants et al. 1982; Okamura et al. 1998).

Timing of tholeiitic basalts, erupted in North China at the beginning of subsidence of the Hubei basin, was constrained within a time interval from 64 to 58 Ma (Zhou et al. 1988). A K–Ar age at 58 ± 2 Ma was measured for Mg-andesite flow cropped in the Birma River, the Amur-Zeya basin. The andesite capped a Mesozoic sedimentary sequence (Rasskazov et al. 2005b). Intrusive bodies and dikes of similar compositions with uncertain ages were found in different areas of East Asia (Sakhno and Moiseenko 2001). The obtained age for Mg-andesites from the Amur-Zeya basin assumes a significance of the Thanetian episode of andesitic magmatism.

Southern Siberia is characterized by development of the Tunka-Eravna rift zone constrained between the Campanian–Maastrichtian and Oligocene (Rasskazov 1994b). In its eastern structural terminus, K–Ar ages of 56 ± 4 Ma, 53.3 ± 1.6 Ma, and 50.6 ± 1.6 Ma were measured for basanites and nephelinites interbedded with Paleocene sediments in the Eravna depression (Ivanov et al. 1995; Rasskazov et al. 2007d). In the western terminus, basalts were found in the Elovskii spur (i.e. within the late Tertiary Tunka rift valley). Here basalts were affected by allitic-type weathering crust formed during the Paleocene climatic optimum. An age of 59 ± 5 Ma was calculated as appropriate to transition of radiogenic argon from semi-sinusoidal to rectangular distribution in large biotites, collected in a profile of a weathering crust from the uplifted southeastern margin of the Siberian craton (Chap. 5). The isotopic closure indicates a time of the area uplift resulted in cessation of weathering processes.

Central Mongolia was enveloped by volcanic events of 65–57 Ma as inferred from K–Ar dating (Ageeva et al. 1988; Genshaft et al. 1990; Genshaft and Saltykovskii 1990; Yarmolyuk et al. 2007; Rasskazov et al. in press). A preceded magmatic gap covers a time span from 71 to 66 Ma. Volcanism in the Ulan-Tsab-Khuduk area took place at the Cretaceous–Paleogene boundary (i.e. 66–65 Ma). In Trans-Altay and Middle Gobi (Dalanzadgad, Under Shil, Mandal Gobi, Altan Shire), basalts erupted between 62 and 57 Ma.

A similar relatively short Pre-Cenozoic magmatic lull was defined in Tien Shan on the basis of $^{40}Ar/^{39}Ar$ dating. Volcanic activity was constrained here within time intervals of 76–71 and 61–60 Ma (Simonov et al. 2008). The only exception is a gabbro intrusion, which was emplaced in the Tuyon basin at 67.3 ± 0.1 Ma (Sobel and Arnaud 2000). This episode was coeval to the main phase of trap magmatism that was constrained in the Deccan Province between 68 and 65 Ma (Courtillot et al. 1988).

12.3.2 Spatial Connections Between the Thanetian and Lutetian Magmatic Events

As a rule, early Tertiary volcanic areas have been affected by both the Thanetian and Lutetian magmatic events (Rasskazov and Taniguchi 2006). Connection between those events is demonstrated by data on volcanic rocks from the East Asian continental margin, western Transbaikal, Central Mongolia, Tibet, and Tien Shan.

Southeastern Sakhalin is of particular interest, in this respect. Here metamorphosed mélange after sedimentary rocks from the collisional Susunai terrane was dated by K–Ar method using white mica separates at 64.5–54.0 Ma (Gouchi et al. 1992). An age of 43.5 Ma was obtained for a sample from the thrust fault cut the southern part of the Susunai terrane (Zharov 2004). The latter falls within a time interval of 49–42 Ma obtained by K–Ar dating of biotite separates from the Aniva intrusion of S-type syncollisional granites (Ivanov et al. 1989). The two time intervals can be attributed to respective accretional phases of the Okhotsk Sea plate to Eurasia.

Both the Thanetian and Lutetian ages were obtained in the southern Sikhote-Alin. In sections of the southern Sikhote Alin (Vanchinskii basin, Zerkalnaya River), basalts were dated at 57 ± 3 Ma, 54.8 ± 1.8 Ma, 47.3 ± 1.8 Ma, and 43.7 ± 1.4 Ma, whereas a rhyolite at 42 ± 3 Ma (Arakelyants et al. 1982; Okamura et al. 1998; Popov et al. 2005). A Rb–Sr isochron age of 46.2 ± 0.5 Ma was obtained for the Shkolnyi dacite extrusion with initial ($^{87}Sr/^{86}Sr$) 0.705 02 ± 0.000 03 at the MSWD value of 1.1 (Rasskazov et al. 2004b). A K–Ar age interval of 46–44 Ma was measured for calc-alkaline andesites from the Giyongsang basin in Korea and for the southern part of the Kuril island arc (Pouclet et al. 1995; Govorov 2002). Events of the latter area were connected with magmatic reactivation in the Proto-Kuril belt accompanied by initiation of subduction along the southern border of the South-Kuril uplift at about 47 Ma.

Numerous Paleocene and Middle Eocene ages were reported for volcanic rocks from the Philippine Sea plate (i.e. Deschamps and Lallemand 2002). A narrow time interval of 47–44 Ma was defined by four K–Ar dates of dredged samples from the Sea of Okhotsk (Yemelyanova et al. 2003). A time interval of 48–44 Ma was reported for basalts of Liyang, North China (Zhou et al. 1988).

In the western Transbaikal, volcanic rocks of the Eravna depression were dated within a time interval of 56–51 Ma and those of the Khilok basin within a time interval of 48–38 Ma (Ivanov et al. 1995; Rasskazov et al. 2007d).

Basaltic eruptions of 65–43 Ma were scattered in Middle and Trans-Altay Gobi of Central Mongolia on an area ca. 140 000 km^2. The Paleocene–Middle Eocene stage was separated from the earlier and later stages by episodes of lava eruptions at 66–65 and ~43 Ma with relatively low potassium contents (0.5–1.5 wt%). A volcanic hiatus between 56 and 49 Ma separated the Thanetian and Lutetian volcanic phases with similar lava compositions; potassium concentrations varied between 0.8 and 1.9 wt% (Fig. 12.6).

$^{40}Ar/^{39}Ar$ dating of volcanic and sub-volcanic rocks in the northern part of the Tien Shan volcanic region, in the vicinity of Lake Issyk-Kul, yielded ages about 61 Ma (Simonov et al. 2008). The younger volcanic episodes were identified in the southern part of the region, in the vicinity of the Tuyon basin, northwest China. In Kirghiz Tien Shan, a biotite from a diabase yielded $^{40}Ar/^{39}Ar$ plateau age at 59.0 ± 1.4 Ma and two basalts showed ages at about 54 Ma. Reliability of these age determinations is confirmed by mutually coincided plateau and isochron ages. Biotite from a lamprophyre showed a plateau age of 46.5 ± 3.8 Ma also consistent with isochron of 46.5 ± 0.1 Ma (Sobel and Arnaud 2000). The southward magmatic shift in time intervals of 76–61 Ma to 59–46 Ma in Tien Shan is comparable with the eastward magmatic shift in time intervals of 66–65 Ma to 62–43 Ma in Central Mongolia.

12.3.3 Geologic Background

Closing of the Neo-Tethys Ocean was finalized by collision between the Indian and Eurasian continents in a time interval of 66–50 Ma (Patriat and Achache 1984; Lee

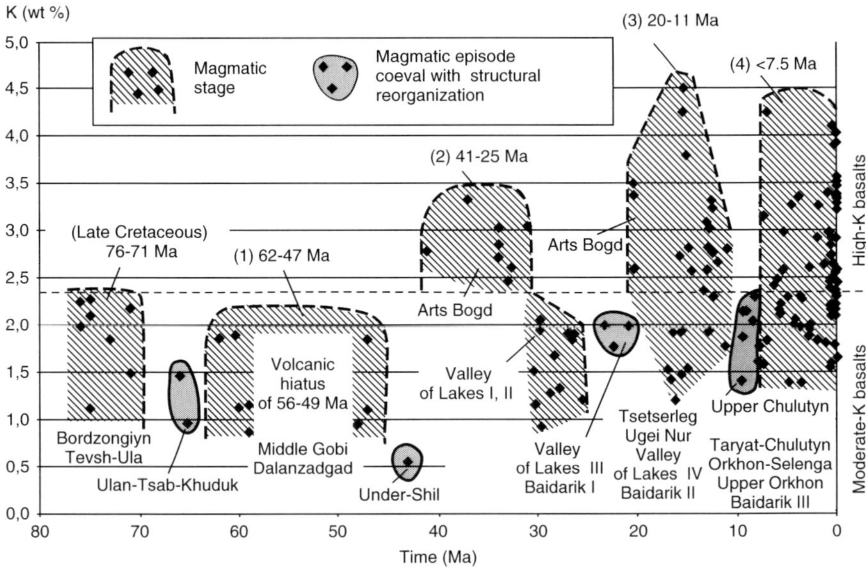

Fig. 12.6 Definitions of principal magmatic stages of Central Mongolia in the latest Cretaceous through Cenozoic as inferred from spatial–temporal variations of potassium in volcanic rocks of basaltic to intermediate compositions. Lavas with relatively low potassium abundances are indicated between the latest Cretaceous and four subsequent Cenozoic magmatic stages and referred to episodes of first-order structural reorganizations. The episode of ~10 Ma is within the late Tertiary time interval dominated by the second-order events (Data source: Rasskazov et al. in press and references therein)

and Lawver 1995; Beck et al. 1995; Butler 1995). In the northwestern Himalaya, the emplacement of the Ladah granitoid batholiths was governed by subduction of the Neo-Tethys plate between 70 and 50 Ma. A cessation of the batholiths growing was interpreted as an indication of a collision-related rupture of the oceanic lithosphere (Weinberg and Dunlap 2000).

A collision between the Pontide (Eurasian) and Anatolide (Gondwanian) blocks resulted in rapid differential uplift between 57.4 ± 2.4 Ma and 47.8 ± 2.4 Ma as was inferred in the East Turkish Pontides from fission-track data on apatite (Boztug et al. 2004). The eastward block extrusion in the northern and eastern flanks of the Indian indenter was delayed relative to initial hitting of Eurasia by India. The earliest tectonic motion with exhumation and cooling rocks below 500°C at 52.6 ± 1.1 Ma was defined by Rb–Sr dating of a muscovite and a biotite from gneisses exposed in the Ailao-Shan Red River shear belt, south China (Zhang and Schärer 1999). The initial stage of the collisional plate interaction of 66–50 Ma was followed by long-lived tectonic and magmatic instability induced into Eurasia.

The beginning of the Thanetian coincided with the strongest global warming of 61–56 Ma that was caused by destruction of gas-hydrates in sediments of the World Ocean due to perturbation of oceanic waters cased by magmatism of the Proto-Icelandic plume in the northeastern Atlantic Ocean (Jolley et al. 2002). Methane was released also due to strong deformations of sediments in an accretional prism of the Alaska gulf in a time interval of 61–56 Ma (Hudson and Magoon 2002). The widespread Thanetian volcanic events in Asia are also indicative for globally increased tectonic activity in the Earth.

In the context of the Earth evolution, the Thanetien corresponded to a transition from a phase of extension to metaphase of compression (comprised several compression phases) (Milanovskii 1995). This transition accompanied plate reorganization, drastically decreasing sea level and increasing frequency of geomagnetic field inversions. The main reorganization in plate motions from the Mesozoic to Cenozoic took place within a post-Thanetian time interval of 53.5–37.5 Ma (Rona and Richardson 1978).

In the Paleocene, the Pacific plate moved northwards with oblique subduction beneath Asia with incidental collisional interaction between volcanic arcs and continental margin (Parfenov et al. 2003). In the Late Cretaceous and early Paleogene, the Sikhote-Alin block rotated counter-clockwise with synchronous opposite rotation of the North China block. These relationships were interpreted as a consequence of a strong Cretaceous–Paleogene stretching continental margin, similar to stretching responsible for Early–Middle Miocene opening Sea of Japan (Otofuji et al. 2003). The end of rotation was defined at 53–51 Ma by K–Ar ages of samples from the Bogopolye Formation. According to these data, the Cretaceous–Paleogene stretching between Sikhote-Alin and North China preceded the main reorganization in plate motions defined by Rona and Richardson (1978) and/or corresponded to its initial phase. The extension might be caused by accretion of the Okhotsk Sea plate to Eurasia (Rasskazov et al. 2004a).

Events responsible for compression in South Eurasia between 57.4 and 47.8 Ma (Boztug et al. 2004) are comparable with those in the southeastern Sakhalin, where accretion of the Okhotsk Sea plate to Eurasia resulted in high-grade metamorphism, constrained between 64 and 54 Ma and subsequent emplacement of S-type granites between 49 and 42 Ma.

12.4 Middle–Late Tertiary through Quaternary: Transition from the First-, through the Second- to the Third-Order Processes

To demonstrate the main regularities of tectonic and magmatic processes in Asia, we consider spatial–temporal distribution of volcanism in two regions: (1) Central Mongolia with adjacent Central Asia as a territory of spatial transition from a region affected by post-collision convergence to a rifted region (Fig. 12.1) and (2) South Primorye as a territory indicative for development of back-arc extension in a continental margin of East Asia. We present a compilation of geochronological data on volcanic rocks and show that temporal evolution of volcanism in Central Mongolia and Tibet was coeval. To explain spatial–temporal particularities of volcanic events in Central Mongolia, we discuss also a possible role of rifting, which was initiated due to extension in the East Asian continental margin. The main purpose of this section is to demonstrate an expression of magmatic processes related to Indo-Asian interaction and to rifting in eruptions of basaltic liquids with varying potassium concentrations. In terms of suggested hierarchical system of Fig. 12.2, these variations will demonstrate transition from the middle to late Tertiary (i.e. from the first- to the second-order events), and from the late Tertiary to Quaternary (i.e. from the second- to the third-order events).

12.4.1 Central Mongolia and Adjacent Regions

Early Tertiary lavas of Central Mongolia are characterized by moderate-K compositions. High-K varieties appear only since the Middle Eocene. A data set of isotopic ages for volcanic rocks with various K abundances for the past 41 Ma was presented by Rasskazov et al. (in press). The moderate- and high-K groups of basalts are partially separated from each other in space and time. For the purpose of assessing relations between these two groups, we examine firstly spatial–temporal distribution of high-K basalts and then their relations with moderate-K ones.

12.4.1.1 Definition of High- and Moderate-K Basaltic Groups

The terms 'high potassic' and 'ultrapotassic' are used to define rocks, which have high contents of K_2O and other incompatible elements, elevated K_2O/Na_2O, and characteristics of relatively primitive basaltic magmas (i.e. high $100Mg/Mg+Fe^{2+}$, Ni, and Cr concentrations etc.). These rocks frequently contain leucite and mica in phenocrysts and groundmass. The ultrapotassic group has been constrained by $K_2O/Na_2O > 2$, $K_2O > 3$ wt%, and $MgO > 3$ wt% (Foley et al. 1987) and potassic feldspathoid-bearing rocks by $K_2O/Na_2O > 1$ with some exceptions (Le Bas 1989).

Cenozoic alkaline basalts from Central Mongolia were referred to a potassic series due to occurrence of sanidine megacrysts (Kepezhinskas 1979). Rocks of some volcanic fields show, however, K_2O/Na_2O ratios less than 1, although typical K-rich compositions with K_2O/Na_2O more than 1 are widespread, for example, in the Ugei-Nur area (Saltykovskii et al. 1984). Concentrations of such trace elements as Ba and Rb, geochemically connected with K, in Quaternary volcanic rocks from Taryat-Chulutyn are comparable to those in moderate-K volcanic rocks with K_2O/Na_2O = 0.47–0.80 from Jingpohu, Northeast China and are lower than in typical potassic rocks from Wudalianchi with K_2O/Na_2O = 1.0–1.7 (Chuvashova et al. 2007). Therefore, rocks from Central Mongolia should be treated as transitional between potassic and non-potassic end-members rather than referred to one of these groups (Fig. 12.7).

Lavas from the western part of the Ugei-Nur volcanic field show Na_2O+K_2O contents between 6 and 10 wt%, SiO_2 = 47–55 wt%, K = 2.8–4.3 wt%, and K_2O/Na_2O = 0.9–1.7. Those from the Tsetserleg volcanic field demonstrate lower total alkalis

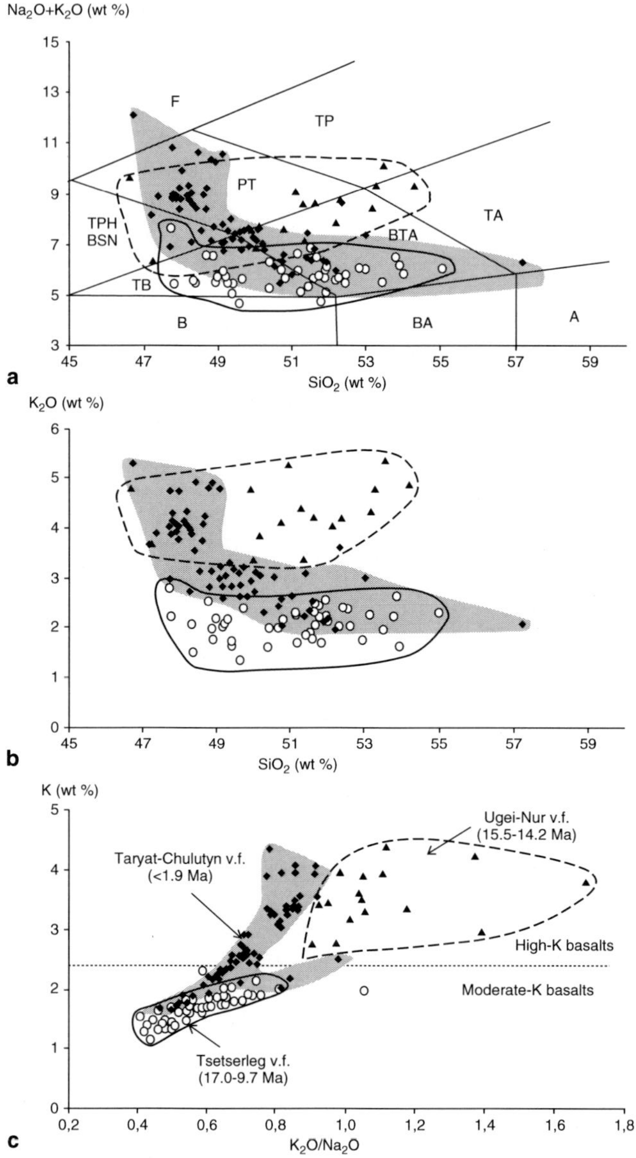

Fig. 12.7 Variations of Na_2O+K_2O vs. SiO_2 (**a**), K_2O vs. SiO_2(**b**), and K vs. K_2O/Na_2O (**c**) in moderate- and high-K basalts from Central Mongolia. Lavas from the Tsetserleg and Ugei-Nur volcanic fields (v.f.) are plotted as representative compositions for moderate- and high-K groups. Both are characterized by wide variations of SiO_2. The Quaternary lavas from the Taryat-Chulutyn volcanic field age shown as demonstrating continuous spectrum from moderate to high-K compositions. Fields of the IUGS classification on the diagram **a**, (Le Bas and Streckeisen 1991): *B* – basalt, *BA* – basaltic andesite, *A* – andesite, *TB* – trachybasalt, *BTA* – basaltic trachyandesite, *TA* – trachyandesite, *TPH*, *BSN* – tephrite, basanite, *PT* – phonotephrite, *TP* – tephriphonolite, *F* – foidite. The diagrams present unpublished data of authors

concentrations (5–7 wt%) with the same range of silica, potassium abundances from 1.0 to 2.0 wt% and variations of K_2O/Na_2O from 0.37 to 0.80.

The volcanic rocks from the Ugei-Nur and Tsetserleg volcanic fields are representative for definition of high- and moderate-K groups, respectively. Lavas from other volcanic fields of Central Mongolia belong to high- or moderate-K groups or reveal variations of K and K_2O/Na_2O overlapping both groups. An example of the latter series is Quaternary lavas of the Taryat-Chulutyn volcanic field, which show a trend from the moderate-K group of the Tsetserleg volcanic field to the high-K group with relatively low K_2O/Na_2O ratio (i.e. 0.8–1.0). Taking into consideration geologic setting of the high- and moderate-K groups in Central Mongolia, we discriminate them by a boundary concentration of K at 2.3 wt%.

12.4.1.2 Distribution of High-K Basalts

Figure 12.8 demonstrates spatial relations between migrating high-K magmatism and late Tertiary orogenic provinces and rift basins. The orogenic provinces of Hangay Highland are comprised into an isometric uplift separated from the East Hangay orogenic province by basins of the Central Mongolia rift zone.

High-K basalts occur mostly along the Noen-East-Hangay locus of volcanic fields-the main magmatic zone that stretches north–south from the Trans-Altay Gobi to East Hangay in distance over 380 km (Fig. 12.8). Its southern and northern portions are as wide as 140 km; the central part does not exceed 80 km. In the Hurmen and Noen volcanic fields of the southern part of the main zone, lavas erupted from 41 to 31 Ma, in the Arts Bogd and Valley-of-Lakes volcanic fields of the middle part from 32 to 30 Ma and from 15.6 to 12.0 Ma, respectively, and in the Harhorin, Upper-Orhon, and Erdene-Tsogt volcanic fields of the northern part from 5.5 to 1.3 Ma. Magmatic processes migrated northward along that zone with average velocity about 1 cm/year.

High-K magmatism, spatially connected with the main zone, occurred at 15.5–14.2, 13–11, 7.5–1.3, and <1.2 Ma. The first episode in Valley of Lakes and Ugei Nur demonstrates a high-K locus extending from the main zone northeastwards. The second episode in Valley of Lakes and Baidarik marks a high-K locus stretching from the main zone to the west–northwest. The third episode exhibits rejuvenated K-magmatism in Ugei-Nur with a subsequent quick southwestward shift to the main zone between 5.5 and 1.3 Ma. At last, the fourth episode expresses northward propagation of high-K magmatism from Ugei Nur to Orhon-Selenga similar to propagation along the Noen-East-Hangay zone.

Note that simultaneous eruptions of high-K basalts in Valley of Lakes and Ugei Nur at 15.5–14.2 Ma was followed by their subsequent shift over 90 km northward from the former area to Erdene Tsogt and from the latter area to Orhon-Selenga.

In Taryat-Chulutyn, eruptions of high-K basalts, commenced at ~7.1 Ma, accompanied rejuvenation of high-K volcanism in Ugei Nur and first appearance of high-K lavas in Tesin Gol. Subsequent high-K volcanic phases in Taryat-Chulutyn at

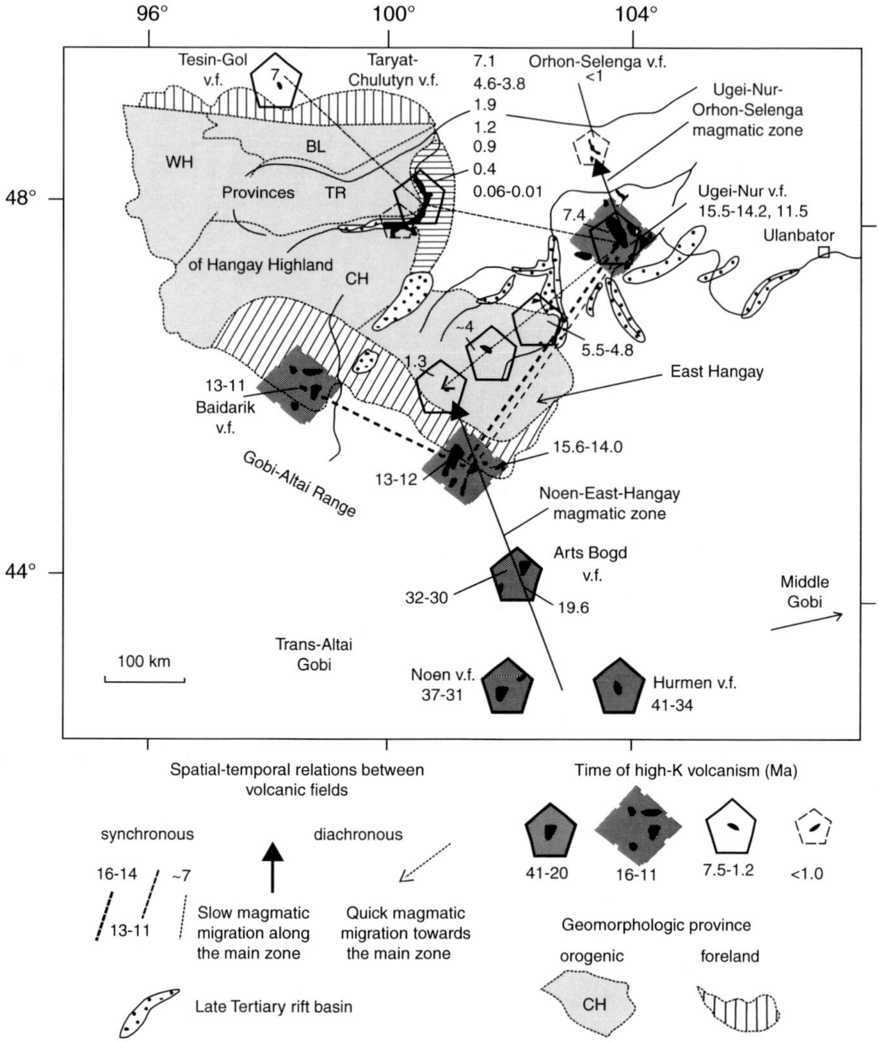

Fig. 12.8 Spatial–temporal distribution of high-K basalts in basic geomorphologic provinces of Central Mongolia. Contours of the geomorphologic provinces in Hangay Highland (*CH* – Central Hangay, *WH* – Western Hangay, *TR* – Tarbagatai Range, *BL* – Bolnai Range) and East Hangay are shown after Korina (1982)

4.6–3.8, ~1.9, and ~1.2 Ma were contemporaneous to southwestward volcanic migration from Ugei Nur to East Hangay. Afterwards, high-K volcanic episodes took place in Taryat-Chulutyn at ~0.9, ~0.4, and 0.06–0.01, when high-K volcanism ceased in East Hangay, but commenced in Orhon-Selenga.

In the past 1.2 Ma, high-K basalts erupted in Taryat-Chulutyn and Orhon-Selenga. Unlike the former locality, where high-K basalts continued erupting since the late

Miocene, high-K magmatic liquids of the latter appeared for the first time. The beginning of these eruptions after cessation of high-K basaltic volcanism in East Hangay demonstrated geodynamic connection between the Noen-East-Hangay and Ugei-Nur-Orhon-Selenga zones provided the Quaternary jump of high-K magmatism from the former, as the main zone of Central Mongolia, to the latter, as the incipient zone of the same rank.

12.4.1.3 Relationships Between High- and Moderate-K Basalts

The initial Middle Late Eocene volcanic interval of 41–31 Ma was exhibited only by high-K basalts in Hurmen and Noen. Afterwards, high-K lavas were followed by those of moderate-K compositions or lavas of high- and moderate-K groups alternated in time. Some areas were flooded also only by moderate-K lavas.

Between 32 and 30 Ma, high-K lavas of Arts Bogd erupted simultaneously with moderate-K lavas that intercalated with sediments of the Shand-Gol Formation in Valley of Lakes. The area of moderate-K basalts located at that time in the northern end of the active high-K part in the main zone (Fig. 12.9). For this unit, Höck et al. (1999) reported here one high-K composition for a sample (Mon10/96).

The Loh Formation, which range in age from 29 to 27 Ma, also included lavas of moderate-K compositions, except two reported samples with elevated potassium contents (Mon09/97 and Mon10/97). However, all the mentioned samples with elevated potassium abundances from Shand-Gol and Loh units are not included by Höck et al. (1999) in a list of reliably dated rocks.

The dated moderate-K lavas from Valley of Lakes span a time interval between 32 and 21 Ma. Compositionally similar lavas from the Ulan Tsab Khuduk and Dzabhan also fall in that time range. All volcanic localities trace the Dzabhan-Valley-of-Lakes zone. Sinistral displacement along the latter resulted in multiple synkinematical echelon-like system of dikes spatially related to the Shand-Gol Formation (Rasskazov and Chuvashova 2007).

Eruptions of high-K lavas in Arts Bogd began a new interval of magmatic activity that lasted from 20 to 10 Ma (Figs. 12.8, 12.9, 12.10). Shift of high-K lava eruptions along the main zone from Arts Bogd to Valley of Lakes between 19.6 and 15.6 Ma was accompanied by initiation of moderate-K volcanism in Tsetserleg. The latter area occupied at that time the northern extension of the active high-K part of the main zone (Fig. 12.10).

The spatial relation between high- and moderate-K magmatic liquids in Valley of Lakes and Tsetserleg repeated the relation between coeval high- and moderate-K lavas of Arts Bogd and Valley of Lakes at 32–30 Ma. Moderate-K eruptions commenced in the west–northwestern Dzabhan-Baidarik-Ulan-Tsab-Khuduk locus at ~16 Ma, just before high-K lavas flooded Valley of Lakes and Ugei Nur along the northeastern locus. Then high-K lavas distributed along the west–northwestern locus in Valley of Lakes and Baidarik in a time interval of 13–11 Ma, while moderate-K magmatism concentrated in the orthogonal locus of Tsetserleg and Hanui (Fig. 12.11).

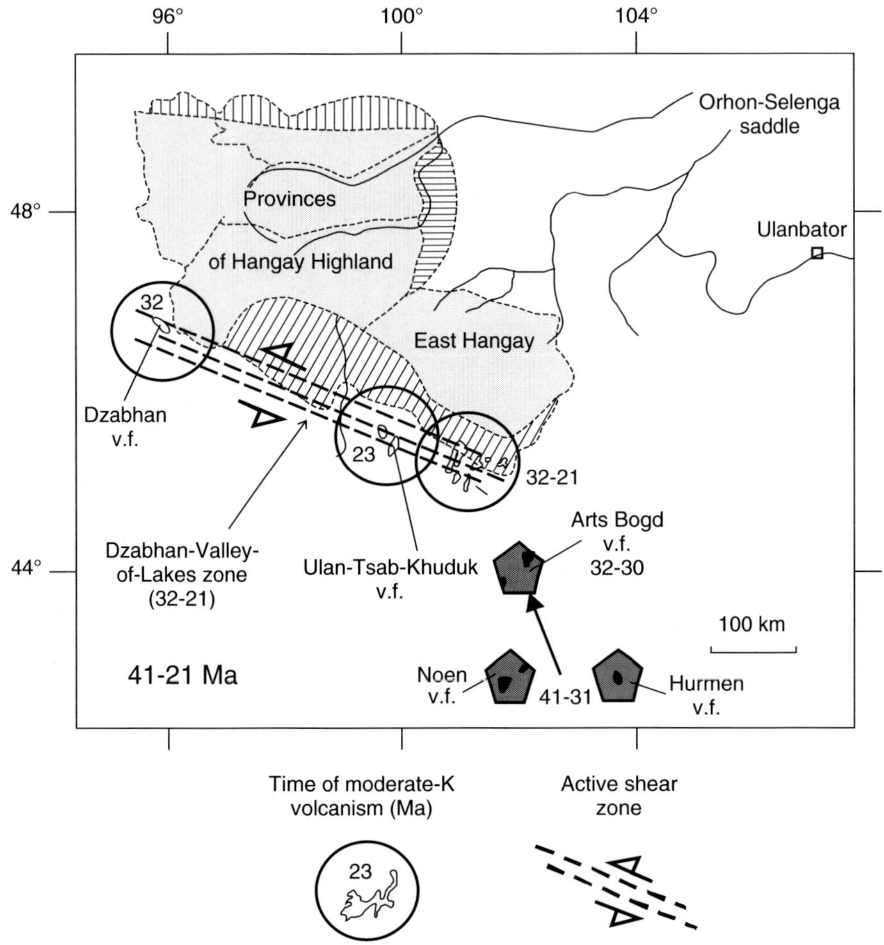

Fig. 12.9 Spatial–temporal distribution of volcanic fields with high- and moderate-K relatively basic geomorphologic provinces of Central Mongolia in a time interval of 41–21 Ma. Legend as in Fig. 12.8. The main zone of high-K magmatism reached Arts Bogd

A time interval from 11 to 8 Ma was characterized by moderate-K basaltic eruptions, separated magmatic intervals of 20–11 Ma and <8 Ma with high-K lavas. A line of active volcanic fields (Upper-Chulutyn, Tsetserleg, and Ugei-Nur) was directed orthogonally relative to the main Noen-East-Hangay zone of high-K magmatism (Fig. 12.12). Commencement of volcanism in Upper Chulutyn marked initiating activity of the right-lateral Chulutyn shear zone, which bounded in the east the main uplift of Hangay Highland. Its activity controlled volcanism both in the Upper-Chulutyn and Taryat-Chulutyn volcanic fields. Eruptions in the latter area began in the Late Miocene at ~7.1 Ma and finalized by the latest Pleistocene–Holocene volcanic phases of 65–50 Ka (disequilibrium ^{238}U–^{230}Th ages) and >8 Ka

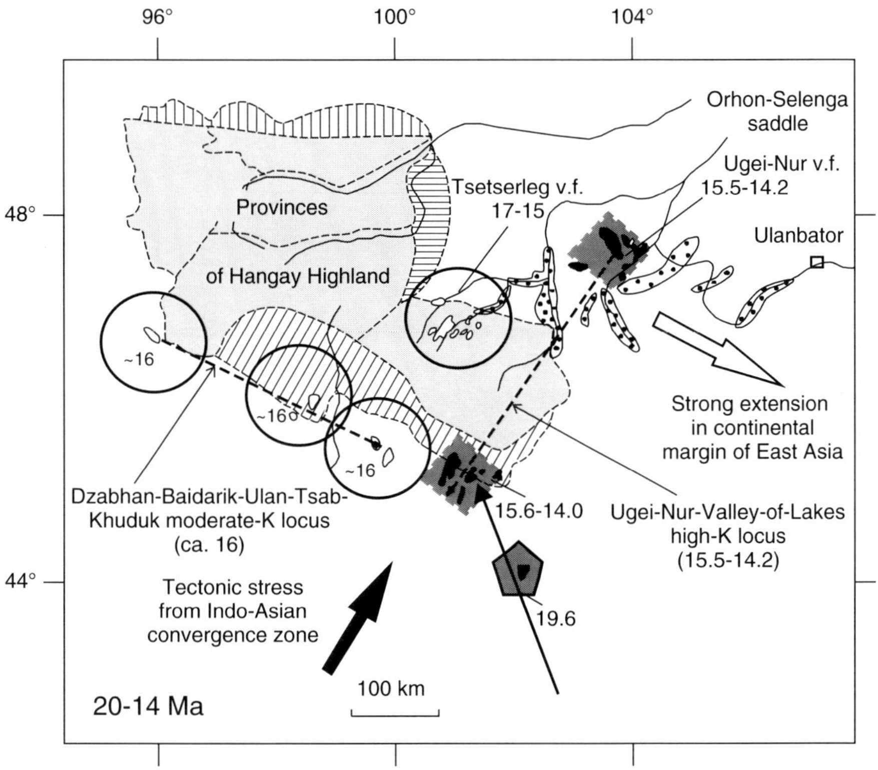

Fig. 12.10 Geometrical relations between volcanic fields with high- and moderate-K lavas of Central Mongolia in a time interval of 20–14 Ma. Legend as in Figs. 12.8 and 12.9. The main zone of high-K magmatism reached Valley of Lakes. Basins of the Central Mongolia rift zone began subsiding

(^{14}C calibrated age) (Devyatkin 2004; Chuvashova et al. 2007; Rasskazov et al. 2008a).

Appearance of high-K basaltic eruptions at ~7.1 Ma accompanied magmatic rejuvenation in a parabolic system of volcanic fields with a common geometric centre at Ugei-Nur. In a time interval of 7.1–6.0 Ma, an axis of the parabola corresponded to a locus of high-K basaltic eruptions in Ugei Nur and Taryat-Chulutyn, while branches extended from the parabolic center by moderate-K basaltic fields to the west by outflanking the uplifted region of Hangay Highland in the south and north (Fig. 12.13). At about 6.0–5.7 Ma, magmatic conditions changed, although eruptions continued in some previously active volcanic fields of the parabola. After extinction of magmatism in Ugei-Nur, the role of parabolic center started to be played by the Chulutyn shear zone. Moderate-K branches of the new parabola extended from that zone eastwards. The northern (Orhon-Selenga) and southern (East Hangay) branches stretched from Taryat-Chulutyn and Upper Chulutyn, respectively (Fig. 12.14).

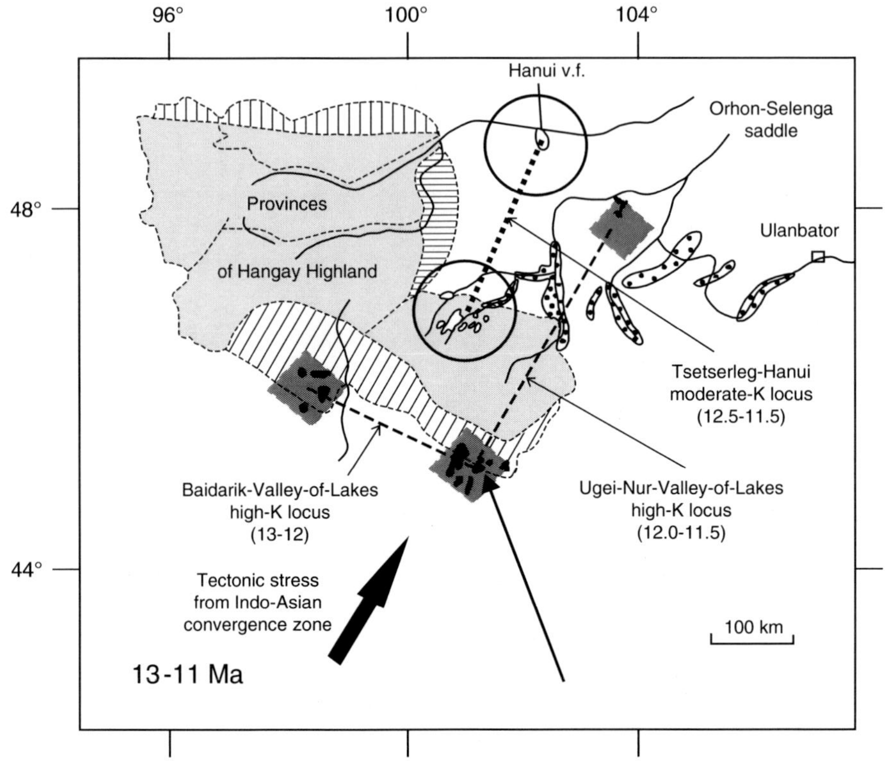

Fig. 12.11 Geometrical relations between volcanic fields with high- and moderate-K lavas of Central Mongolia in a time interval of 13–11 Ma. Legend as in Figs. 12.8 and 12.9. The locus of high-K lavas is directed orthogonally to the one of moderate-K compositions

In the past 1.2 Ma, moderate-K magmatism inherited the previous volcanic activity of the parabolic pattern, while high-K magmatism redistributed from Erdene Tsogt to Orkhon-Selenga (Fig. 12.15). The final moderate-K eruptions in Upper Orhon occurred at ~68 Ka as defined by disequilibrium ^{238}U–^{230}Th dating (Rasskazov et al. 2008a). Well preserved surface morphology of moderate-K lava flows in Orhon-Selenga indicates that eruptions finalized here less than 100 thousand years ago.

From the general analysis of spatial–temporal distributions of high- and moderate-K basalts, we infer that evolution of magmatism in Central Mongolia was governed mostly by the Noen-East-Hangay zone from 41 to 10 Ma with subsequent transition of magmatic activity to the parabolic magmatic pattern. The latter style of activity was maintained in the further magmatic evolution, though influence of the Noen-East-Hangay zone increased and resulted in migration of high-K magmatism from Ugei Nur to East Hangay between 7.4 and 1.3 Ma (Figs. 12.6, 12.11, 12.12). Afterwards, a role of the main control of high-K magmatic activity was transferred from the Noen-East-Hangay to the Ugei-Nur-Orhon-Selenga zone.

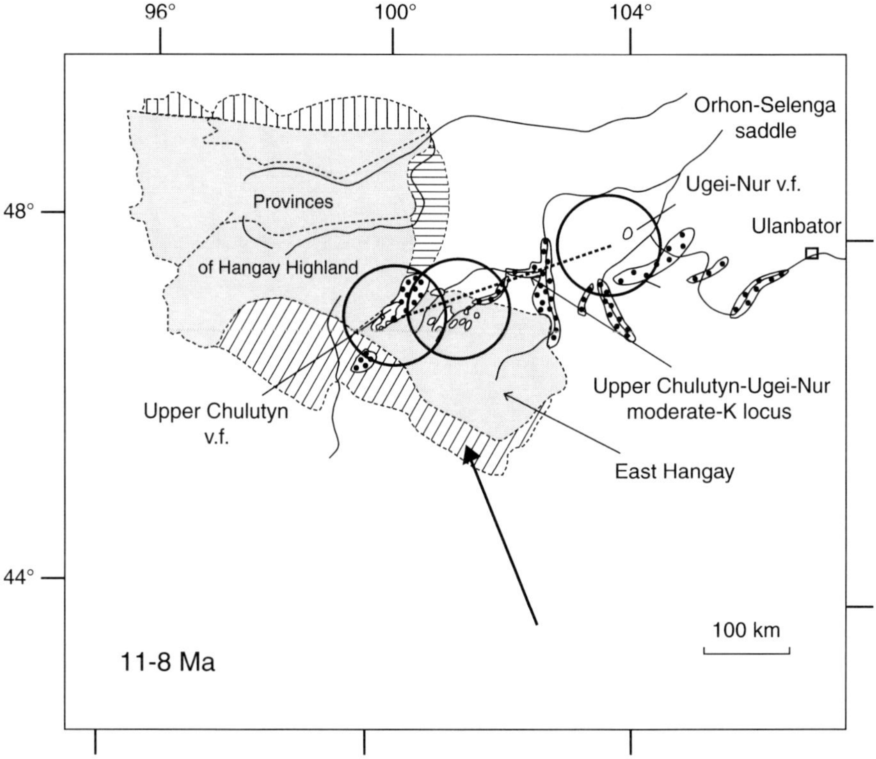

Fig. 12.12 Distribution of active volcanic fields of Central Mongolia in a time interval of 11–8 Ma. Legend as in Figs. 12.8 and 12.9. All lavas are moderate-K

12.4.1.4 Convergence-Related Magmatism: Comparisons with Tibet

In the Lhasa terrane of Southern Tibet, the granite batholith Gangdes was emplaced between 94 and 42 Ma, voluminous basaltic to silicic volcanic rocks of the Linzizong Formation were accumulated near Lhasa from 65 to 49 Ma and in southwestern Tibet from 62 to 37 Ma. High- and ultra-K lavas of basaltic to intermediate compositions erupted afterwards at 24–22 and ~13 Ma, while those with lower potassium concentrations occurred at 17–15 Ma. Intermediate to silicic volcanism took place from 26 to 8 Ma (Copeland et al. 1995; Ding et al. 2003; Hou et al. 2004; Guo et al. 2007; Mo et al. 2007).

In the Qiangtang terrane of Central Tibet, Cenozoic basaltic magmatism began simultaneously with magmatism of Southern Tibet, but its evolution was different. In the western part of the terrane, basalts erupted in a time interval of 65–44 Ma. In its eastern part, granites intruded from 51 to 39 Ma, simultaneously with initiation of basaltic eruptions, which occurred mostly between 41 and 32 Ma with shift to the western part of the terrane in a time interval of 31–24 Ma. In the Songpan-Ganzi terrane of Northern Tibet, the first magmatic events took place between 19

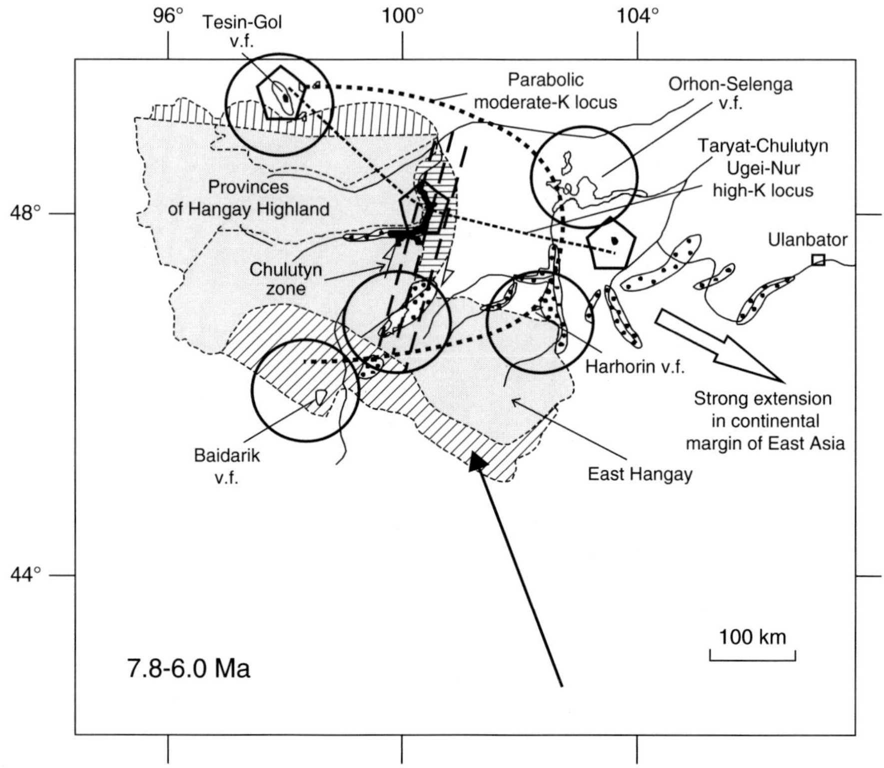

Fig. 12.13 Geometrical relations between volcanic fields with high- and moderate-K lavas of Central Mongolia in a time interval of 7.8–6.0 Ma. Legend as in Figs. 12.8 and 12.9. The high-K locus is an axis of a parabola formed by moderate-K lavas. The branches of the parabola are directed westward by outflanking the uplifted region of Hangay Highland in the south and north

and 17 Ma. Magmatism of intermediate through silicic compositions occurred at ~19 Ma, between 13 and 5 Ma and in the past 3 Ma (Arnaud et al. 1992; Otofuji et al. 1995a; Turner et al. 1996; Deng et al. 2000; Roger et al. 2000; Ding et al. 2000, 2003; Jolivet et al. 2003; Williams et al. 2004) (Figs. 12.16, 12.17).

Comparisons of data on Central Mongolia and Tibet reveal temporally consistent changes of magmatic processes expressed in variations of potassium abundances.

Emplacement of the pre-collision batholith Gangdes and eruptions of low-K basalts in Tibet correlate with eruptions of moderate-K basalts in Central Mongolia. We comprise the coeval liquids of basaltic to intermediate compositions by a common corridor that extends from the western part of the Qiangtang terrane through Gobi towards Northeast China and adjacent part of Russian Far East (Fig. 12.16) and suggest that the corridor reflects deformations of Asia caused by simultaneous collision-derived compression of its southern margin and back-arc extension of the eastern margin. The latter process was governed by a rollback mechanism operated

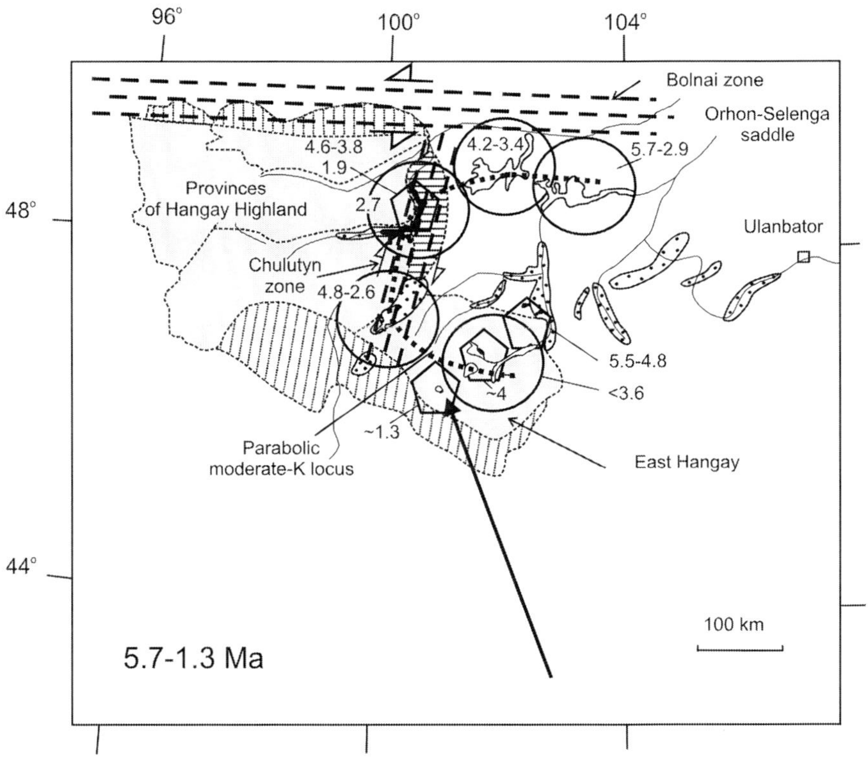

Fig. 12.14 Geometrical relations between volcanic fields with high- and moderate-K lavas of Central Mongolia in a time interval of 5.7–1.3 Ma. Legend as in Figs. 12.8 and 12.9. Fields of high-K basalts in East Hangay are shifting towards the Noen-East-Hangay main zone up to the northern terminus of the latter at Erdene Tsogt. Taryat-Chulutyn and Upper Chulutyn mark, respectively, the northern (Orhon-Selenga) and southern (East Hangay) branches of the moderate-K basaltic parabola

in the early Tertiary and repeated again in the Early–Middle Miocene with opening of the Sea of Japan (Otofuji et al. 1994, 1995b; Rasskazov and Taniguchi 2006).

Structural reorganization between 45 and 40 Ma—the proposed Middle Eocene transition from a ‘soft’ collision to a ‘hard’ post-collision convergence (Mo et al. 2007)—resulted in increasing potassium abundances in basaltic liquids and drastic spatial redistribution of volcanic activity both in Tibet and Central Mongolia. A new magmatic corridor extended from the eastern-central part of the Qiangtang terrane to the southern part of the Siberian craton in a time range of 41–21 Ma. We speculate (1) that Central Mongolia became a focus of magma-generating deformations developed between the Indian indenter and Siberian cratons and (2) that a temporally created tectonic stress was transferred from the Indo-Asian boundary through a low-velocity weakened layer at a level appropriate in depth to the

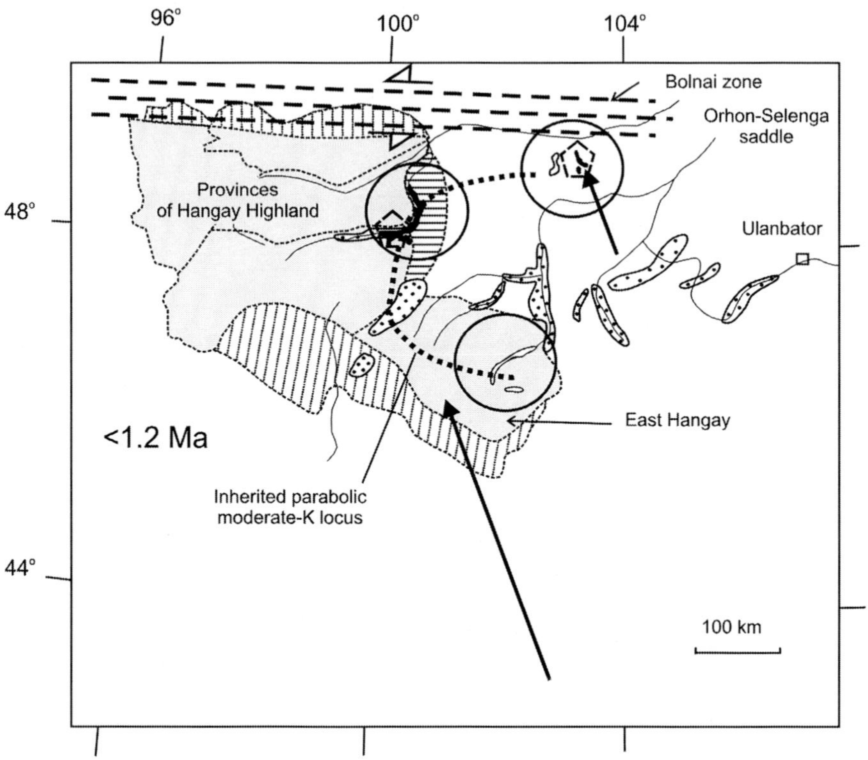

Fig. 12.15 Geometrical relations between volcanic fields with high- and moderate-K lavas of Central Mongolia in the past 1.2 Ma. Legend as in Figs. 12.8 and 12.9. Spatial distribution of volcanism is inherited from preceded time interval

Lehman discontinuity beneath the Siberian craton (Rasskazov et al. 2007a). The discontinuity is well defined here at depth ~210 km (e.g. Pavlenkova et al. 2002).

Similar temporal evolution of arc-like magmatic zones in Tibet and Central Mongolia was expressed by synchronous northwesterly-migrating volcanism in both areas at about 32–30 Ma, when volcanic eruptions shifted from the eastern part of the Qiangtang terrane to its western part (Deng et al. 2000; Ding et al. 2003; Williams et al. 2004) and from Trans-Altay Gobi to the Dzabhan-Valley-of-Lakes zone, respectively.

In time intervals of 20–10 and <10 Ma, volcanism of Central Mongolia was exhibited by both high- and moderate-K lava compositions. In the former interval, a new magmatic corridor was created with westward shift of magmatism in Tibet and was directed along the southeastern edge of the Siberian craton. That change is consistent with strengthening extension at the East Asian margin that resulted in opening of the Sea of Japan. In the latter interval, the Tibetan magmatic area remained spatially fixed simultaneously with a relative westward shift of magmatic area in Central Mongolia.

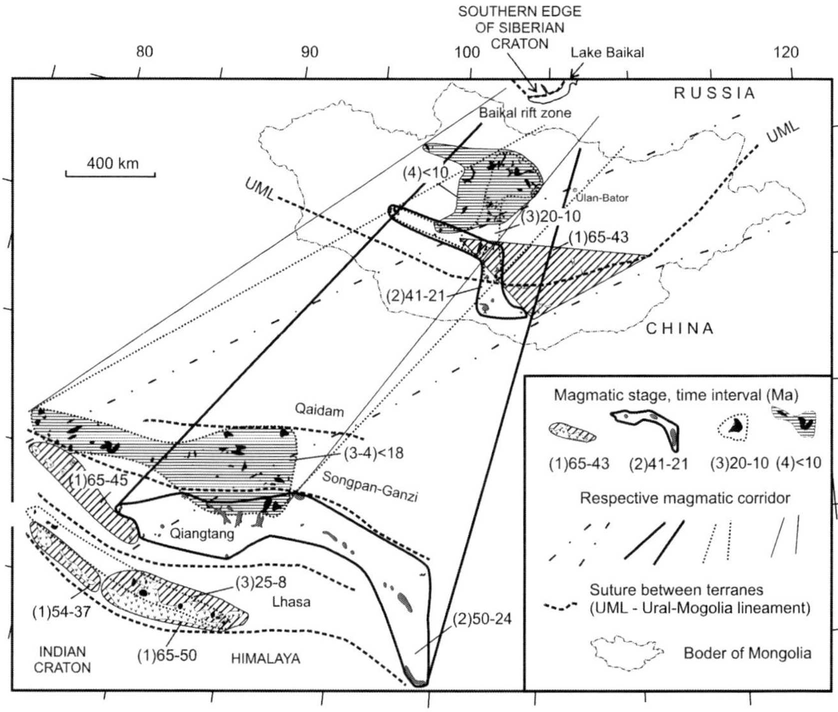

Fig. 12.16 Spatial correlation of Cenozoic magmatic evolution in Tibet and Central Mongolia. Magmatic corridors of different time intervals in Tibet join coeval volcanic fields in Central Mongolia and Tibet (Qiangtang and Songpan-Ganzi terranes)

12.4.1.5 Mantle Levels of High- and Moderate-K Magmatic Sources

Central Mongolia is underlain by the Mongolia-Okhotsk high-velocity slab located in the mantle depth more than 200 km (Bijwaard et al. 1998; Van der Voo et al. 1999b). A high-resolution seismic tomography model from surface S-waves (Yanovskaya and Kozhevnikov 2003) has shown existence under Central Asia of a vast low-velocity region in an interval of depths 50–200 km. This region, spatially coincided with area occupied by Cenozoic basalts in Central Mongolia and Sayan Mountains (Russia), was defined as the Sayan-Mongolia low-velocity domain (Rasskazov et al. 2003b). The image of the domain is consistent with a seismic velocity cross-section, compiled on records of exchanged SV-waves from MOBAL–2003 experiment (Mordvinova et al. 2007). Along the whole length of the cross-section, from the southern margin of the Siberian craton and southwestern Baikal rift zone to Central Mongolia, the lower and upper limits of the Sayan-Mongolia domain are traced at the same levels as in the previous S-wave model. A depth of the crust–mantle boundary corresponds to estimates of about 45 km based on study of deep-seated xenoliths from the Quaternary Shavaryn-Tsaram volcano, Taryat-Chulutyn (Kopylova et al. 1995). The domain is underlain by the mantle with normal and elevated velocities (Fig. 12.18).

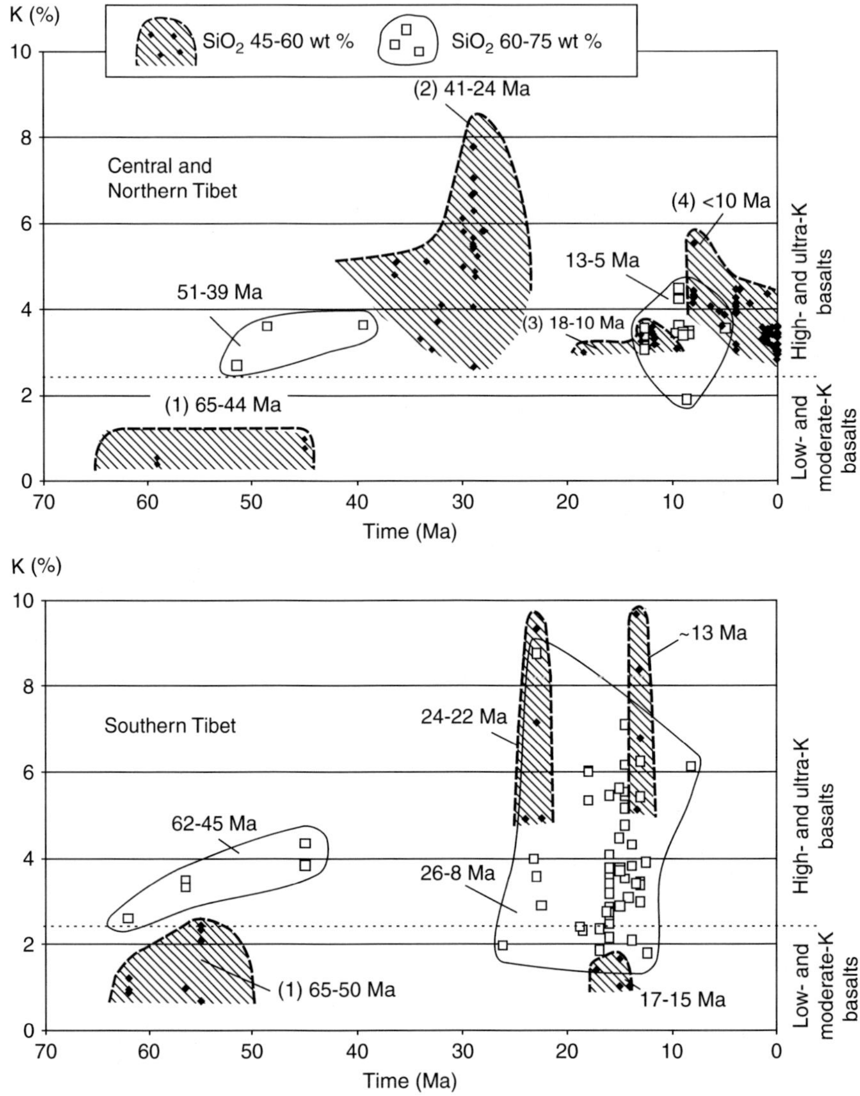

Fig. 12.17 Stages of Cenozoic magmatic evolution inferred from variations of potassium concentrations in volcanic rocks from basaltic to silicic compositions in Tibet (Data sources: Arnaud et al. 1992; Copeland et al. 1995; Deng et al. 2000; Ding et al. 2000, 2003; Guo et al. 2007; Hou et al. 2004; Jolivet et al. 2003; Mo et al. 2007; Otofuji et al. 1995a; Roger et al. 2000; Turner et al. 1996; Williams et al. 2004)

The cross-section displays details of the domain interior. Below a foot of the crust, there is a wedge-like low-velocity region bounded by a system of high-velocity steps lowered from the north to the south. The steps restrict low-velocity lenses under the station ARSH (Tunka rift valley of the Baikal rift zone) by depth 100 km, under

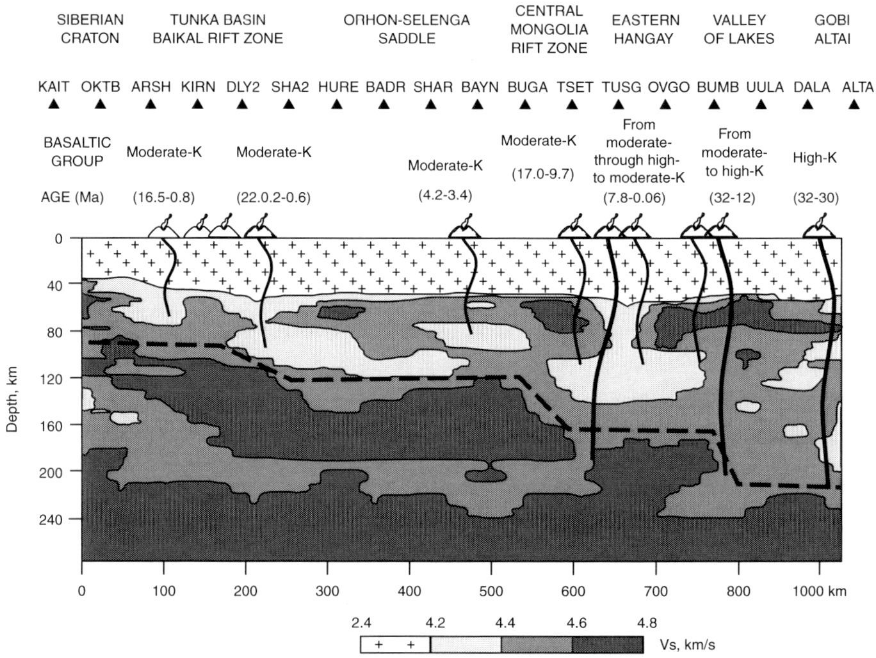

Fig. 12.18 Deep (120–200 km) and shallow (120–80 km) levels of sources for high- and moderate-K basaltic liquids on the simplified 2D velocity section (Mordvinova et al. 2007). The section is compiled using a receiver function (P-to-S-waves) for data of the MOBAL–2003 profile in depth interval 0–270 km on a model from 60 layers (interval of minimization 5–35 c, interpolation on a grid with 30 columns and 90 rows). A system of high-velocity steps (bounded at the top by a dashed line) is getting lower from 80–100 km beneath the Baikal rift zone (stations ARSH–DLY2) to ~200 km beneath Valley of Lakes and Gobi Altay (stations BUMB–ALTA). Location of the stations from BADR to ALTA in the southern and central parts of the profile MOBAL–2003 is shown in Fig. 12.19

the stations SHA2–BAYN (Orhon-Selenga saddle) by depth 140–120 km, under the stations TUSG–OVGO (East Hangay) by depth ~170 km, and under the stations BUMB–DALA (Valley of Lakes, Gobi Altay) by depth of the domain bottom. Under East Hangay, between stations BUGA and OVGO, a block with elevated speed occurs in a depth interval of 210–170 km, i.e. occupies the bottom of the low-velocity domain. A similar block with elevated velocity occurs at a depth interval of 180–130 km beneath the Orhon-Selenga saddle and separates a low-velocity layer at the bottom of the domain from the wedge-like low-velocity region.

The East Hangay block of elevated speed is overlain by a large low-velocity lens defined at a depth interval of 140–80 km. To the south, a region with irregular low-velocity patches is extended between 220 and 80 km beneath Valley of Lakes and Gobi Altay. To the north, some large low-velocity lenses are stretched beneath the Orhon-Selenga saddle and Baikal rift zone. These lenses exhibit a common low-velocity level constrained mostly in a depth interval from 120 to 80 km.

To describe spatial–temporal magmatic evolution in Central Mongolia, we have selected potassium concentration as the main characteristic of volcanic rocks. The main reason is that potassium contents in mantle liquids are indicative for depth estimations of melting processes. Sources of high-K liquids, erupted in Gobi Altai and Valley of Lakes, spatially correspond to the region with irregular low-velocity patches extended from the bottom to the top of the domain, unlike those of moderate-K liquids, erupted in East Hangay, Orhon-Selenga saddle and Baikal rift zone and expressed by shallow low-velocity lenses in the level of 120–80 km.

High activity of potassium characterizes the upper mantle of both old cratons and younger fold regions, being reflected in occurrence of potassium-bearing clinopyroxenes in xenoliths from kimberlites and inclusions in diamonds, as well as in diamond-bearing pyroxene-garnet-carbonaceous rocks of ultrahigh-pressure metamorphic complexes, tectonically moved from the mantle to earth surface. Elevated potassium concentrations of clinopyroxenes were recorded in experiments at high pressures and temperatures (Sobolev 1974; Tsuruta and Takahashi 1998; Perchuk et al. 2002; Shatskii et al. 2006). Natural rocks with potassium-bearing clinopyroxenes, constrained by pressures of 5–7 GPa (Safonov et al. 2005), appropriate to depths 120–180 km. That level appears to be the most probable for origin of high-K basaltic liquids in Central Mongolia, whereas more shallow melting results in generation of moderate-K liquids.

Depth of melting depends on degree of heating adiabatically raised mantle material. In a case of high heat capacity, initial melting occurs at a deeper level. The lower heat capacity, the more shallow level should be reached by ascending hot buoyant material for producing partial melting (Fukuyama 1985). We have shown that the beginning of the main Noen-East-Hangay zone (Gobi Altay) is characterized only by high-K magmatic liquids, but its middle part (Valley of Lakes) is exhibited by eruptions of moderate-K lavas with subsequent transition to high-K compositions. Hence, initial magmatism under Gobi Altay is produced in conditions of high heat capacity, sufficient for generation of deeper high-K magmas. Magmatic evolution in Valley of Lakes demonstrates initial melting at low heat capacity by producing moderate-K partial melts in shallow mantle, but subsequent increasing heat capacity resulted in melting of high-K magmas at deeper levels. The velocity image of the mantle region under Valley of Lakes in a depth interval of 220–80 km reflects, therefore, deeper magmatic generation, whereas shallow low-velocity lenses in the mantle beneath East Hangay, Orhon-Selenga saddle, and Baikal rift zone might demonstrate consequences of adiabatically rising mantle material with relatively low heat capacity.

12.4.1.6 Convergence- Versus Rift-Related Magmatic Processes

Orogenic provinces of Hangay Highland and East Hangay with the adjacent part of the Orhon-Selenga saddle are aseismic (Fig. 12.19). A lack of earthquakes is explained by crustal heating, which prevents accumulation of tectonic strains and reduces effective elastic thickness of crust up to 10 km and less (Bayasgalan et al. 2005). Anomalous heating is expressed also by occurrence of multiple thermal springs with temperatures

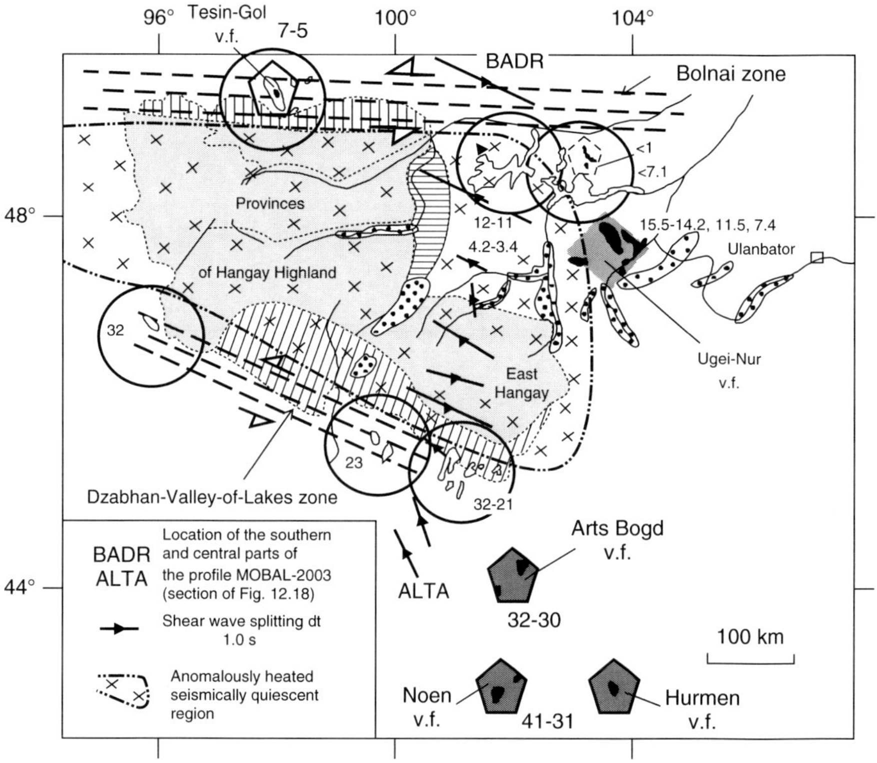

Fig. 12.19 Schematic presentation of successive magmatic reactivation of high- and moderate-K basaltic magmatism at the southern and northern boundaries of the aseismic anomalously heated region in Central Mongolia. Legend as in Figs. 12.8 and 12.9. The southern boundary was reactivated in a time interval of 41–21 Ma due to Indo-Asian convergence, whereas the northern boundary reactivated in the past 15.5 Ma, when the convergence processes were complicated by rifting contemporaneous to extension at the East Asian continental margin. A contour of the aseismic region is shown after a map compiled by Dugarmaa and Schlupp (2003). Data on the mean SKS splitting *dt* are shown after Barruol et al. (2008)

up to 89°C (Pissarsky et al. 2003). The GPS-data show concentration of present-day deformations on periphery of the aseismic area with decreasing deformation velocities toward its central part and irregular directions of contraction and extension axes (San'kov et al. 2005). We connect magmatic evolution, high heat flow, and spatial pattern of present-day crustal deformations with superimposed convergence and rifting processes.

High-K magmatism was produced in the main Noen-East-Hangay zone between 41 and 31 Ma due to increased heat capacity in a deep-seated magma generating region. Moderate-K magmatism, predominated in a time interval of 11–8 Ma, was controlled by the Upper-Chulutyn-Tsetserleg-Ugei-Nur locus trending orthogonally to the main zone (Fig. 12.12) and was favored by shallow melting that required minimal heat capacity of magmatic processes. Temporal coincidence of the Middle–Late

Eocene initiation of high-K magmatism in Tibet and Central Mongolia emphasizes a leading role of the Indo-Asian convergence, whereas a lack of high-K magmatism in the latter region means minimal influence of convergence processes. The latter conditions existed during structural reorganization in the Indo-Asian zone (Molnar 2005). At the time of the structural reorganization, the only volcanic phase of moderate-K lava eruptions occurred in Ugei-Nur, where high-K lavas prevailed. Long-term volcanic activity in Tsetserleg, lasted from 17.0 to 9.7 Ma and ceased simultaneously with a launch of long-term volcanic activity in Upper-Chulutyn, lasted from 9.6 to 2.1 Ma. Before the structural reorganization, magmatic activity was centered at Valley of Lakes. The structural reorganization resulted in initiating activity of the right-lateral Chulutyn shear zone and in creation of the parabolic magmatic pattern.

It seems plausible that high-K magmatism reflected melting caused by increasing role of convergence processes at the lower part of the Sayan-Mongolia low-velocity domain, at a level 120–210 km, whereas moderate-K liquids resulted from diminishing effect of convergence processes and melting in the upper part of the domain, in a depth interval of 80–120 km. The development of shallow low-velocity lenses appears to be responsible for origin of the Central Mongolia rift zone.

Tectonic stress results in melting of the mantle, which begins in a weakened locus, where tangential strains concentrate and cause lateral displacement by producing deep-focused earthquakes. Such a weakened portion of the mantle is deformed, elongated, and propagated linearly with speed of seismic waves (Spencer 1981). High-K melting in the main Noen-East-Hangay zone, caused by convergence, was developed in Central Mongolia under an angle of ca. 50–55 grades relative to maximal contraction of the lithosphere between the Indian indenter and Siberian craton. Optimal regime of melting in the tangential zone, however, was changed due to extension of the lithosphere in the East Asian continental margin or accelerated Indo-Asian convergence. In the former case, magmatic processes propagated from the main zone along the principle compression axis (i.e. north–northeastward). In the latter case, additional compression of the mantle resulted in lateral motion of a material. Such an effect was expressed by orthogonally (i.e. west–northwestward) migrating magmatism from the main zone (Chuvashova and Rasskazov 2008). This kind of mantle flow is deducted also from splitting of teleseismic shear waves (SKS phases) collected in the MOBAL–2003 experiment (Barruol et al. 2008). Figure 12.19 shows sufficient seismic anisotropy consistent with elongation of major orogenic provinces in Central Mongolia. Unlike a prominent shear wave splitting beneath East Hangay, anisotropic effect becomes negligible beneath basins of the Central Mongolia rift zone, probably, because of relaxation of tectonic strains around low-velocity lenses that underlay the rift zone at the shallow mantle level.

The initial Middle Eocene–Early Miocene magmatic stage, demonstrated by a regime of optimal convergence responsible for high-K basaltic eruptions along the Noen-East-Hangay main zone in a time interval of 41–31 Ma, was followed by accelerated convergence that resulted in lateral flow of mantle material and transition to moderate-K magmatism in the Dzabhan-Valley-of-Lakes zone in a time interval of 32–21 Ma. Similarly, an optimal convergence regime started the Miocene magmatic stage by eruptions in Arts Bogd (high-K basalts of 19.6 Ma) and

Tsetserleg (moderate-K basalts of 17 Ma). Afterwards, the normal convergence conditions changed to conjugate convergence and extension processes that resulted in orthogonal arrangements of high- and moderate-K magmatic loci between 16 and 11 Ma. A phase of accelerated convergence was expressed by lateral motion of mantle material along the Dzabhan-Valley-of-Lakes zone at about 16 Ma and accompanied high-K basaltic eruptions at the eastern terminus of the latter. A phase of extension was responsible for high-K basaltic eruptions along the Ugei-Nur-Valley-of-Lakes locus at 15.5–14.2 Ma with extinction of magmatic activity along the moderate-K part of the Dzabhan-Valley-of-Lakes zone. Then, a 13–11 Ma phase of accelerated convergence was pronounced again by high-K basaltic eruptions in the Dzabhan-Valley-of-Lakes zone simultaneously with extension expressed by the orthogonal Tsetserleg-Hanui locus of moderate-K basalts (Figs. 12.6, 12.7).

If a northward shift of high-K magmatism from Ugei-Nur to Orhon-Selenga is of the same nature as the magmatic shift along the main Noen-East-Hangay zone (Fig. 12.8), high-K magmatism in Orhon-Selenga marks an optimal convergence regime of the past 1 Ma.

The anomalously heated aseismic region of Central Mongolia is bounded on the south and north by asynchronous but dynamically similar zones (Fig. 12.19). Volcanic fields with moderate-K lavas are shifted northward and westward relatively to those with high-K lavas both in the main Noen-East-Hangay and incipient Ugei-Nur-Orhon-Selenga zones. In the latter, high-K basaltic eruptions of Ugei Nur at 15.5–14.2 Ma were followed by moderate-K ones in Hanui at 12.5–11.5 Ma, in Tesin Gol at 7–5 Ma, in Hanui and Orhon-Selenga in the past 7.1 Ma. Moderate-K basalt distribution along the Bolnai shear zone is similar to volcanic scattering along the Dzabhan-Valley-of-Lakes one, so both magmatic loci appear to originate due to lateral flow of mantle material caused by accelerated convergence.

If high-K magmatism in Ugei-Nur was promoted by pulses of stretching in the East Asian continental margin at 15.5–14.2 and ~7.4 Ma, migration of high-K magmatism from Ugei-Nur to the Noen-East-Hangay zone in a time interval of 7.4–1.3 Ma might indicate relative decreasing the imposed stretching effect. High-K magmatism, caused by extension pulses, occurred at latitude between $47^{\circ}50'$ and $48^{\circ}40'$ (Fig. 12.20) and took place simultaneously with subsidence of basins in the Central Mongolia rift zone.

We infer that high-K convergence-related magmatism migrated along the main Noen-East-Hangay zone towards the rift zone developed in Central Mongolia in the late Tertiary. The initial rift-related magmatism occurred in Ugei Nur, which, in turn, was a starting point for development of convergence-related events of the incipient Ugei-Nur-Orhon-Selenga zone. Evolution of the convergence-caused magmatism was complicated by an extension pulse resulting in a quick movement of high-K basaltic magmatism from the main Noen-East-Hangay zone to an area of spatially superimposed convergence- and rift-related processes, while the convergence-caused magmatism continued in the main Noen-East-Hangay zone. Melting regions of both the main long-lived and incipient zones were enlarged principally by westward propagation of moderate-K basaltic eruptions consistent with lateral flow of mantle material in conditions of accelerated convergence.

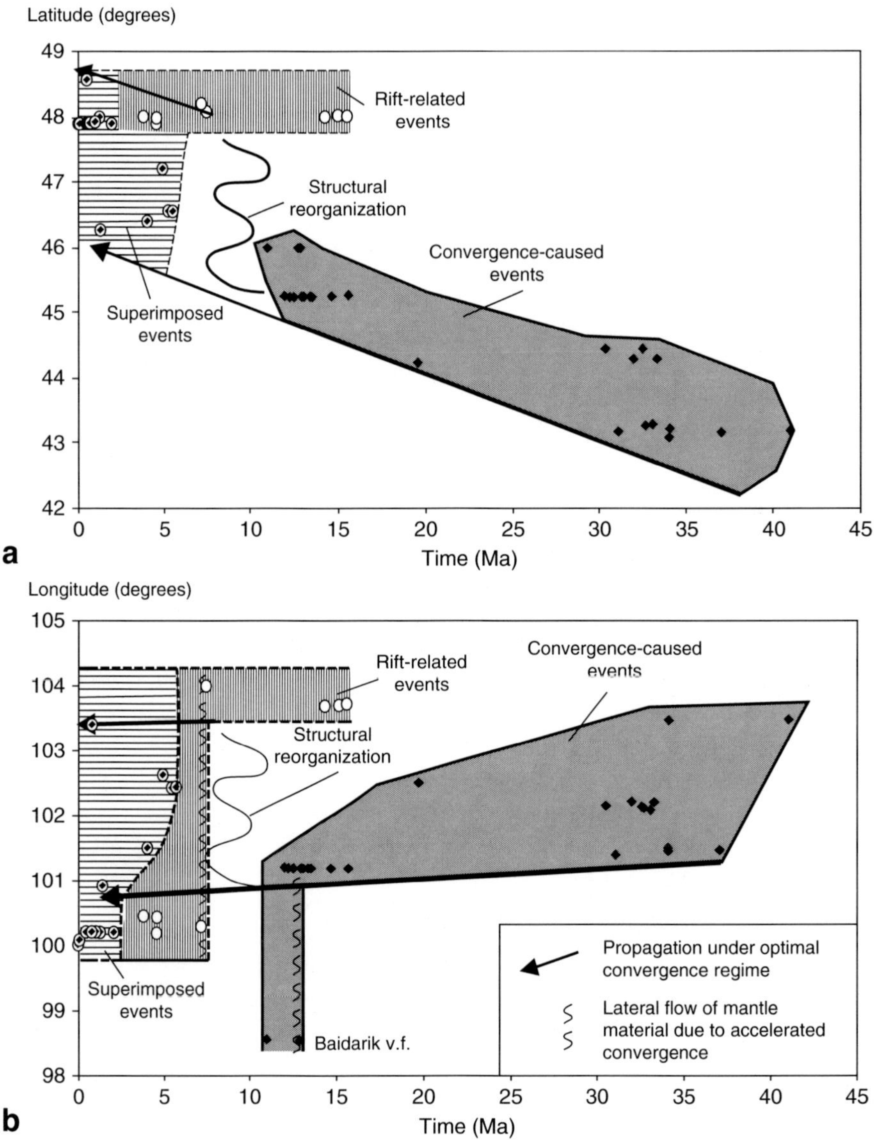

Fig. 12.20 North–south (**a**) and east–west (**b**) temporal transition from convergence-caused through rift-related to superimposed events of high-K magmatism in Central Mongolia

12.4.1.7 Cyclic Variations of Potassium

As depth antipodes, convergence- and rift-related processes result in spatial–temporal separating of the deeper (i.e. convergence-related high-K) and shallow (i.e. rift-related moderate-K) liquids, as shown in Fig. 12.18. This relationship is complicated, however, by occurrence of high-K basalts in the Central Mongolia rift zone. In a time interval of 41–21 Ma convergence-caused high-K liquids of the first 10 Ma changed to moderate-K ones of the subsequent 10 Ma. The conjugated convergence- and rift-related high- and moderate-K magmatic phases alternated.

Figure 12.21 demonstrates opposite temporal variations of potassium concentrations in convergence-related basalts from the southern part of the Central Mongolia volcanic region and rift-related basalts from its northeastern part. In general evolution of convergence-related magmatism, potassium maxima occurred in lavas erupted at ~37, ~19.6, 13–11, ~5.5, ~1.3 Ma alternated with its minima at ~30, ~16, ~10, and 3.6–3.4 Ma. Potassium minima of convergence-related magmatism are close to potassium maxima of rift-related magmatism at ~15.5 and ~3.6–3.4 Ma. Mutually consistent extremes of potassium concentrations in convergence- and rift-related basalts at 13–11 Ma and ~10 Ma preceded and accompanied reorganization of magmatic system. The potassic maximum of rift-related basalts in the northeastern volcanic region at ~7.4 Ma has no appropriate convergence-related counterpart in its southern portion.

In the conjugated convergence- and rift-related processes, the longest interval of high-K magmatism accompanied initiation of rifting between 15.6 and 11.0 Ma.

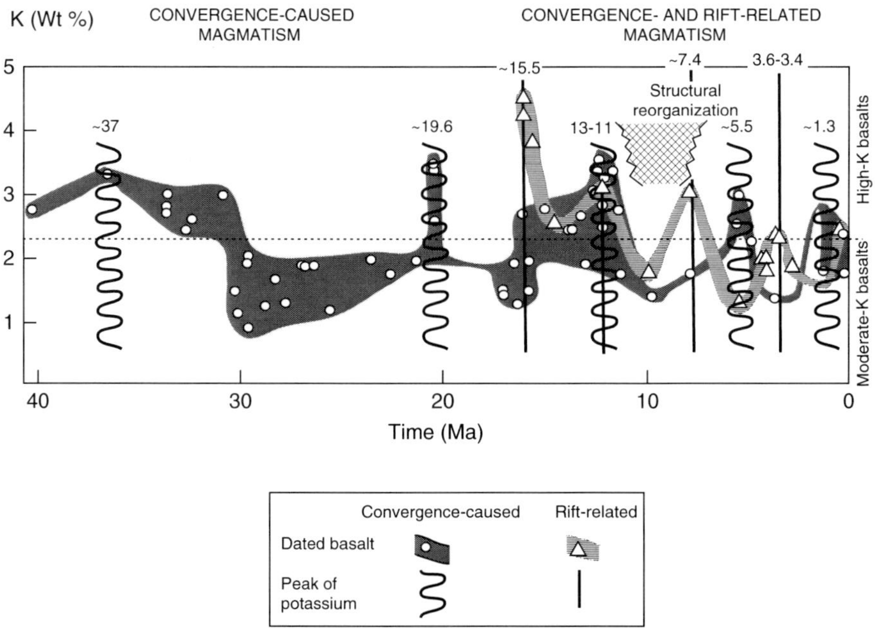

Fig. 12.21 Temporal transition from convergence-caused magmatism to coeval convergence- and rift-related magmatism developed with opposite potassium variations in the southern and northeastern parts of the Central Mongolia volcanic region (lavas from the Chulutyn zone are not plotted). Change of quasi-periodicity at about 20 Ma demonstrates transition from the first-order middle Tertiary to the second-order late Tertiary magmatic events (Fig. 12.2)

A short high-K volcanic phase in Arts Bogd at ~19.6 Ma was followed by lavas erupting in Valley of Lakes at ~15.6 Ma with concentration of potassium ~2.7 wt%. Afterwards, the latter area was flooded by lavas intermittently erupted up to 12.0 Ma with variable potassium concentrations (from 2.3 to 3.3 wt%) and was accompanied by lava eruptions in Baidarik at 13–11 Ma with higher potassium abundances (3.4–3.6 wt%). In the latter area, this volcanic phase was anticipated and followed by phases of moderate-K lavas at ~15.6 and ~7.0 Ma, respectively. On the contrary, a phase of moderate-K lavas at ~9.6 Ma in Ugei Nur was anticipated and followed by phases of high-K lavas at 15.5–14.2 and ~7.4 Ma, respectively (Fig. 12.22).

Individual volcanic fields show moderate-K lavas erupted over 7.0–7.5 m.y. intervals. In Tsetserleg, this kind of volcanic activity is constrained between 17.0 and 9.7 Ma with relative increasing potassium concentrations at 15.6–15.5 Ma, consistent with initial high-K lava eruptions in Valley of Lakes and Ugei Nur. Such shift of lava compositions demonstrates a connection between high- and moderate-K magmatic processes.

In a time interval of 10–2 Ma, variations of potassium in basalts from Upper Chulutyn and Taryat-Chulutyn were mutually consistent and revealed cyclic processes with quasi-periodicity of 2.5 Ma. In the latter area, quasi-periodical phenomena were

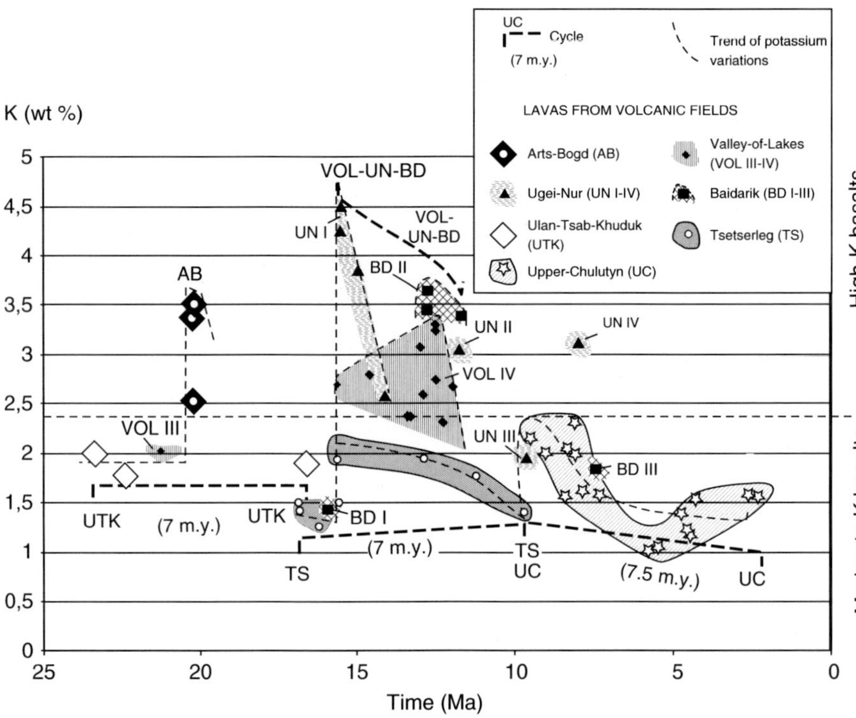

Fig. 12.22 Temporal relations of volcanic activity with high- and moderate-K lavas in a time interval of 23.4–2.1 Ma. High-K compositions are exhibited by the common volcanic interval of 15.5–11.0 Ma in Valley of Lake, Ugei-Nur, and Baidarik (VOL–UN–BD)

expressed by eruptions of high-K lavas at ~7.1 and 4.6–3.8 Ma in background of moderate-K basalt eruptions in the former area. The 9.6–7.4 Ma lavas of the Upper Chulutyn predated eruptions in Taryat-Chulutyn and those of 5.5–4.0 and 2.6–2.1 Ma marked final phases of the cycles with minimal potassium concentrations.

Change of magmatic conditions in the past 2 Ma resulted in extinction of volcanism in Upper Chulutyn and in four new cycles of high-K lava eruptions in Taryat Chulutyn with quasi-periods of 0.3–0.7 m.y. The two latest cycles exhibited, beside high-K, moderate-K lava compositions (Fig. 12.23).

Quasi-periodic variations of potassium in lavas from Upper Chulutyn and Taryat-Chulutyn reflect alternating convergence- and rift-related processes. Between 7.1 and 2.7 Ma lavas of Taryat-Chulutyn were derived from two mantle sources (Fig. 12.24). High-K basalts show isotopically-depleted (Nd, Sr) compositions with wide variations of K/Ta and La/Ta at K/La = const. Such variations are characteristic of basalts from axial basins in continental rift zones (Rasskazov et al. 2008b). Moderate-K lavas have less depleted isotopic signatures than high-K lavas and elevated K/Ta ratios with relatively low La/Ta ratios, originating from slab fragments within the Sayan-Mongolia low-velocity domain. Character of cyclic potassium variations changed in Taryat-Chulutyn at 1.9 Ma, when lavas revealed elevated $^{87}Sr/^{86}Sr$ ratios (0.7052–0.7053). From ~1.2 Ma to the Holocene, basalts showed a narrow range of $^{87}Sr/^{86}Sr$

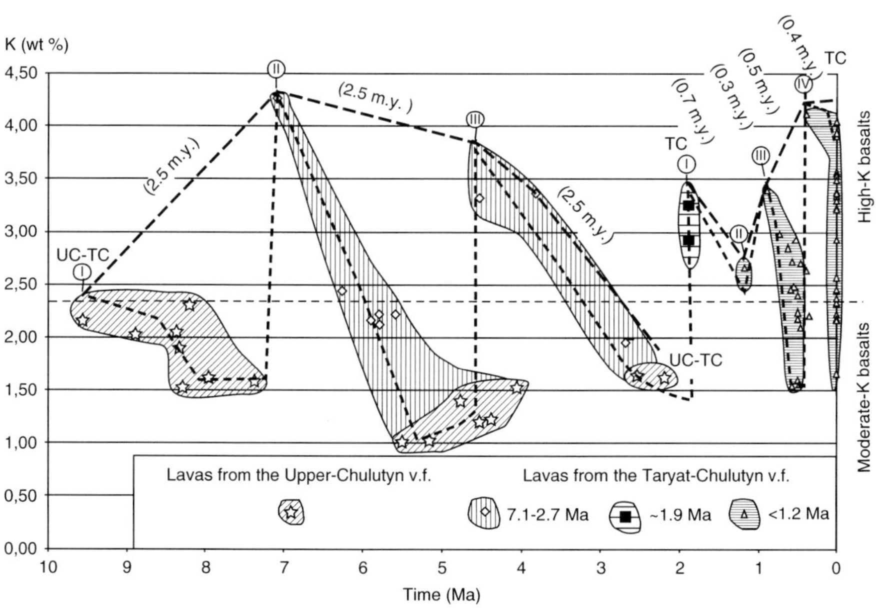

Fig. 12.23 Transition from eruptions in Upper Chulutyn and Taryat-Chulutyn (9.6–2.1 Ma) with consistent variations of potassium (*UC–TC cycles I–III*, quasi-periodicity 2.5 Ma) to eruptions in Taryat-Chulutyn (*TC cycles I–IV*, quasi-periodicity 0.3–0.7 Ma). Symbols as in Fig. 12.22. Change of quasi-periodicity about 2 Ma demonstrates transition from the second-order Tertiary to the third-order latest Pliocene–Quaternary magmatic events (Fig. 12.2)

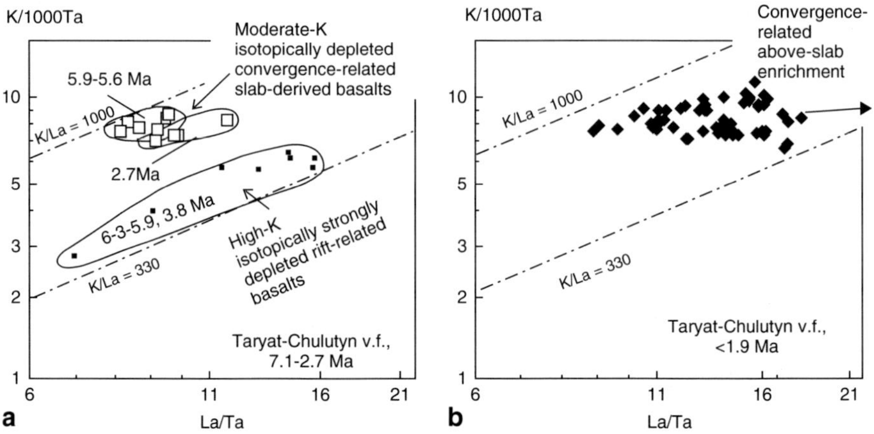

Fig. 12.24 Variations of K/1000Ta vs. La/Ta in convergence- and rift-related basalts of Taryat-Chulutyn from 7.1 to 2.7 Ma (**a**) and <1.9 Ma (**b**). Panel **a**, demonstrates groups of high-K rift-related compositions with K/La = const and moderate-K convergence-related compositions with elevated K/1000Ta and relatively low La/Ta. Panel **b**, shows convergence-caused enrichment of mantle sources with relative increase of La/Ta that is characteristic for above-slab post-collision convergence (Rasskazov et al. 2008b). Change of trace-element abundances at 1.9 Ma is consistent with a hierarchical transition from the second-order Tertiary to the third-order Quaternary magmatic processes. Potassium concentrations were measured by flame photometry, La and Ta by ICP MS technique as described by Rasskazov et al. (2004b)

ratios (0.7046–0.7047) with Sr concentrations from 2000 to 625 ppm (Chuvashova et al. 2007). In those basalts, La/Ta ratios increase without sufficient variations of K/Ta that is characteristic for melts from an enriched above-slab source of post-collision convergence in Central Mongolia. Similar component was identified in the East Hangay lavas, which erupted in a time interval of the past 3.6 Ma (Rasskazov et al. 2008b). Such type of magmatic sources was initiated here 2–3 Ma earlier, than in Taryat-Chulutyn.

12.4.2 South Primorye and Adjacent East Asia

Strong extension in an active continental margin of East Asia is expressed in the Sea of Japan opening. Timing and mechanisms of extension is under discussion. A back-arc extension has been suggested to begin in a time interval from 32 to 15 Ma (Tamaki et al. 1992; Shuto et al. 1993; Ohki et al. 1994a, b; Jolivet et al. 1994; Pouclet et al. 1995; Okamura et al. 1998). Middle–late Tertiary magmatism in the southwest of Primorye was examined generally as a post-subduction process related to the Sea of Japan opening following Late Cretaceous subduction of Pacific plates beneath Asia (Martynov et al. 2001). Geochronological data obtained for Southeast Primorye, located at the continental margin in an opposite side of Sea of Japan relative to Northeast

Honshu, revealed, however, evidence of the earlier extension-related intraplate-type magmatic events commenced at about 38 Ma (Rasskazov et al. 2004b).

On the one hand, a boundary between the Pacific plate and Eurasia is often referred to transform type (Natal'in 1993; Khanchuk 2000; Parfenov et al. 2003). This kind of interplate kinematics prevents calc-alkaline magmatism usually related to plate subduction (Maruyama and Seno 1986). On the other hand, the major driving forces responsible for plate motions are created due to pull-down effect of subducting slabs (Forsyth and Uyeda 1975). In this respect, subduction is considered as the main factor operating at interplate convergent zones, strike-slip faulting (transform motion) being a secondary phenomenon complicated structural relations between lithospheric plates.

12.4.2.1 Volcanic Sequence

Middle–late Tertiary volcanic and volcano-sedimentary sequence involves: (1) Narva Formation (rhyolites and dacites), (2) Zaisan Formation (Middle–Upper Eocene calc-alkaline basalts and andesites), (3) Klerk stratum (Upper Eocene high-Ti basalts and trachyandesites), (4) Kraskino Complex (Lower Oligocene rhyolites and dacites), (5) Slavyanka Complex (Upper Oligocene basalts and andesites), (6) Sineutesovsky Formation (Late Oligocene–Lower Miocene coal-bearing sediments capped by basaltic lavas), and (7) Shufan Formation (Middle Miocene through Pliocene basaltic lavas) (Figs. 12.22, 12.23). This stratigraphic sequence is based on results of geologic mapping, paleontological records in sediments, and isotopic dating of volcanic rocks by the K–Ar, $^{40}Ar/^{39}Ar$, and Rb–Sr methods.

Volcanic rocks of the Narva Formation, dated at ~46.2 Ma, were mentioned in the previous section as the unit related to the Lutetian event predated the early Tertiary structural reorganization. The Zaisan Formation was traced in the Kraskino basin in drill holes above sediments of the Paleocene–Lower Eocene Nazimov Formation and Eocene–Oligocene flora-bearing part of the Hasan Formation, but below lavas of the Kraskino volcano-plutonic complex.

Lavas of the Klerk peninsula are moderately alkaline basalts with elevated abundances of TiO_2 (1.5–2.1 wt%), Al_2O_3 (16.1–17.7 wt%), P_2O_5 (0.55–0.92 wt%) and moderate concentrations of MgO (1.8–6.2 wt%). Trace element ratios, such as Ba/Nb, Ba/La, Ce/Pb etc. are typical for intraplate basalts. These lavas are coeval with calc-alkaline lavas of the Zaisan Formation and represent a specific stratigraphic unit referred to as 'Klerk stratum' (Rasskazov et al. 2004b).

Two samples of high-Ti basalts from the Klerk peninsula yielded whole rock $^{40}Ar/^{39}Ar$ plateau ages of 37.7 ± 1.3 Ma and 34.4 ± 1.0 Ma. One more $^{40}Ar/^{39}Ar$ plateau age of 38.1 ± 1.3 Ma was obtained for a high-Mg basaltic dike cropped in the Lukin cape (sample Sh–00–97, see location on Fig. 12.24). The dike was the latest in a rhyolite–dacite–basalt sequence. The age obtained is comparable within analytical error with the older age of the high-Ti basalts from the Klerk peninsula. No dating of dacites and rhyolites was performed in a K–Ar isotope system because these rocks could lose radiogenic argon due to heating by younger basaltic dikes.

Rhyolites from the Kraskino Complex yielded the whole rock Rb–Sr isochron age of 33.5 ± 1.1 Ma with initial $^{87}Sr/^{86}Sr$ of 0.704 67 ± 0.000 03 and the MSWD value of 0.7. This unit was formed after high-Ti basalts of the Klerk peninsula (Figs. 12.25 and 12.26).

The Slavyanka Complex includes two stratigraphic members. The lower one consists of agglomerate tuffs with rare basaltic lava flows. The upper member exhibits trachyandesitic, andesitic, and dacitic lavas as well as trachydacitic and dacitic plagues. An age of the upper member was constrained by Rb–Sr isochron at 22.9 ± 0.3 Ma obtained for two K-feldspar separates and their host rocks of the Nerpa dacite

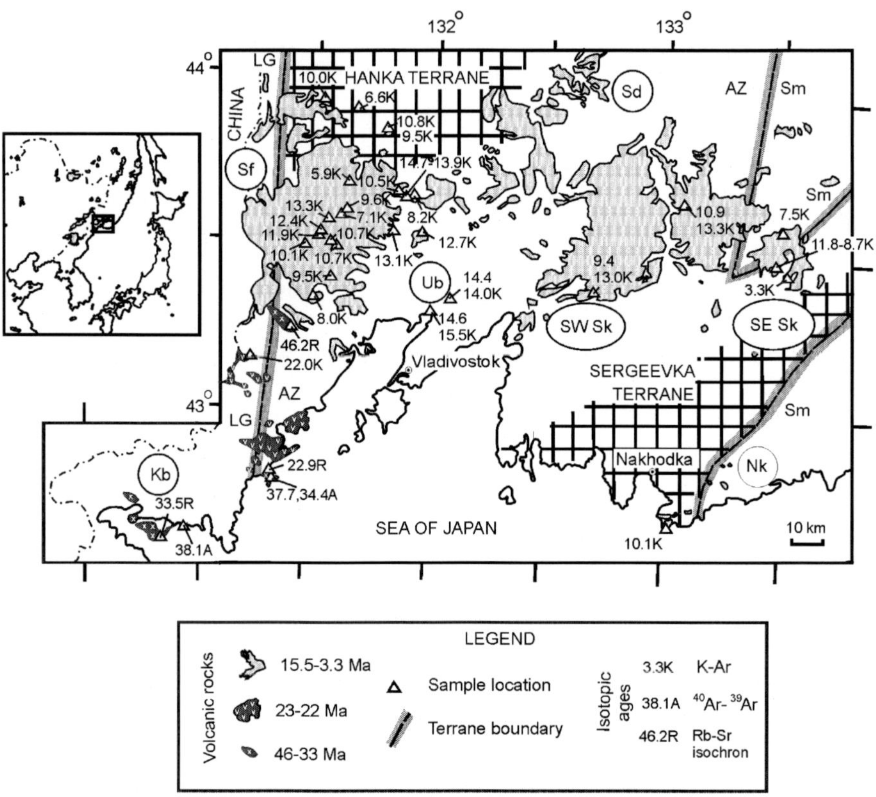

Fig. 12.25 Spatial–temporal distribution of middle–late Tertiary volcanic rocks in South Primorye. Volcanic fields: *Sf* – Shufan, *Sk* – Shkotov, *Nk* – Nakhodka, *Sd* – Sandugan, *Ub* – Uglovka basin, *Kb* – Kraskino basin. Units of 46–33 Ma: Narva Formation (46.2 Ma), Klerk stratum (38–34 Ma), Zaisan Formation (38–33 Ma), and Kraskino Complex (~33–32 Ma). Units of 23–22 Ma: Slavyanka Complex and Sineutesovsky Formation. Units of 15.5–3.3 Ma: two low-volume and two voluminous lava generations of the Shufan Formation. Tectono-stratigraphic terranes: *LG* – Laoelin-Grodekov terrane, *AZ* – Arsenev zone of the Hanka terrane, *Sm* – Samarka terrane. (Data sources: Okamura et al. 1998; Rasskazov et al. 2003g, 2004b; Popov et al. 2005; Sakhno et al. 2004; Martynov et al. 2002; Chashchin et al. 2007; Maksimov and Sakhno 2008; Brandt et al. 2009)

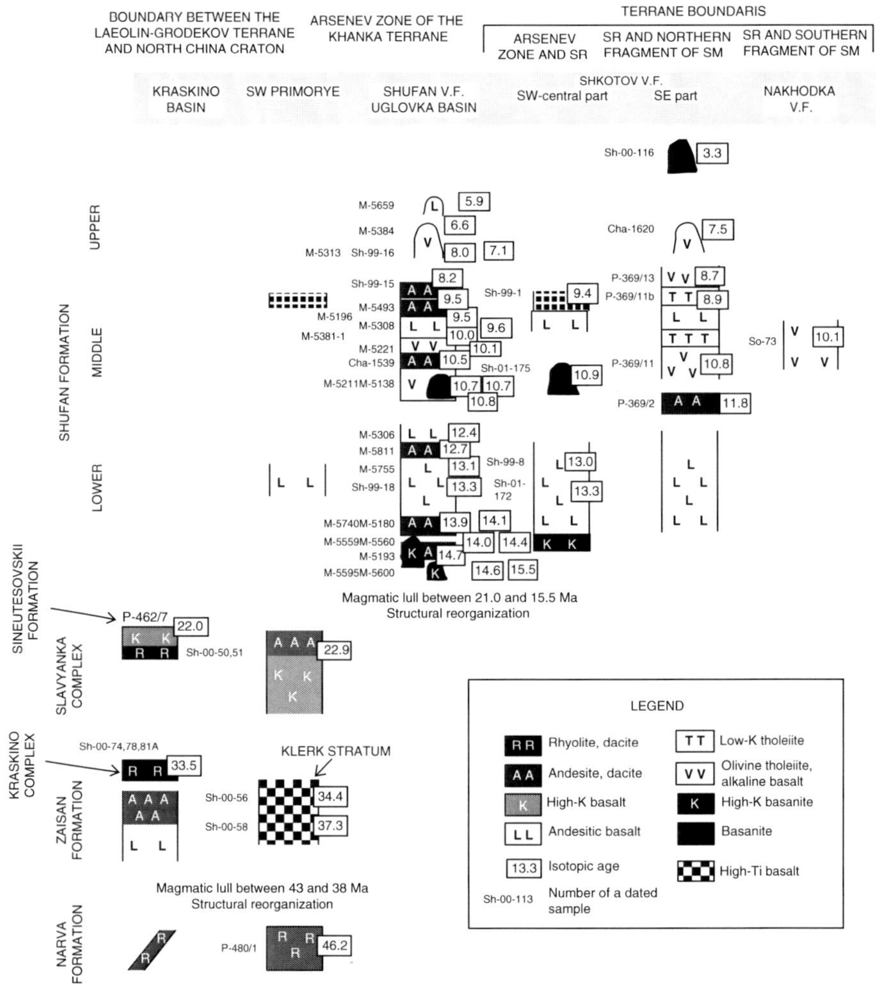

Fig. 12.26 A sequence of the Middle through Upper Tertiary volcanic rocks in South Primorye. Locations of dated samples are shown in Fig. 12.25

dome with initial strontium isotope ratio 0.704 16 $\pm$ 0.000 02 and MSWD 0.8. An age of the lower (basaltic) member of the Slavyanka Complex was not measured because of possible distortion due to heat impact from the younger evolved magmatic liquids.

Late Tertiary volcanic activity was expressed in four volcanic fields: Shufan, Shkotov, Sandugan, and Nakhodka (Fig. 12.24). The Shufan volcanic field occupies a margin of the uplifted basement of the Pre-Cambrian Hanka terrane, the subsided Arsenev zone, and adjacent area of the Laoelin-Grodekov terrane, which is situated largely in Northeast China. The Sandugan volcanic field is trending northeastward along the boundary between the Arsenev zone and uplifted basement of the Khanka terrane. The Shkotov volcanic field lies mainly within the Arsenev zone and partly at a margin of the Samarka terrane. The Nakhodka volcanic field is located at the southeastern boundary of the Sergeevka terrane.

From K–Ar dating of lavas with different compositions, we distinguish in evolution of late Tertiary magmatism of Primorye three time intervals: (1) 15.5–11.8 Ma, (2) 10.9–8.2 Ma, and (3) 8.0–3.3 Ma.

The oldest magmatic unit is exhibited by low-volume high-K basanites that occur at the foot of a lava section in the southwestern part of the Shkotov volcanic field and in the Uglovka basin (Fig. 12.27). The coeval trachyandesites of the Baranovskii volcano were dated at 14.7 Ma (Rasskazov et al. 2001, 2003g; Sakhno et al. 2004;

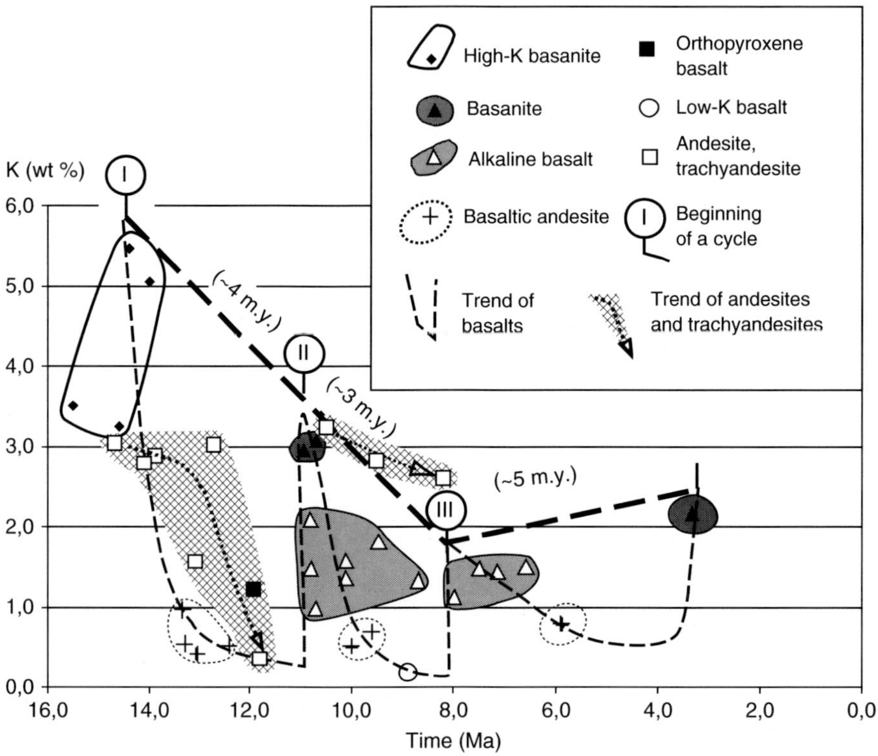

Fig. 12.27 Temporal variations of potassium in late Tertiary volcanic rocks from South Primorye between 15.5 and 3.3 Ma. Three cycles are defined by elevated concentrations of potassium in basanites and alkaline basalts of 15.5–14.0 Ma, 10.9–9.5 Ma, 8.6–6.6 Ma, and ~3.3 Ma. These episodes are separated from each other by those of low-K basaltic andesites of ~13.3–12.4 Ma, 10.0–9.5 Ma, and ~5.9 Ma. Andesites and trachyandesites of 14.7–12.7 Ma contain about 3 wt% of potassium that was decreasing up to 0.35 wt% by 11.8 Ma. This trend is delayed for 1.0–1.5 m.y. relative to the one of decreasing potassium contents from 3.25–5.46 wt% in high-K basanites of 15.5–14.0 Ma to 0.43–0.97 wt% in basaltic andesites of 13.3–12.4 Ma. The lowest K content (0.35 wt%) in the 11.8 Ma andesite is comparable with the lowest K content in the 8.9 Ma low-K basalt (0.18 wt%). Both dated samples belong to a group with the most depleted isotopic signatures ($^{87}Sr/^{86}Sr$ as low as 0.703 51) (Rasskazov et al. 2003g)

Maksimov and Sakhno 2008). The upper part of the unit consists of basaltic andesites that have flooded a vast area and created large volcanic shields mostly at 13.3–13.0 Ma. This phase of eruptions was accompanied by trachyandesites and andesites erupted at 12.7 and 11.8 Ma.

The second unit, also voluminous, is represented by lavas with varied compositions; by basanites and alkaline basalts of 10.9–10.7 Ma at the foot, basaltic andesites, andesites, and olivine tholeiites of 10.5–9.5 Ma in the middle part, and low-K tholeiitic basalts, Ti-rich megaplagiophyric basalts, and trachyandesites of 9.4–8.2 Ma at the top.

The third unit is low-volume lava flows and small extrusions of olivine tholeiites, alkaline basalts, and basaltic andesites with basanites at the end. Some extrusions contain deep-seated inclusions (Rasskazov et al. 2003g; Chashchin et al. 2007).

The late Tertiary magmatic stages correspond to three cycles of potassium variations. High-K basanites erupted only at the beginning of the first cycle, andesites and trachyandesites in the first and second cycles, and basaltic andesites in all three cycles. Basanites, alkaline basalts, and olivine tholeiites, which possess typical intraplate trace-element signatures (Rasskazov et al. 2003g), dominated in the second and third cycles. Low-K tholeiites marked the final episode of the second cycle.

12.4.2.2 Correlations

Similar to Central Mongolia, South Primorye is characterized by northward shift of the late Tertiary Shufan, Shkotov, Nakhodka, and Sandugan volcanic fields relative to the Middle Tertiary fields of Southeast Primorye (Figs. 12.22, 12.23). So, a question arises about possible links between events of these two regions that belong to time intervals of 43–21 Ma and <20 Ma.

Late Eocene through Oligocene lavas occur in East Asia, beside Primorye, in East Sikhote Alin, Sakhalin, Central Hokkaido, Honshu, Korea, Sea of Japan, and Northeast China. Although spatially shifted from the Middle Tertiary volcanic rocks, those dated back to the late Tertiary are widespread practically in the same regions. In this section, we emphasize particularities of magmatic evolution within time intervals of 43–21 and <20 Ma in East Asia for comparisons with convergence- and rift-related magmatic processes in Central Mongolia.

Time Interval of 43–21 Ma

Structural reorganizations at 43 and 21 Ma are reflected in liquid compositions not only in Central Mongolia but also in East Asia. The former episode is represented by dacitic eruptions in the Kizi area, north Sikhote Alin (Rasskazov et al. 2003e). There was a magmatic lull along the whole length of the East Asian continental margin between 43 and 38 Ma (Deschamps and Lallemand 2002; Rasskazov and Taniguchi 2006). We assume also that a transition from high- to moderate-K lava compositions at $\sim$32–30 Ma in Central Mongolia has an adequate expression in variations of lava compositions in East Asia.

Before opening Sea of Japan and Tatar strait, the Eocene–Oligocene sediments deposited in basins of the Hasan-Amur rift zone trending southwest–northeast from the Kilchzhu-Menchkhon and Kesson basins in Korea to the Buhtyanka and Maslovka basins in vicinity of the Amur mouth (Popov et al. 2001). Subsidence of the Kraskino basin, located in the southern part of the rift zone, was accompanied by basaltic magmatism of the Klerk stratum and Zaisan Formation with subsequent silicic magmatism of the Kraskino Complex. In a time interval between 38 and 32 Ma lavas erupted in both the southern and northern parts of the rift zone (i.e. in southwest Primorye and northern Sikhote Alin). Within the same time interval, volcanism occurred in the northern and southern parts of the Sakhalin island as well as in the Okushiri island.

Silicic magmatic activity was defined between 33.5 and 31.7 Ma by K–Ar dating of unaltered biotite separates from granites of the Hamada cauldron located in the Inner Zone of Southwest Japan. Here the Paleogene cauldrons were arranged along a line with intervals of about 20 km (Imaoka et al. 2001). Granites and volcanic rocks of I-type magnetite series of ca. 30 Ma were considered as the youngest events in the San-in belt (Iizumi et al. 2000). In the central San'in region, K–Ar and fission-track ages of the middle and late Paleogene volcano-plutonic complexes span a time interval from 41 to 26 Ma (Matsuura 1989).

In Central and Northern Sikhote Alin, an interval of K–Ar ages between 38 and 25 Ma was reported by Tatsumi et al. (2000) for volcanic rocks with uniform "volcanic arc" geochemical signature ($K/Y > 300$, $K/Nb > 900$). This interval includes K–Ar ages presented by Okamura et al. (1998) for lavas from the Eocene–Oligocene Siziman Formation: two olivine basalts dated at ca. 37 Ma, five olivine and aphyric basalts of 36–34 Ma and two olivine basalts of 32–29 Ma. Three olivine and olivine-clinopyroxene basalts with ages between 25 and 21 Ma were described by Okamura et al. (1998) as basalts of the Kizi Formation, dated back to the Lower–Middle Miocene. Two basalts of 24.4–21.1 Ma from the Sovgavan volcanic plateau showed lower K abundances of 0.63 and 0.53 wt%. The sample Yu–770 from the Amur mouth area is characterized by elevated potassium content (1.42 wt%), which is in a range of K concentrations in basalts from the Siziman Formation (i.e. 1.07–2.14 wt%). This sample yields an age of 24.8 ± 0.8 Ma corresponding to the end of the 38–25 Ma age interval reported by Tatsumi et al. (2000). Data reported by Martynov (1999) show that the sample Yu–S122/7, dated at 24.4 Ma, is isotopically (Nd, Sr) comparable with lavas of the Siziman Formation rather than those of the Kizi Formation.

Andesitic basalts in the vicinity of Lake Kizi—a stratotype area of the Kizi Formation—show potassium contents largely in a range of 0.82–2.16 wt%, comparable with the dated basalts of the Siziman Formation. Lavas with lower K concentrations of 0.61–0.89 wt% have been found only in a volcanic sequence of the Arbat mount. Taking into account possible time-dependant potassium variations in the area (i.e. the elevated concentrations in basaltic lavas of 38–24 Ma and lower abundance in the sample Yu–19 of 21.1 Ma), low-K andesitic basalts from the lower and upper parts of the Arbat mount sequence were selected for $^{40}Ar/^{39}Ar$ incremental measurements (Rasskazov et al. 2003e). A groundmass of the stratigraphically older sample yielded a plateau age of 28.80 ± 0.10 Ma consistent with isochron ages produced by steps with two different contaminating components: isochron age of lower temperature steps at 29.71 ± 1.79 Ma, $^{40}Ar/^{36}Ar$ of the component 237.0 ± 13.0 (MSWD 1.3) and isochron age of high temperature steps at 28.53 ± 0.10 Ma, $^{40}Ar/^{36}Ar$ of contaminating component at 292.6 ± 3.1 (MSWD 2.1). Also, two contaminating argon components were identified in feldspar separates from the younger andesitic basalt with ages of lower temperatures steps at 27.32 ± 0.28 Ma, $^{40}Ar/^{36}Ar$ of contaminating component at 297.2 ± 4.2 (MSWD 3.2) and high temperature steps at 27.29 ± 0.53 Ma, $^{40}Ar/^{36}Ar$ of contaminating component 232.6 ± 7.3 (MSWD 1.8). Although no plateau was obtained, identical ages of high and lower temperature steps of the dated plagioclase demonstrate a satisfactory preservation of K–Ar isotopic system and reliability of the age determination.

The results obtained by $^{40}Ar/^{39}Ar$ method show ages of 29–27 Ma for low-K andesitic basalts from the Arbat sequence, which are in a range of the Siziman Formation (i.e. between 38 and 25 Ma). Taking into consideration no ages older than 24 Ma in the middle and northern parts of the Eastern Sikhote Alin, we connect the 21.1 Ma low-K basalts with structural reorganization resulted in concentration of subsequent (late Tertiary) volcanic activity of the Eastern Sikhote Alin within the Sovgavan, Nelma, and Bikin volcanic fields.

Geochronological studies of the middle Tertiary volcanic rocks were focused also on the western shore at the northernmost and southernmost parts of the Sakhalin island showed volcanic activity of 39–37 Ma, ~31 Ma, and 26–25 Ma.

Volcanic rocks of the northern part (Schmidt peninsula) are represented by the Upper Eocene–Oligocene Matchigarka Formation. This unit overlays the Cretaceous sedimentary and volcano-sedimentary deposits and was capped by the Upper Tertiary sediments. The lower part of the Matchigarka Formation is sedimentary, the upper is predominated by lavas varying in composition from latites to andesites and trachyandesites. The Upper Eocene–Oligocene age of sediments has been determined on mollusks and foraminifers (Gladenkov et al. 1999). The biostratigraphic data are consistent with K–Ar ages of 37 ± 2 Ma (andesite) and 30 ± 2 Ma (trachyandesite). Besides volcanic rocks of the Matchigarka Formation, there are sub-volcanic tephrite intrusions of the Espenberg Complex (Rozhdestvenskii 1988; Sal'nikov 2004). The intrusions yielded K–Ar dates at ca. 26–25 Ma (Sal'nikov 2004; Rasskazov et al. 2007e). Similarly, andesites dredged in the Okhotsk Sea from the Kashevarov were uplifted near the Schmidt peninsula, yielding K–Ar ages of 39 and 25 Ma (Lelikov et al. 2001; Yemelyanova et al. 2003).

In the southern part of Sakhalin, the Middle Tertiary volcanic rocks occur in the Arakai and Kholmsk Formations. The former is dated back to the Upper Eocene–Oligocene, the latter to the Upper Oligocene–Lower Miocene. Four K–Ar dates were obtained in a range from 39 to 26 Ma for volcanic rocks from the Arakai Formation. Dates of 38.7 ± 1.2 Ma and 30.7 ± 0.7 Ma were reported for lavas sampled in outcrops (Okamura et al. 1998). The younger dates were obtained for volcanoclastic material (Takeuchi 1997). The latter age determinations were performed on two parallel aliquots. One sample yielded dates of 27.1 ± 0.9 Ma and 28.0 ± 0.9 Ma overlapped within analytical error, another one showed two identical values at 26.0 ± 0.8 Ma. In the Holmsk Formation, a volcanic layer was recorded in the middle part of the section exposed in the Lutoga basin. These rocks underwent secondary hydrothermal alteration with resetting of K–Ar dates (Rasskazov et al. 2005b).

Particularities of magmatic evolution at 23–20 Ma, ~15.5 Ma, and 10–8 Ma are emphasized by specific variations of initial Sr-isotope compositions in basaltic-intermediate volcanic rocks from the Northeast Japan Arc and onshore areas of the Tatar strait (Figs. 12.28, 12.29). Similar coeval variations of Sr, Pb, and Nd isotope signatures have been found in the Baikal Rift System (Rasskazov et al. 1999a, 2002a, c, 2005a).

Taking into account close geologic correlations between the Sakhalin and Hokkaido terranes in the Mesozoic and early Tertiary, we consider data on the middle–late Tertiary rocks from the southwestern part of the latter in comparison with the western Sakhalin and from the central-eastern part with the Eastern Sakhalin. In Central Hokkaido, magmatism lasted from ca. 43 Ma to 10 Ma. The Hidaka zone was intruded by calc-alkaline granites of the Eastern Chain at 43–41 Ma, which are spatially connected with the 49–41 Ma Aniva intrusion of the Southwestern Sakhalin. The Western Chain of the Hidaka zone was active from 36 to 17–16 Ma. Temporal variations of magmatic rocks are not constrained properly. On a geologic basis, it was suggested that basic liquids were followed by silicic. Magmatic rocks represented a wide range of compositions of the Hidaka magmatic Complex with both tholeiitic and calc-alkaline affinities (Maeda 1990).

A time span of K-rich alkaline magmatism with emplacement of absarokites, shonkinites, and syenites in western Kamchatka was constrained at 35–31 Ma by $^{40}Ar/^{39}Ar$ dating, although K–Ar measurements yielded a range of dates from 41 to 21 Ma. This magmatism postdates the initial Late Eocene magmatic episode of 38–34 Ma in East Asia. Eruptions of K-rich liquids of 35–31 Ma accompanied the final phase of the early-middle Paleogene subduction of the Okhotsk Sea plate beneath Western Kamchatka and transition to the late Tertiary geodynamics (Perepelov et al. 2003).

In the Philippine Sea plate, K–Ar and $^{40}Ar/^{39}Ar$ ages from 38 to 36 Ma were reported for volcanic rocks from its western margin near the Luzon Arc and for plagiogranites at the northern end of the Palau-Kyushu Ridge. Comparable K–Ar ages were measured also for boninites, tholeiitic basalts, and dacites from the Bonin islands (Deschamps and Lallemand 2002).

In Southeast China, Rb–Sr ages between 37 and 34 Ma were obtained for granites. An age of 35.0 ± 0.3 Ma was measured by U–Pb method on zircons from Jinping alkali granites emplaced into the Ailao-Shan Red River shear belt (Zhang and Schärer 1999). This granite magmatism was coeval to the initial magmatic pulse of 38–35 Ma in the Sea of Japan region.

K–Ar ages of 35.9 ± 1.2 Ma and 38 ± 1 Ma were reported for possible initial events at Fanshi (northern part of the Circum-Ordos (Shansi) Rift System, China) and Khilok (Transbaikal), respectively (Xu et al. 1996; Ivanov et al. 1995). For the former area, besides an age of 35.9 ± 1.2 Ma, three ages of 30–28 Ma, two ages of 26–24 Ma, and one of 22 Ma were determined for alkali olivine basalts, which can be comprised into a volcanic interval of 36–22 Ma. The dated samples show moderate

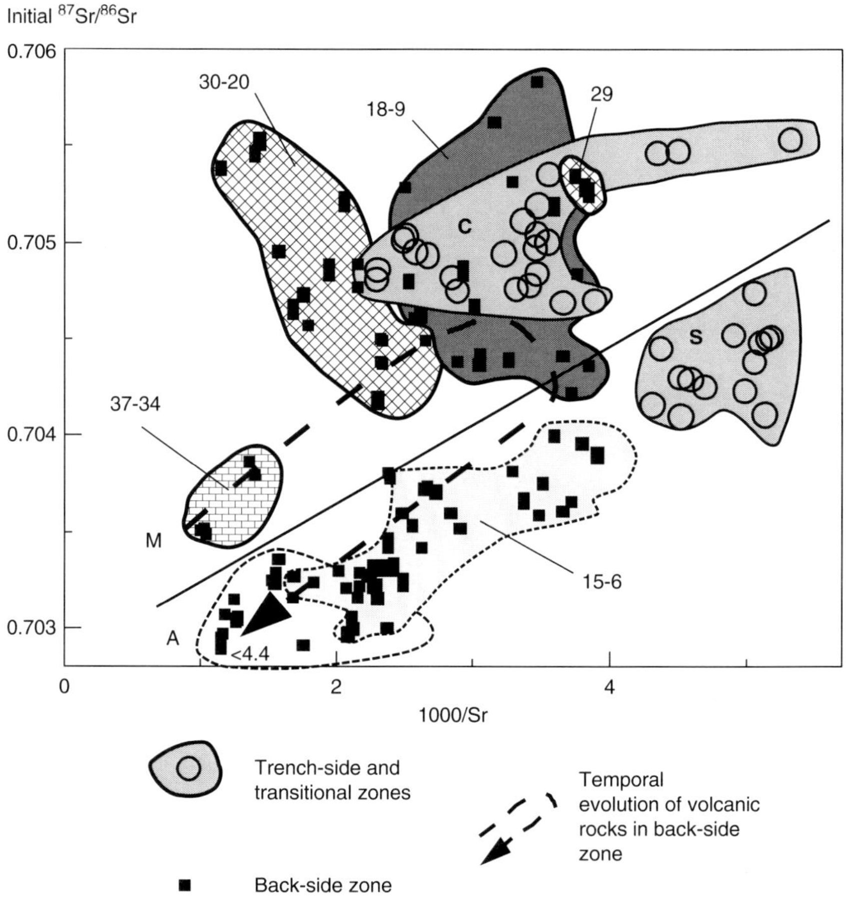

Fig. 12.28 Initial $^{87}Sr/^{86}Sr$ versus 1000/Sr in middle–late Tertiary volcanic rocks of basaltic through intermediate compositions ($SiO_2 < 60$ wt%) from trench-side—transitional and back-side zones of the Northeast Japan Arc. Volcanic rocks from the trench-side and transitional zones that are younger than 23 Ma are referred to the first-order late Tertiary cycle. *C* and *S* are volcanic material of this zone, probably, derived from the mantle wedge of the continental margin and subducting slab, respectively. For volcanic rocks from the back-side zone, ages are shown in Ma. Temporal change of strontium isotope compositions is directed from the isotopically depleted component (*M*) that is characteristic of an extended continental margin to the asthenospheric-like depleted component (*A*). Compositions that locate above and below the *heavy line* are produced from a mantle wedge and oceanic slab, respectively. Transition from the former to the latter is exhibited by the 18–9 Ma lavas from the mantle wedge source and by the 15–6 Ma lavas from the slab source. (Data sources: Shuto et al. 1993; Ohki et al. 1994b; Rasskazov et al. 2005b and references therein)

potassium contents in an interval of 0.9–1.5 wt%. Only Yuxian basalts of ca. 29 Ma reveal relatively high potassium concentrations (K = 2.3–2.7 wt%).

In Inner Mongolia (northern margin of the North China block), ages in a range between 32.5 Ma and 21.5 Ma were measured for low-potassium quartz- and olivine

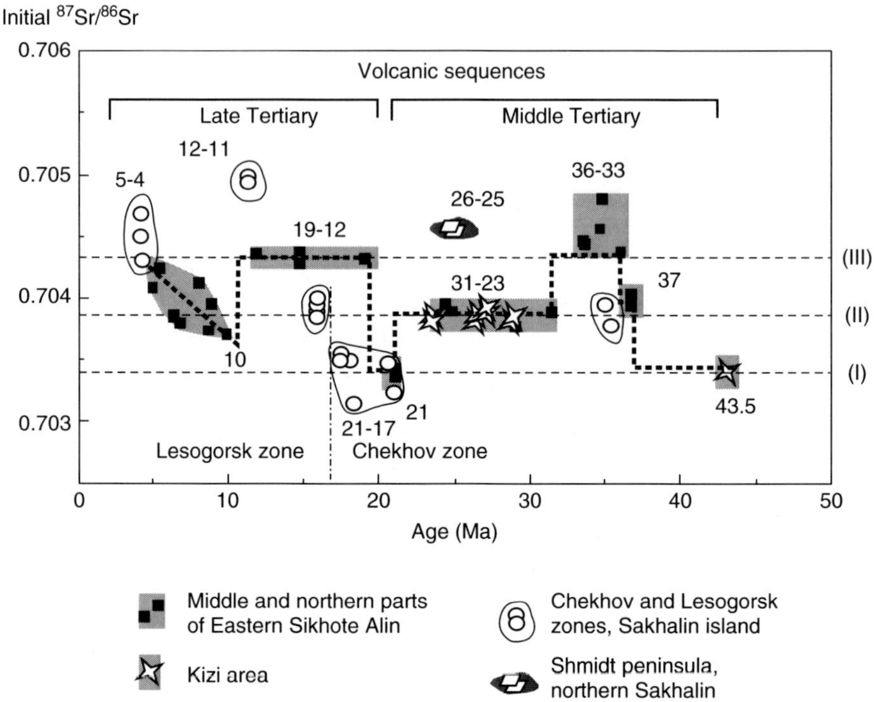

Fig. 12.29 Temporal variations of initial $^{87}Sr/^{86}Sr$ ratios in middle–late Tertiary volcanic rocks of basaltic through intermediate compositions (SiO_2 < 60 wt%) from the on-shore area of the Tatar strait (Rasskazov et al. 2006). The variations are shown by a *dotted line*. Strongly depleted, moderately depleted, and enriched sources are marked by numbers in brackets (*I*), (*II*), and (*III*). Magmatism of the middle Tertiary cycle commenced at ~43 Ma and finalized at ~21 Ma by compositions from the source (*I*) with subduction-type trace-element signatures. The late Tertiary cycle marked a transition to magmatism with trace-element signatures corresponding to withinplate setting. The 19–11 Ma lavas originated from separated sources (*I*), (*II*), and (*III*). Since about 10 Ma, role of material from the source (*II*) was persistently decreasing with increasing role of material from the source (*III*). (Data sources: Esin et al. 1994; Okamura et al. 1998; Martynov 1999; Rasskazov et al. 2003e, f, g, 2005b, 2006, 2007d)

tholeiites (K = 0.13–0.48 wt%) (Zheng et al. 2002). Some analyzed samples revealed elevated potassium abundances (0.5–1.1 wt%), but because of poor freshness, these samples might yield distorted dates. Among the low-K basalts, the fresh samples showed ages of 28.9–28.3 Ma, 25.0 ± 0.6 Ma, and 22.9 ± 1.0 Ma. No relative stratigraphic control was presented to show reliability of this data set, however. These results demonstrate eruptions of low-K basalts in the Oligocene–earliest Miocene. Comparable ages were reported in interval of the Oligocene–earliest Miocene for tholeiites from Northeast China and Inner Mongolia by Zhou et al. (1988): 31 Ma (Yitong Hengtoushan), 28–24 Ma (Dunhua), 22 Ma (Yongji), and 21 Ma (Yitong Xigianshan).

Time Interval of <20 Ma

Geochronological data on middle–late Tertiary magmatic events in East Asia were reported in recent papers and were discussed in monographs (Rasskazov et al. 2000b, 2005a, 2007b; Sakhno 2001; Khanchuk 2006; Rasskazov and Taniguchi 2006). For details, a reader is referred to these publications. Here we focus on particularities of magmatic evolution in East Asia related to structural reorganizations and episodes of relative northeastward shift of volcanism in Central Mongolia at ~15.5 Ma and ~7.4 Ma.

To correlate subduction-related and intraplate magmatic events in Asia, we have adopted a temporal sequence of events that Watanabe (2002) suggested for the southwest Hokkaido and have placed additional boundaries, which we found to be important from isotopic evolution of magmatic sources in the back-side zone of the Northeast Japan Arc. The sequence has quasi-periodicity of 2–3 Ma with initial points of time intervals at: (1) 23 Ma, (2) 20 Ma, (3) 18 Ma, (4) 15 Ma, (5) 12.1 Ma, (6) 9 Ma, (7) 6.2 Ma, (8) 4.4–3.6 Ma, and (9) 1.5 Ma (Rasskazov and Taniguchi 2006) (Fig. 12.30). A correlation gives an opportunity for discussion of relationships between tectonic and magmatic expressions of structural reorganizations.

The initial and final late Tertiary events in Primorye (Fig. 12.27) are correlated with boundaries of 15.0 and 4.4–3.6 Ma and the lowest K concentrations with boundaries of 12.1, 9.0, and 6.2 Ma in the Northeast Japan Arc. Starting events of the second and third cycles (10.9–10.7 and ~8.0 Ma, respectively) fall between boundaries of the Northeast Japan Arc. But these boundaries are consistent with those of Central Mongolia, where an interval of 11.0–7.5 Ma is constrained as the time of structural reorganization.

Transition from the middle to late Tertiary events was exhibited by sparse eruptions at about 23 Ma in Central Asia and widespread volcanic activity between 22 and 17 Ma in a vast region between the Sea of Japan and Eastern Sayan. In the latter area, volcanism, commenced in the Oligocene between 28 and 23 Ma, did not coincide spatially with the subsequent late Tertiary volcanism. The 28–23 Ma lavas were compared in terms of highly alkaline compositions and depletion of normative *di* with rocks of Late Cretaceous pipes that were dated by the $^{40}Ar/^{39}Ar$ method in Hakassia between 79 and 72 Ma (Rasskazov 1993; Bragin et al. 1999; Malkovets et al. 2003). In this respect, both the Cretaceous and Oligocene events in Eastern Sayan and adjacent areas were considered as events preceding the late Tertiary magmatic processes that related to the development of the Baikal Rift System (Rasskazov 1993; Rasskazov et al. 2007b).

Reconstructions in the Sea of Japan region yield evidence on extension that resulted firstly in incipient rifting and subsidence initiated in Primorye between 38 and 34 Ma and in Japan at ca. 32–30 Ma with transition to fast spreading and arc rotation between 21 and 15 Ma. Southwest Japan quickly rotated at about 15 Ma. This tectonic phase is considered as responsible for the Sea of Japan opening. It was accompanied by change of isotopically enriched liquids from mantle wedge to those from isotopically depleted subducting slab (Fig. 12.28) (Jolivet et al. 1994; Rasskazov et al. 2004b, 2005b; Rasskazov and Taniguchi 2006 and references therein).

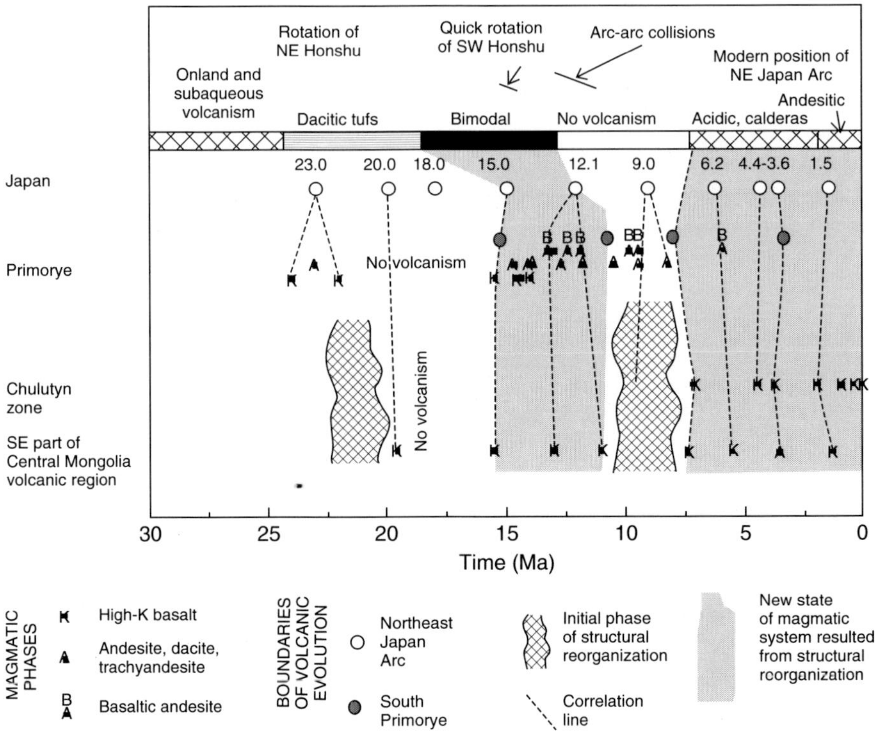

Fig. 12.30 Correlation of the late Tertiary magmatic events in the Central Mongolia and Sea of Japan regions. In the SE part of Central Mongolia and Chulutyn zone, initial phase of structural reorganization is exhibited by moderate-K basalts of ~21 and 11.0–7.5 Ma, respectively. In Primorye, the latter time interval corresponds to the second cycle of eruptions commenced by basanites and finalized by basaltic andesites and low-K basalts (Fig. 12.27). Note that andesites occur in both regions. (Sources: Jolivet et al. 1994; Ohki et al. 1994b; Rasskazov et al. in press; this work)

We have hypothesized that the middle–late Tertiary magmatic evolution in Primorye reflected dynamics of oceanic slab flexures subducted from Pacific beneath East Asia. Three segments of the currently active Pacific slab, divided by flexures, descend beneath the overriding Okhotsk Sea plate, Eurasia, and Philippine Sea plate (Fig. 12.1). The Hokkaido-Amur slab flexure, located between the two former units, stretches landward from a junction between the Japan and Kuril trenches below Southwest Hokkaido to the southern end of the Middle Amur basin. The Honshu-Korea slab flexure, located between Eurasia and Philippine Sea plate, goes from a junction between the Izu Bonin and Japan trenches bellow Central Honshu and the southeastern edge of the Korean peninsula. The directly subsided Hokkaido-Amur flexure is trending N60°W in accordance with the long-term Pacific-Eurasia convergence, unlike the Honshu-Korea one originated after a quick clockwise turn of Southwest Japan at ca. 15 Ma due to retreat of the Japan-Izu-Bonin junction.

The Hokkaido-Amur slab flexure coincides spatially with the high-velocity divider separating the Okhotsk Sea and Philippine Sea low-velocity domains (Fig. 12.1). These domains occur above the northeastern and southwestern slopes of the Hokkaido-Amur slab flexure, respectively. Taking into account possible connection of the latter with landward propagation of the Pacific slab, we speculate that these low-velocity anomalies could reflect different conditions of convective instabilities above the northeastern and southwestern slopes of the slab flexure. Processes above the former slope were affected by a retreat of the Japan-Izu-Bonin junction and subsequent relaxation, whereas those above the latter slope were generated in accordance with general late Tertiary rollback effect of arc system but were not influenced by subsequent complications introduced by relaxation around the Honshu-Korea slab flexure. In other words, the northeastern slope of the slab flexure preserved subduction features (i.e. should maintain subduction-related magmatism), unlike the southwestern one that underwent adaptation to new geodynamic conditions. These relations are plausible for progressive magma generation along a zone dynamically related to convective instability above the Okhotsk Sea segment of the Pacific slab. The zone is supposed to be extended from the back-arc Kuril basin through South Sakhalin and Tatar strait to the mainland of Asia, where one can expect magmatic activity, spatially related to deeply subducted portions of the currently active Pacific slab beneath Far East Russia and adjacent Northeast China.

Alternatively, magmatic evolution in South Primorye (Fig. 12.27) shows transition from an initial anomalous state of high-K magma generating processes, which accompanied the main phase of opening the Sea of Japan (rotation of Southwest Japan with retreat of the Japan-Izu Bonin junction), through an intermediate state in a deep magmatic system, underwent adaptation to new conditions followed the opening, to its final state that satisfy to the new geodynamic conditions. It is noteworthy that the Philippine Sea plate behaved as an independent unit drifting in the Pacific in the early and middle Tertiary until ca. 21 Ma, when it accreted to Eurasia with commencement of its subduction beneath the latter at ca. 10 Ma (e.g. Deschamps and Lallemand 2002). So, the new geodynamic conditions in South Primorye might be temporally connected with this kind of structural transformations.

We interpreted the Middle Tertiary magmatism of Southwest Primorye as a result of convective instability in the mantle wedge above the Honshu-Khingan slab flexure presumably active during drifting of the Philippine Sea plate before its accretion to Eurasia. The post-accretion structural reorganization was expressed by a magmatic lull in South Primorye between 21 and 15 Ma with subsequent rejuvenation of magmatism of 15.5–12.4 Ma accompanied landward propagation of the Hokkaido-Amur slab flexure with retreat of the Kuril-Japan junction and spreading of crust in a pull-apart back-arc basin of Sea of Japan.

The 21 Ma episode of low-K basalts in Central Sikhote Alin postdated high-K basalts of the Sineutesovsky Formation in the southwest Primorye is followed by a magmatic lull between 21.0 and 15.5 Ma. Interestingly that the 15.5–14.0 Ma high-K basanites from the Uglovka basin, which are coeval to the main phase of Sea of Japan opening, show K/1000Ta and La/Ta comparable with these ratios in the post-collision basalts from the North Tibet (Rasskazov et al. 2008b). Hence, this feature again emphasizes the specific role of the 15.5–14.0 Ma magmatic phase both for the Central Mongolia and Sea of Japan regions.

We have defined a moderate-K magmatic interval of 11.0–7.5 Ma as indicative for a structural reorganization resulting in transition from a linear- to parabolic-type volcanic pattern in Central Mongolia. For a while, the Indo-Asian convergence had no influence on magmatic processes with subsequent new launch of high-K magmatism at 7.4–7.1 Ma. A similar assumption is valid for the Northeast Japan Arc that rotated in the Late Oligocene and Miocene, but possessed a single locus corresponding to the modern arc since ca. 8 Ma (Ohki et al. 1994b). A collision between the Izu-Bonin and Honshu arcs in a time interval of 14–12 Ma resulted in a comparatively neutral stress regime with weak pulses of compression and extension lasting up to time intervals of 7–4 Ma. A compression stress field was distinctly expressed after 4 Ma, when the most isotopically depleted liquids erupted in a back-side zone of the Northeast Japan Arc (Jolivet et al. 1994; Ohki et al. 1994a; Sato 1994; Rasskazov et al. 2004b, 2005b; Rasskazov and Taniguchi 2006 and references therein).

In Primorye, andesites erupted in the late Tertiary between 14.7 and 8.2 Ma, whereas moderate-K basanites and alkaline basalts followed between 10.9 and 3.3 Ma. The former reflected magmatic processes in a mantle wedge above the Pacific slab, the latter showed transition to intraplate conditions. Similar to Central Mongolia, the contemporaneous andesitic and alkaline basaltic magmatism reflects here relaxing previous tectonic state between 10.8 and 8.2 Ma and the subsequent alkaline basaltic magmatism corresponds to a launch of a new tectonic regime at $\sim$8.0 Ma.

The tectonic change attained expression approximately at the same time in practically all of Asia as was argued by Molnar (2005). In this respect, motions are indicative also in the Baikal Rift System that consists of extensional and transtensional segments with diachronous development expressed in spatial–temporal evolution of volcanic activity. In the central part of the rift system, simultaneous activity of the mutually overlapping transtensional Tsypa-Muyakan and extensional Barguzin-North Baikal segments favored for voluminous eruptions in the Vitim volcanic field between 14.4 and 9.0 Ma. The subsequent magmatic lull lasted from 9.0 to 5.2 Ma, when minor eruptions occurred only in the Namaru volcano at 7.5–6.5 Ma. The new voluminous magmatic phase of 5.2–2.8 Ma was governed by reactivation of the Tsypa-Muyakan segment (Rasskazov 1996; Rasskazov et al. 2000b, 2007d). The magmatic lull in the Vitim volcanic field was coeval with a tectonic phase responsible for a change of sources for sedimentary material that deposited on the underwater Academician ridge in the central part of Lake Baikal. In a 600 m deep drill hole, the sediments of the lower and upper strata consist of material brought from the Barguzin and Upper Angara basins, respectively (i.e. from the southeastern and northern parts of the lake). The sediment sources drastically changed at 7.65–7.09 Ma (Kashik and Lomonosova 2006). In the southwestern part of the rift system, where high mountain ridges commenced uplifting approximately at the same time, lavas of 8.7 Ma occupy a top of the Munku-Sardyk ridge and those of 6.9 Ma fill up river valleys as deep as 400 m (Rasskazov et al. 2000c).

12.5 Summary: Magmatic Geodynamics in Asia

We have considered late Phanerozoic magmatic evolution of Asia in a frame of a hierarchic system, which involves five levels. After the main Phanerozoic structural reorganization commenced at ca. 260–250 Ma, the most important tectonic processes occurred at about 140 Ma (closure of the Mongolia-Okhotsk Ocean), between 116 and 107 Ma (accretion of the Kema terrane), between 65 and 50 Ma (Indo-Asian collision and accretion of the Okhotsk Sea plate), at about 43 Ma (global plate reorganization), and between 23 and 20 Ma (accretion of the Philippine Sea plate to Eurasia). We have identified numerous coeval magmatic episodes as evidence on common geodynamic factors operating on a scale of the whole Asia.

To estimate influence of Indo-Asian interaction and rifting on magmatic processes, we have examined spatial–temporal relations between high- and moderate-K Cenozoic basalts in such key regions as Tibet, Central Mongolia and Primorye. The latter region is located on the opposite side of the Sea of Japan relative to the Northeast Japan Arc, so magmatic evolution here is indicative for deciphering geodynamic conditions of the Pacific slab subduction beneath East Asia and accompanied processes responsible for strong extension of the continental margin that resulted in opening the Sea of Japan. Central Mongolia is located between regions affected by convergence and rifting and, similar to Primorye, is underlain by low-velocity domains and fragments of high-velocity slabs. Due to interaction between India and Asia, the lithosphere and sub-lithospheric heterogeneous mantle appeared to be affected by sub-horizontal breaks and relative motion of mantle layers. We propose that magmatic processes in the Sayan-Mongolia low-velocity domain were initiated under the leading role of a weakened level of 200–210 km, traced beneath the Siberian craton as the Lehman discontinuity. A keel of the craton lithosphere could play a role of an obstacle for convergence-caused northward motion of shallow mantle material, whereas below the level 200–210 km, mantle material could move beneath the craton (Fig. 12.31).

Spatial–temporal evolution of Cenozoic magmatism in Central Mongolia is consistent with evolution of magmatism in Tibet and includes a stage of 66–43 Ma, launched by Indo-Asian collision, and three stages of post-collision convergence of 41–21, 20–10, and <10 Ma. The collision of India with the southern margin of Asia, and coeval stretching of its eastern margin, was accompanied by moderate- and low-K basaltic magmatism in areas between those margins. The post-collision high-K basaltic eruptions began only when the lithosphere and sub-lithospheric mantle underwent contraction between the Indian indenter and Siberian craton.

Due to accommodation of Indo-Asian convergence in Central Mongolia, high-K basaltic magmatism was generated along a tangential (north–south) zone with northward migration in an interval of the past 41 Ma with average velocity of about 1 cm/year. Intermittent late Tertiary extension of the lithosphere in the East Asian margin resulted in activity of high-K magmatism in time intervals of 15.5–14.2 and <7.5 Ma in the Central-Mongolia rift zone that developed simultaneously with the Baikal rift zone. Detailed correlations of tectonic and magmatic events in these rift zones will be considered elsewhere.

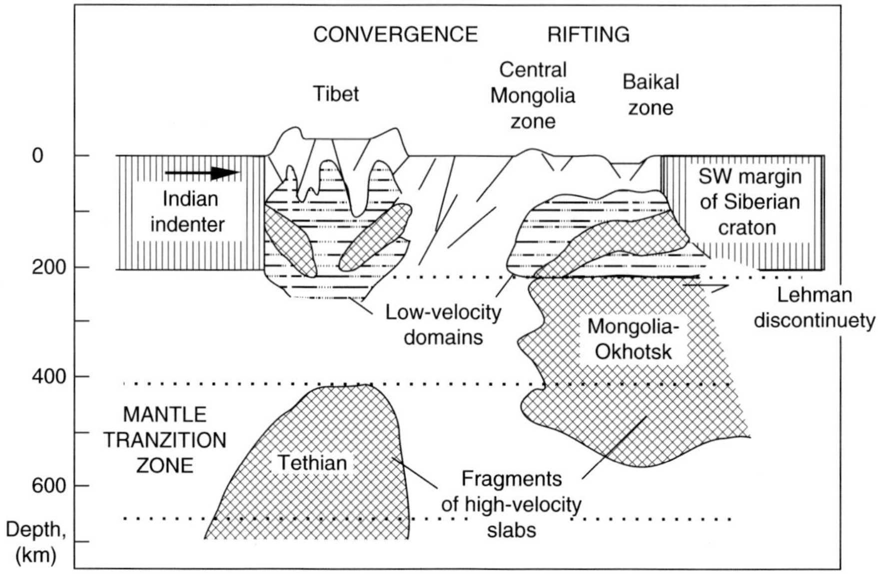

Fig. 12.31 Schematic north–south cross-section of the main deep mantle structures between the Indian indenter and Siberian craton. Central Mongolia occurs between regions affected by long-term convergence and rifting. The Tethyan and Mongolia-Okhotsk slab fragments are shown after tomographic images (Bijwaard et al. 1998; Van der Voo et al. 1999a, b). Shallow slab fragments under Tibet are assumed from models of magmatic evolution (Roger et al. 2000; Ding et al. 2003). A large low-velocity anomaly under Tibet (i.e. a dome of width 250–300 km rising from depth more than 300 km up to 150–200 km) is indicated according to high-resolution seismic tomography (Wittlinger et al. 1996). Vertical scale of the section is exaggerated

High-resolution seismic images and petrologic constraints lead to the conclusion that high- and moderate-K basaltic magmatism was controlled by melting processes at the mantle levels of 120–200 and 80–120 km, respectively. The convergence-caused processes of 41–21 Ma resulted firstly in deeper and then in shallow magmatic liquids. Afterwards, high- and moderate-K magmatic phases alternated. The longest interval of convergence-caused high-K magmatism accompanied initiation of rift-related processes between 15.6 and 11.0 Ma, whereas moderate-K magmatism lasted in different volcanic fields for 7.0–7.5 m.y. in time intervals of 23.4–16.0 Ma, 17.0–9.7 Ma, and 9.6–2.6 Ma. Convergence- and rift-related magmatic phases showed opposite variations of potassium contents at ~15.5, ~5.5, 3.6–3.4, and ~1.3 Ma, but consistent increasing potassium between 13 and 11 Ma and decreasing between 11 and 8 Ma. The latter magmatic episode corresponded to structural reorganization of the magmatic system with transition to alternating convergence- and rift-related magmatic pulses with quasi-periodicity 2.5 m.y. In the past 2 Ma magmatic events repeated more frequently—every 0.3–0.7 m.y.

We emphasize the most important magmatic changes in the Central Mongolia and Sea of Japan regions in time intervals of 23–20, 15.5–14.0, 11.0–8.0, and 2.0–1.2 Ma, which characterize the second-order events in the proposed hierarchical system. Strong collision-caused compressive stresses appeared to affect the lithosphere of Central Mongolia and the southwestern Baikal Rift System by short pulses expressed in volcanic episodes with quasi-periodicity about 0.3–0.5 Ma of the past 2 Ma. These magmatic processes represent the third-order events. Comparisons of volcanic peaks within the continent with those near the eastern and southern plate boundaries showed that the volcanic episodes were not coeval in a time interval of 2.0–1.5 Ma but were synchronized in the past 1.2 Ma. This change was interpreted as evidence of a transition from the independent to tightly connected interplate and withinplate processes.

Post-collision magmatism of Central Mongolia was dominated firstly by convergence processes and then by alternating convergence- and rift-related ones with final activity of mantle sources originating in conditions of convergence-caused above-slab enrichment. Present-day anomalous heating of crust in Central Mongolia exhibits an integrated thermal effect of convergence and rifting.

For understanding the present-day situation, we have analyzed timing of the latest Pleistocene through Holocene magmatic eruptions, which reflect fourth- and fifth-order magmatic processes, and have found regular temporal relations of the volcanic events in Central Mongolia, Northern Transbaikal, and Northeast China, where these events might be triggered by strong tectonic pulses propagated from the Indo-Asian collision zone. The maximal tectonic stress corresponded to the structural reorganization of the volcanic zone in Udokan, Northern Transbaikal at 8.78 $\pm$ 0.26 Ka. In Taryat of Central Mongolia, the east–west volcanic line of the volcanoes Odnobokii, Listvennichnyi, and Sosnovyi was active at ca. 70–50 Ka, i.e. before the reorganization, while the eruptions at the north–northeast line of the volcano Khorgo occurred either at the time of the structural reorganization or a little bit after (not later than 7.71 $\pm$ 0.81 Ka calibrated age). In Northeast China, the Holocene volcanism was initiated at 8.740 $\pm$ 0.40 Ka, i.e. simultaneously with the structural reorganization in Udokan. Frequency of eruptions began increasing in Northeast China at about 2.5–2.2 Ka, i.e. at the time of the latest eruption in Udokan.

Asia exhibits regions with potentially possible volcanic eruptions in the nearest future. The fifth- and fourth-order events imply increasing volcanic activity in Northeast China and Korea between 2100 and 2300 A.D. and an eruption in Udokan about 2050 A.D.

Chapter 13
Late Phanerozoic Magmatic Evolution of North America and Northeast Africa: Comparisons with Asia

Comparisons of the Baikal, Rio Grande, and East African intracontinental rifts have demonstrated similarities of the rifted continental regions in terms of structural features, high heat flow, seismicity, thinned lithosphere, low-velocity sub-lithospheric mantle domains, and accompanied magmatism (Florensov 1948; Logatchev 1977; Lipman et al. 1989; Rasskazov 1993; Barry et al. 2000). We consider continental rifting within Eurasia, North America, and Africa as manifestation of global effects produced by lithospheric plate interaction and/or induced by deep mantle plumes, and we analyze spatial–temporal magmatic evolution in the rifted continental regions, as a principle indicator of different geodynamic factors.

13.1 North America

Late Phanerozoic geodynamic models for the western continental margin are based on: (1) stepwise shift of intermediate to silicic magmatism from continental margin landwards between 120 and 37 Ma, (2) change of intermediate–silicic to basaltic magmatism between 37 and 18 Ma, and (3) predominating basaltic magmatism in the past 17–15 Ma. The stages were distinctly defined in the southern Basin and Range province (Seager et al. 1984) (Fig. 13.1).

The 36 Ma change from Laramide shortening to mid-Tertiary extension coincided with the end of early volcanism of intermediate composition (lower Datil Group) and the beginning of bimodal volcanism (upper Datil Group and younger rocks) in the Mogollon-Datil volcanic field of the southern Basin and Range province (Cather 1990). A major change of stress azimuth by $\sim 90°$ occurred in Mexico at 33 ± 2 Ma and in the western United States at 30 ± 2 Ma (Bird 2002). In the northern part of the Great Basin, crust began stretching about 32–30 Ma (Axen et al. 1993).

An early structural development of the Rio Grande rift and Basin and Range province was characterized by low-angle faulting that commenced about 30 Ma and lasted 10–12 m.y., whereas a later development was dominated by high-angle normal faulting. The change of tectonic style was accompanied by transition from caldera-generated intermediate–silicic to basaltic magmatism (Baldridge et al. 1984; Seager et al. 1984; Lipman 2007). In studies of 1970s–1980s, the Rio Grande rift

DOI 10.1007/978-90-481-2999-7_13,

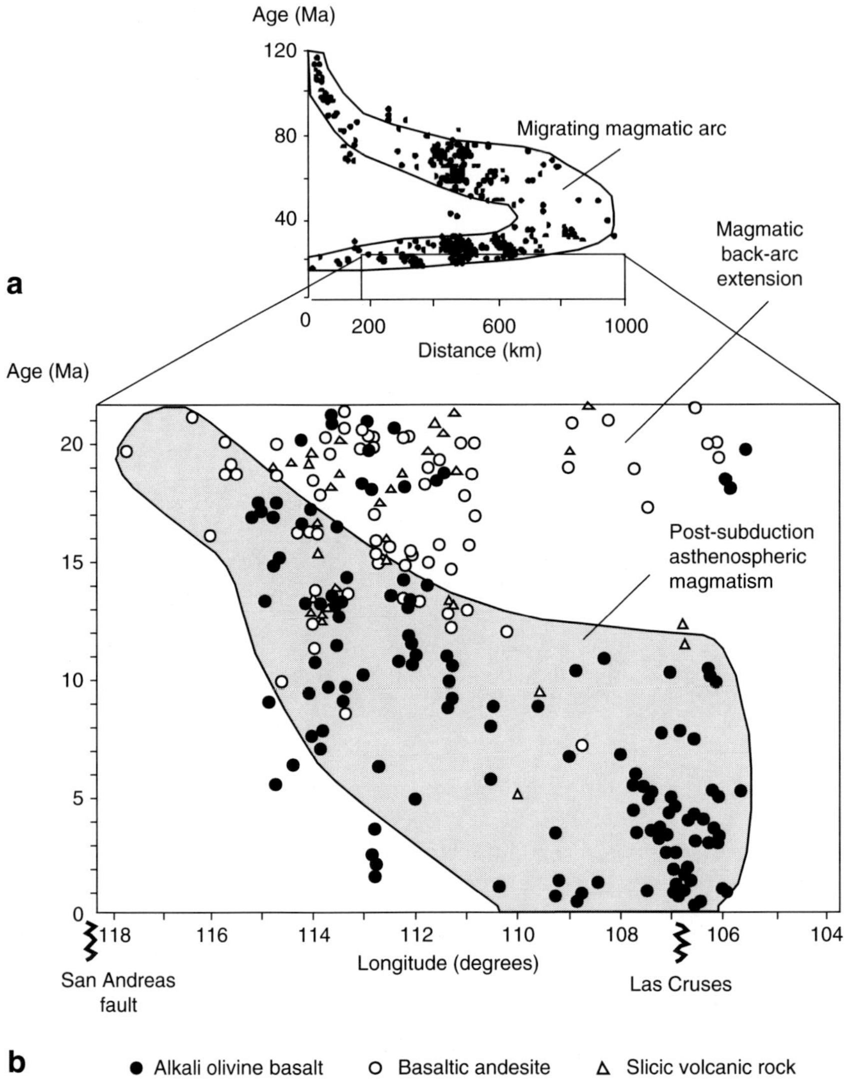

Fig. 13.1 Cretaceous–Tertiary migration of a magmatic arc (**a**) and distribution of radioisotope ages of volcanic rocks less than about 21 Ma across the southern Basin and Range province, mainly south of lat. 35°N (**b**). A *gray field* in **b** shows temporal eastward increasing of the "slab-free" region exhibited by eruptions of the depleted asthenospheric liquids. Compositions that fall above this trend have originated in back-arc setting. Modified after Coney and Reynolds (1977) and Seager et al. (1984)

was considered as intensively extending in the northern part since about 27–25 Ma and in the southern part since about 31–28 Ma (Lipman and Mehnert 1975; Chapin 1979; Baldridge et al. 1980). Observations were limited, however, by magmatic areas. From studies of sedimentation, it was inferred that rift basins subsided, firstly (between 21 and 10 Ma), due to weak WSW–ENE extension and then (in the past 10–8 Ma) due to stronger WNW–ESE extension (Ingersoll 1988; Ingersoll et al. 1990).

13.1.1 Magmatic Migration of 120–37 Ma

Granite batholiths were emplaced in the western part of the Great Basin in the Late Jurassic through Early Eocene. U–Pb zircon ages of the Sierra Nevada batholith ranges from 165 to 80 Ma, in some intrusions of its eastern part from 220 to 200 Ma (Chen and Moore 1982). Ages of the Peninsula batholith fall in a time interval of 125–85 Ma. U–Pb and K–Ar ages of both batholiths are mutually consistent. In terms of Nd, Sr, and Pb isotope composition of rocks, the batholiths resemble arc volcanic rocks (De Paolo 1981a, b).

Magmatism, related to Laramides, migrated from a continental margin (California and Nevada) through northern and central parts of the Great Basin up to the future Rio Grande rift. A magmatic hiatus occurred between 70 and 55 Ma. For example, the 71–70 Ma granodiorites of south Arizona were separated from the ~36 Ma rhyolites (Spencer et al. 1995). In the Slave province of Canada, kimberlite pipes of the Lac de Gras field were emplaced firstly from 75 to 70 Ma and then from 56 to 48 Ma (Heaman et al. 2003). In the area of the future northern Rio Grande rift, late Mesozoic magmatism took place in a time interval of 70–65 Ma and afterwards commenced only about 37 Ma (Lipman et al. 1978; Lipman 2007).

Andesites, dacites, absarokites, and shoshonites of supergroup Absaroka erupted in Yellowstone after the magmatic hiatus, in a time interval of 52–43 Ma (Smedes and Prostka 1972). Minettes, phonolites, leucitites, syenites, and shonkinites distributed in Montana, to the north and northeast of Yellowstone, in a time interval of 53–47 Ma. Carbonatites, mica peridotites, and kimberlites emplaced in the northern part of Wyoming in a time interval of 51–48 Ma. Then magmatism of the Salt Lake City area commenced between 44 and 32 Ma (Stewart and Carlson 1976). Volcanism developed sporadically in the Cascades between 55 and 42 Ma with concentration in the Cascade magmatic arc in a time interval of 40–36 Ma (e.g. Swanson et al. 1989).

13.1.2 Transition from Intermediate–Silicic to Basaltic Magmatism of 37–18 Ma

In the Great Basin, southern Basin and Range province, and northern Rio Grande rift, volcanic rocks of 37–18 Ma sufficiently varied in composition. Firstly, trachyandesite and trachyrhyolite lavas and welded tuffs dominated in caldera complexes. Then, compositions of volcanic rocks changed to mostly basaltic with an essential role for alkaline basalts.

Trachyandesites, trachyrhyolites, and minor basalts erupted in calderas of the San Juan volcanic field in a time interval of 37–23 Ma with possible final events of 18 Ma (Lipman et al. 1978; Lipman 2007). Similar volcanic activity occurred in the

Mogollon-Datil volcanic field in a time interval of 38–31 Ma (Dungan et al. 1989; Cather 1990). Low-volume latite magmatism took place in the Espanola basin about 38–34 Ma (Baldridge et al. 1980).

Andesites and rhyolites began erupting in the northern part of the Great Basin about 43 Ma, concentrated in its northern and central parts until 30 Ma, and finalized in an area between the Great Basin and Basin and Range province in a time interval of 22–18 Ma. The intermediate to silicic volcanism shifted across these regions southwards. Two stretching belts in the Great Basin extended north–northeast in a time interval of 23–15 Ma (Axen et al. 1993). In the southern Basin and Range province, similar volcanism shifted northwestwards in a time interval of 33–22 Ma (Fig. 13.2).

In southwestern Arizona, volcanic activity renewed about 25 Ma after a hiatus lasting 10 m.y. Mostly silicic volcanic rocks of 25–22 Ma changed to mostly basaltic of 21–19 Ma and younger (Spencer et al. 1995). Similar temporal change of volcanic rock compositions occurred in the Mogollon-Datil volcanic field (Hawkesworth et al. 1995). In the central Colorado Plateau, volcanic activity was dominated by minettes concentrated within the Navajo volcanic field in a time interval of 28–19 Ma (Laughlin et al. 1996) (Fig. 13.3).

The 29.5–28.4 Ma Treasure Mountain welded tuffs of the San Juan volcanic field are underlain by trachyandesites and trachyrhyolites of the Conejos Formation and

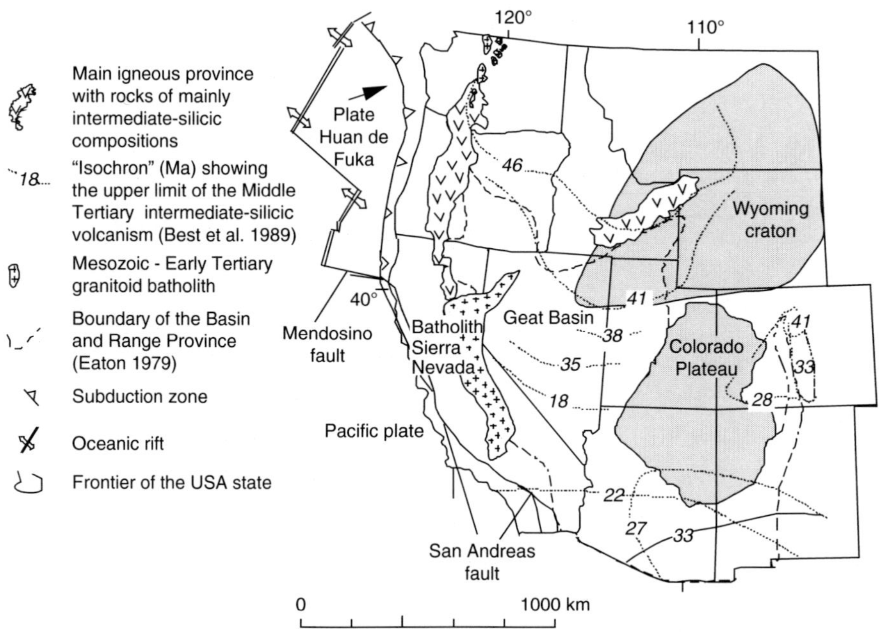

Fig. 13.2 Spatial distribution of the main late Phanerozoic intermediate–silicic magmatic units in the Western United States. Modified after Eaton (1979), Luedke and Smith (1983), Aldrich and Laughlin (1984)

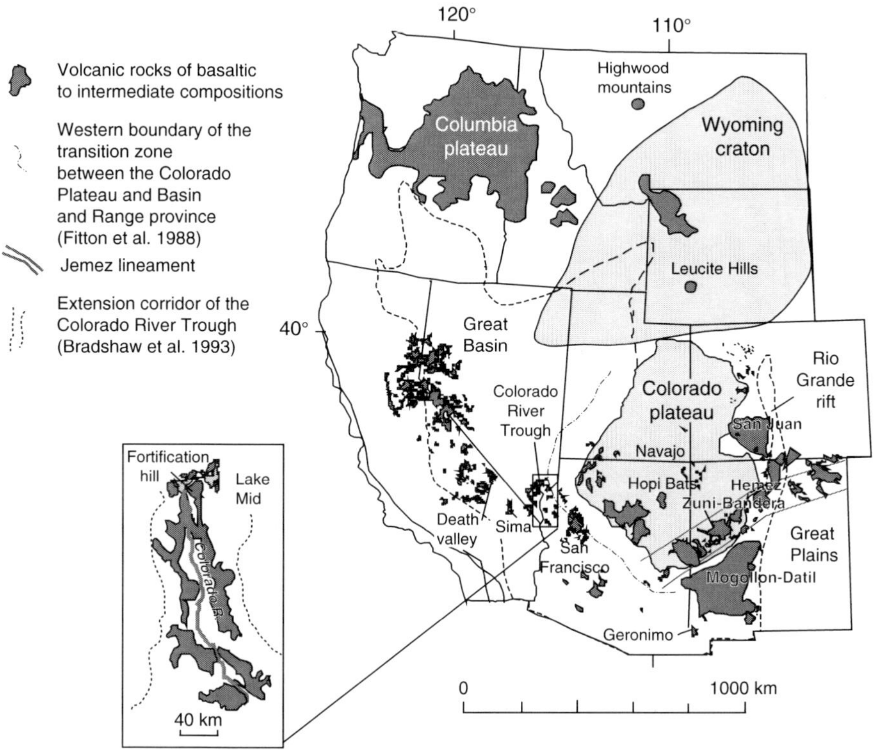

Fig. 13.3 Spatial distribution of the main middle–late Tertiary volcanic fields in the Western United States. Some elements of a legend as in Fig. 13.2. Modified after Luedke and Smith (1983), Aldrich and Laughlin (1984)

are capped by basaltic andesites, alkaline basalts, and andesites of the Hinsdale Formation. Minor dacites of the San Luis basin and geochemically distinct low-K basanites of the Espanola basin erupted 25–22 Ma and 25–18 Ma, respectively (Dungan et al. 1989; Gibson et al. 1993; Rasskazov et al. 2010).

13.1.3 Predominating Basaltic Magmatism of the Past 17–15 Ma

Basaltic volcanism was distributed mainly in the western part of the Great Basin, near the border between Nevada and California, in a time interval of 17–6 Ma with peak of activity between 17 and 12 Ma (Best et al. 1989). Eruptions became rare in the past 8 Ma, but extremely variable compositionally. In central California, trachybasalts and basaltic andesites erupted in the San Hoakin volcanic field between 11 and 8 Ma (in the Kern volcanic field between 13 and 10 Ma) and basanites, trachybasalts, trachyandesites, leucitites, and minettes about 3.5 Ma (Farmer et al. 2002).

In the Death Valley, located on a border between the Great Basin and southern Basin and Range province, intensive stretching of crust took place between 12 and

5 Ma. Firstly, hypersthene-normative and then, nepheline-normative basalts erupted in time intervals of 12–8 Ma, 6.5–4.0 Ma, and <1 Ma (Asmeron et al. 1994). Areas of stretching in Lake Mid and Death Valley were separated from each other by a block that was not affected by extension (Fig. 13.3). Alkaline basalts of Lake Mid erupted 11–10 Ma and 6–4 Ma and those of the Colorado River Trough (south of Lake Mid) mainly in a time interval of 22–10 Ma with rare subsequent events (Bradshaw et al. 1993). Crust underwent stretching here up to 100% between 22 and 9 Ma. North of Lake Mid, lavas of basaltic to intermediate compositions erupted in time intervals of 17–12 Ma and 10–5 Ma (Feuerbach et al. 1993).

Alkaline basalts and trachyandesites erupted in the Mogollon-Datil volcanic field between 19 and 11 Ma with the latest event of 0.9 ± 0.2 Ma and in the southern Rio Grande rift in time intervals of 21–13 Ma and 7–6 Ma (Davis and Hawkesworth 1995).

In the Colorado Plateau, magmatism developed after a long magmatic lull, since ~9 Ma. Basalts marked mostly activity of a transition zone near the border between the Colorado Plateau and Basin and Range province (Alibert et al. 1986).

After long-lasting weak magmatism, basaltic eruptions enveloped the southeastern Basin and Range province in a time interval of 9–3 Ma (Seager et al. 1984). In the past 8 Ma, basanites, alkaline basalts, and tholeiitic basalts erupted in other areas of the southern Basin and Range province: on volcanic fields Sima, southeast California (8–3 Ma and <1 Ma), Colorado River Trough (6–5 Ma), Zuni Bandera and San Francisco, lineament Jemcz (<1.4 Ma and <5 Ma, respectively), Palomas, Sierra de la Uvas, and Potrillo in the southern Rio Grande rift (12–4 Ma and <0.5 Ma) (Bradshaw et al. 1993; Seager et al. 1984; Menzies et al. 1991; Farmer et al. 1995).

Basalts erupted in the Lobato area of the Jemez volcanic field in a time interval of 14–9 Ma (Lipman and Mehnert 1975; Aldrich and Deither 1990) and mostly basalts and andesites (Hinsdale Formation) in the northern Rio Grander rifted in a time interval of 18–5 Ma. Tholeiitic basalts of the Servilleta Formation flooded the San Luis basin 4.5–3.5 Ma simultaneously with emplacements of trachyandesite-trachyrhyolite extrusions. Subsequent eruptions of trachyandesites, hawaiites, and transitional basalts lasted in the San Luis basin and Tusas mountains between 2.2 and 0.24 Ma (Lipman 1969; Dungan et al. 1986, 1989).

13.1.4 Geodynamic Models

Atwater (1970) proposed that the oceanic lithospheric plate Farallon, which existed between the Pacific and North American plates to the south of the system of Mendocino faults in the Mesozoic and early–middle Tertiary, was consumed by subduction beneath the North American continent between ca. 20 and 10 Ma. Subduction processes were expressed by a temporal shift of magmatic arc from a continental margin landwards for a distance over 1000 km in a time interval of 120–20 Ma (Coney and Reynolds 1977). This migration might have been caused by (1) stepwise shallow slab subduction (Christiansen and Lipman 1972) or (2) change of subduction angle (Spencer and Reynolds 1989; Spencer et al. 1995). The proposed models for tectonic and magmatic evolution of the continental margin stress an important role of temporal change from the Farallon plate subduction to dextral displacement of the Pacific plate relative to North America.

The temporal change from intermediate–silicic to mainly basaltic magmatism accompanied significant (up to 200%) stretching of crust and upwelling of asthenospheric material beneath a territory extending from southern California to the Rio Grande rift. Basaltic volcanism that was distributed mainly south of lat. 35° N in a time interval of 21–15 Ma resulted from back-arc extension (Seager et al. 1984) (Fig. 13.1). Such magmatic change was due to cessation of the Farallon plate subduction about 29 Ma and reactivation of the San Andreas transform fault (Atwater 1970; Seager et al. 1984).

Some authors (Atwater 1970; Eaton 1979; Bohannon and Parsons 1995) connected a crust stretching in the Basin and Range province with interaction between the East Pacific rise and North American plate south of the California peninsula in a time interval of 29–25 Ma. Others (Axen et al. 1993; Dickinson 2002) emphasized distinction in geodynamics of those regions that occurred in the Late Oligocene–Middle Miocene to the north and to the south from the Mendocino fault system, respectively, in the Great Basin and southern Basin and Range province. A transition zone occurred in the southwestern Colorado Plateau. Crustal extension in the northern Basin and Range province accompanied northward migration of the Mendocino fault system, whereas the southern part reflected cessation of the Farallon plate subduction and fast moving of a triple junction to the south (Glazner and Supplee 1982). Respectively, basaltic volcanism marked formation of a slab window beneath the transition zone in a time interval of 28–18 Ma. From distribution of P and S seismic-wave velocities, the upwelling mantle appears to have associated itself with the sinking Farallon slab fragment, located in the transition mantle zone beneath the Colorado Plateau. The ascending convective flow has removed a foot of the lithosphere from below the northern Rio Grande rift with entraining of this lithospheric material in the eastern downwelling beneath the Great Plains (Gao et al. 2004).

In some hypotheses, subduction was not examined as a leading geodynamic factor, however. It was proposed, for instance, that extension and volcanism in the Western USA was controlled by occurrence of convective cells in the sub-lithospheric mantle. According to this model, an extensive stretching in the Basin and Range province could have been caused by asthenospheric convection that provided hydrodynamic forces for drag of the lithospheric foot (Eaton 1979). It was suggested also that subduction had no influence on compositions of Oligocene volcanic rocks in the northern part of the Rio Grande rift. Subduction-like geochemical signatures of basalts and evolved rocks were interpreted as a consequence of mantle metasomatism initiated in conditions of lithospheric extension (Gibson et al. 1993).

13.2 Northeast Africa

In the Mesozoic and early–middle Tertiary, the African continent was affected by sporadically distributed tholeiite, alkaline, kimberlite, and carbonatite magmatism (Kampunzu et al. 1986; Reid et al. 1990). The Rukwa, Rungwe, and Chilwa areas of the future Western rift, for instance, were occupied by the 136–96 Ma carbonatites accompanied Late Jurassic–Early Cretaceous rifting (Pentel'kov and Voronovskii 1977).

Spatial–temporal interpretation of the middle–late Tertiary magmatism in rifts of Northeast Africa was based mainly on results of dating by the conventional K–Ar method (e.g. Logatchev 1977). In recent papers, these data were supplemented by $^{40}Ar/^{39}Ar$ ages, which brought new understanding of rift-related processes. For instance, according to former K–Ar dating, 2–3 km thick tholeiitic lava stratum in the northern Ethiopian rift was constrained within a time interval of 50–10 Ma. By incremental $^{40}Ar/^{39}Ar$ dating in combination with paleomagnetic data, the age range of the whole lava sequence was bracketed within a short episode of 30 $\pm$ 1 Ma (Hofmann et al. 1997).

We have measured $^{40}Ar/^{39}Ar$ ages for thirteen samples of volcanic rocks from the Rungwe province of Tanzania, that fall mostly within a time interval of the past 9 Ma recorded for volcanic rocks from this area by Ebinger et al. (1989). Some dated samples, however, have marked a previously unknown magmatic episode in the Western branch of the East African Rift System. From this data, we have proposed a modified scenario for spatial–temporal evolution of rift-related processes in Northeast Africa (Rasskazov et al. 2003a).

13.2.1 Earliest Volcanic Events in the Western Rift

Initiation of rift-related processes in the Western rift was constrained in previous publications by the earliest Late Miocene sediments with fauna in the Malawi basin, southern part of the rift (Ebinger et al. 1989) or from the Late Cretaceous K–Ar dates of volcanic rocks from South Kivu, northern part of the rift. The Cretaceous–Paleogene dates were not confirmed by subsequent K–Ar and $^{40}Ar/^{39}Ar$ age measurements. Lavas from the Virunga volcanic field of 12.6 $\pm$ 0.7 Ma were considered to be the oldest reliably dated unit from the Western rift (Ebinger et al. 1989; Pasteels et al. 1989; Rasskazov 1993). The older unit was found in the Rungwe province (Fig. 13.4).

Phonolite domes were mapped in the Usangu basin as the Quaternary "young extrusives" due to sharply expressed morphology (Harkin 1960). A phonolite from the Machioto Hill (Mwasihutu) dome showed, however, an $^{40}Ar/^{39}Ar$ age of 17.3 $\pm$ 0.1 Ma. To confirm this age value, we dated the phonolites by the Rb–Sr method. Two data points of rocks from the Wuzi dome with elevated Rb/Sr ratios (Ru–15 and Ru–15A) marked a line with a slope appropriate to an age of 17.03 $\pm$ 0.03 Ma with initial $^{87}Sr/^{86}Sr = 0.704\,73 \pm 0.000\,09$. Together with a phonolite sample from Ngalupala (Ru–14), plotted near this line, the Wuzi samples yielded an isochron age of 17.01 $\pm$ 0.01 Ma with initial $^{87}Sr/^{86}Sr = 0.704\,80 \pm 0.000\,05$ and MSWD = 0.83 (Fig. 13.5).

Unlike phonolites from the Wuzi and Ngalupala domes, those from other domes—Chaulimisa Hill (Ilongo after Harkin (1960)) and Machioto Hill (samples Ru–12 and Ru–13)—are shifted above the isochron line of 17.03 Ma. Respectively, in

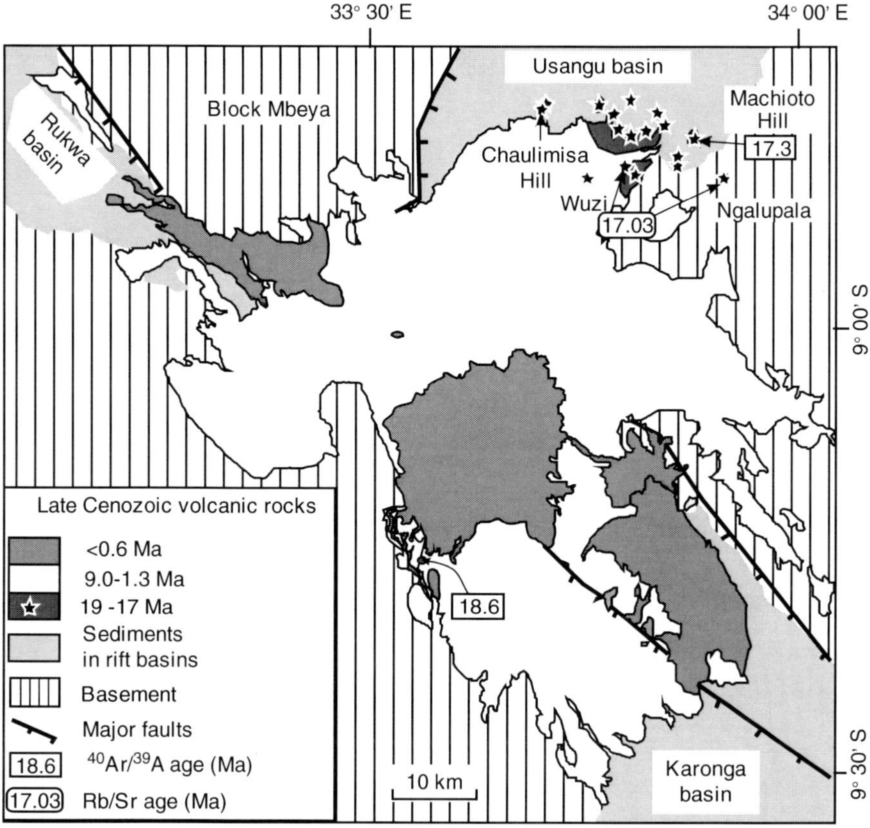

Fig. 13.4 Volcanic units of the Rungwe province, Western rift. In the Usangu basin, phonolite domes and flows are shown by *stars* and *dark-gray fields*, respectively. The unit of 19–17 Ma exhibits olivine tholeiite flows, as well as phonolite flows and domes trending northeast–southwest, along the Usangu basin. The unit of 9.0–1.3 Ma—lavas and edifices with variable (from basaltic to intermediate) compositions—forms an isometric volcanic field centered at a structural junction between the Usangu, Rukwa, and Karonga basins. The unit of the past 0.6 Ma represents a sequence started by high-Mg olivine melanephelinites and demonstrates focusing of magmatism along the Rukwa and Karonga basins. Geochemical data on volcanic rocks from the Rungwe province were reported by Ivanov et al. (1998)

coordinates of initial $^{87}Sr/^{86}Sr$ versus 1/Sr, these samples are distributed along an ordinate axis demonstrating contamination of phonolite liquids by a ^{87}Sr-rich component. The $^{40}Ar/^{39}Ar$ age of 17.3 ± 0.1 Ma was obtained for the most contaminated phonolite Ru–13 from the Machioto Hill dome. The dated contaminated phonolites were followed by uncontaminated ones within ~0.3 m.y.

Relatively old $^{40}Ar/^{39}Ar$ age of 18.6 ± 1.0 Ma was measured also for a groundmass of olivine tholeiite Ru–21 located in the southwestern continuation of the former Usangu basin (Fig. 13.4). The dated olivine tholeiite was sampled from a lava flow that occupied a watershed between Kiwira and Mbaka rivers. In a road cut, the flow

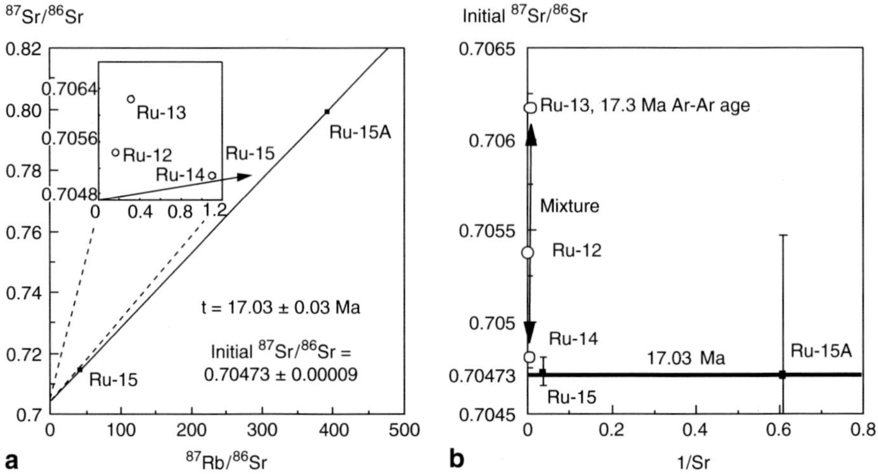

Fig. 13.5 Rb-Sr isochron diagram (**a**) and relation between initial $^{87}Sr/^{86}Sr$ and 1/Sr (**b**) for phonolites from the Usangu basin. On an insert to panel **a**, phonolite samples Ru–12, Ru–13, and Ru–14 lie higher than the isochron line of samples Ru–15 and Ru–15A from the Wuzi dome. Errors of initial $^{87}Sr/^{86}Sr$ increase with increasing $^{87}Rb/^{86}Sr$ in samples Ru–15 and Ru–15A

is exposed as a fragment within a mature weathering crust. The massive basaltic blocks as large as 2 m were affected by weathering in periphery without alteration in inner parts. The olivine tholeiites belong to a group of rocks demonstrating the lowest $^{87}Sr/^{86}Sr$ ratios (~0.704 70), which were considered as characteristic of the partly melted foot of the lithospheric mantle affected by fluids rising from the asthenosphere at the initial stage of rifting.

13.2.2 Hypotheses

The initial phase of rifting is usually defined by the earliest sediments in basins or earliest rift-related lava eruptions, because both subsidence of rift basins and magmatism reflect thermo-mechanical thinning of the lithosphere.

Middle–late Tertiary magmatism and rifting in the African continent was considered to be caused by forty different plumes (Burk and Wilson 1972), by a single (i.e. Ethiopian) plume (Ebinger and Sleep 1998) or by two (i.e. Afar and Kenyan) plumes (Rogers et al. 2000). It was proposed also that stretching in the Read Sea–Aden rift branch was initiated about 30 Ma due to collision between Africa and Eurasia (Baker et al. 1996).

Authors of the mono-plume hypothesis have assumed existence in East Africa of a single centre of deep thermal activity expressed by voluminous magmatism of the Ethiopian plateau in a time interval of 45–35 Ma. According to this hypothesis, spatial–temporal evolution of magmatism and rifting in Africa was governed by a mechanism of radial spreading of hot material from a plume source. Timing of initial eruptions was based on results of both K–Ar and $^{40}Ar/^{39}Ar$ dating (Ebinger et al. 1993), but in fact, only a volcanic episode of 36.90 ± 0.05 Ma was reliably dated by the $^{40}Ar/^{39}Ar$ method. The interval of dates from 45 to 35 Ma was obtained by the conventional K–Ar method. Because of probable distortion of K–Ar ages of Paleogene volcanic rocks, the latter time interval requires confirmation by $^{40}Ar/^{39}Ar$ data. The mono-plume hypothesis was not supported by subsequent high-precision dating of volcanic rocks in Ethiopia, where voluminous lava strata accumulated in the northern part of the Ethiopian plateau about 30 Ma (Hofmann et al. 1997), i.e. later than those of the southern part of the plateau.

13.2.3 Bilateral along-Axis Propagation of Rift-Related Magmatism

The Ethiopian and East African plateaus are separated from each other by the saddle-like Turkana downturn of a relief. Initial magmatism in the Turkana saddle ~37 Ma (or probably ~45 Ma) indicates the earliest rift activation of the most weakened chain in the East African Rift System. On the one hand, the saddle corresponds spatially to a series of transverse (trending northwest–southeast) lineaments and shear zones (Surma etc.). On the other hand, within the saddle, the north–south trending Kenyan (Gregory) rift basins join north–northeast – south–southwest trending basins of the Ethiopian rift (Fig. 13.6).

The onset of volcanism in the Turkana basin corresponds to impact of the mantle plume. Sr-isotope ratios of both Tertiary and Quaternary lavas here are lower than values inferred for either of the two plumes proposed by Rogers et al. (2000), whereas Nd-isotope ratios of the Turkana lavas overlap those of the present-day Afar plume rather than the proposed Kenyan plume (Furman et al. 2004).

Northward propagation of rifting and magmatism from the Turkana saddle was accompanied by voluminous melting in a large thermal anomaly developed along the Ethiopian rift in the central part of the Ethiopian plateau ~30 Ma. Voluminous basalt–rhyolitic magmatism of Yemen, where a volcanic stratum as thick as 2.5 km accumulated in a time interval of 31–26 Ma (Baker et al. 1996; Menzies et al. 1997), was preceded by uplift and erosion of an area along the Red Sea rift initiated about 34 Ma (Omar and Steckler 1995).

South of the Turkana saddle, in the Kenyan rift, minor alkaline magmatism with carbonatites, that predated rifting, began 35–33 Ma. More voluminous magmatic processes, accompanied by basin subsidence and erosion of a foot of the lithosphere, took place only since 23 Ma (e.g. Logatchev 1977). Trace-element signatures of erupted alkaline basalts were of the OIB-type. Unlike the Ethiopian rift that centered on the Ethiopian plateau, the Kenyan rift developed on the periphery of the East African plateau.

Taking into account occurrence of the Middle Miocene fauna in alluvial sediments of the Albert basin and extrapolating thickness of accumulated sediments, Petterson showed that rifting in the northern part of the Western rift began in the Early Miocene (Ebinger et al. 1989). Similarly, phonolites and olivine tholeiites from the Rungwe

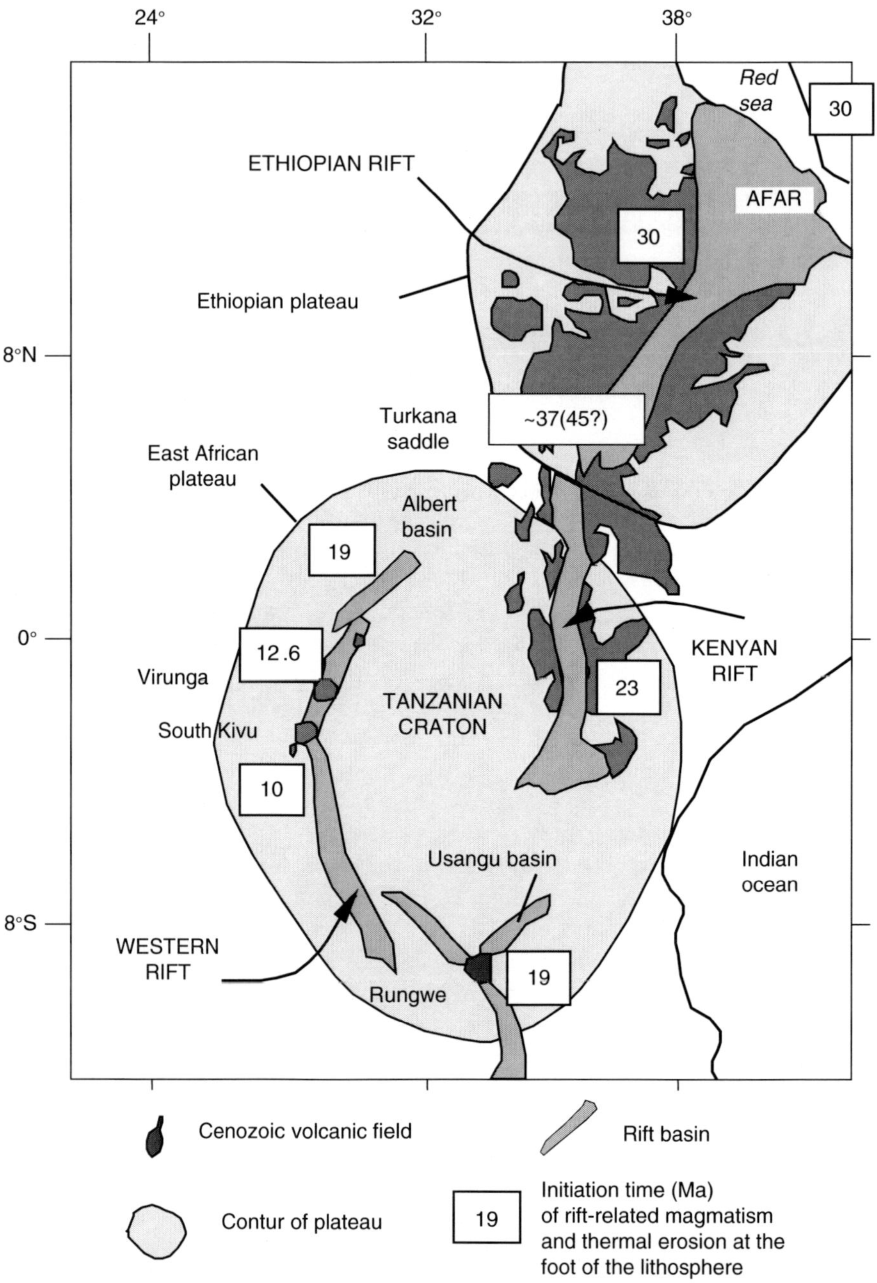

Fig. 13.6 Spatial–temporal propagation of magmatism and thermal erosion of the lithosphere in Northeast Africa. For the Albert basin, the initial time of sedimentation is shown. See explanations in the text

volcanic field accompanied the Early Miocene initiation of rifting in the southern part of the Western rift. In fact, the volcanic episode of 19–17 Ma in the Usangu basin was coeval with the beginning of sedimentation in the Albert basin. Therefore, the southern and northern parts of the rift began developing approximately simultaneously, in the Early Miocene, with subsequent propagation from the north and south to its central part.

13.2.4 Geodynamics of the East African Plateau

From gravimetrical data collected along an equatorial profile, a present-day thickness of the lithosphere beneath the Tanzanian craton does not exceed 100 km with relative decreasing in the Kenyan and Western rifts, respectively, to 38 and 65 km (Hay et al. 1995). Teleseismic tomography has revealed under the Kenyan rift a low-velocity anomaly of P and S waves, which extends from a foot of the lithosphere to a depth of 200 km. The asthenospheric uplifts of the Kenyan and Western rifts are connected to each other beneath the Tanzanian craton in a depth interval of 100–200 km, but are separated in the shallow mantle. Seismic tomography shows a large low-velocity mantle region dipping from the East African plateau southwards, beneath the southern margin of Africa (Ritsema and Heijst 2000).

Occurrence of diamond-bearing kimberlites Mwadui with a zircon track age of 52 ± 2 Ma in the central part of the Tanzanian craton (Heaman et al. 2003; Stiefenhofer and Farrow 2004) indicates that the Paleocene lithosphere was here as thick as 200 km. At present, the East African plateau is supported dynamically by asthenospheric convection that resulted in erosion of a foot of the lithosphere (Ebinger et al. 1989).

Spatial–temporal distribution of magmatism and rifting implies propagation of thermal erosion of the lithosphere from the Turkana saddle firstly southwards, along the Kenyan rift, and then westwards, along the Western rift. The erosion began beneath the Rungwe volcanic field and Albert basin in a time interval of 19–17 Ma, being delayed beneath the Virunga and South Kivu volcanic fields until a time interval of 12.6–10.0 Ma (Fig. 13.6).

The Late Miocene through Quaternary volcanic activity in the northern and southern parts of the Western rift was not coeval at the beginning but became more coherent in the late Quaternary. For instance, in the South Kivu volcanic field, near Bukavu, volcanic activity was dominated by fissure eruptions of three stages (Kampunzu et al. 1986; Pasteels et al. 1989; Furman and Graham 1999). The earliest activity of 10.0–7.5 Ma that occurred prior to tectonic motions was limited to the East Kivu basin and the southern part of the Idjwi island. The second-stage volcanism of 7.5 and 4.0 Ma, which likely corresponded to the start of basin subsidence (Furman et al. 2004), occurred in both the East and West Kivu basins. The third-stage (Quaternary) volcanism exhibited lavas that erupted primarily along the West Kivu border fault system trending northeast–southwest. In the Rungwe province, the 9.2–1.3 Ma lavas and edifices with variable (from basaltic to intermediate) compositions created an isometric volcanic massif centered at a structural junction between the Usangu,

Rukwa, and Karonga basins. This unit included three age groups: (1) trachytes of ∼9.2 Ma strongly contaminated by crustal material with initial $^{87}Sr/^{86}Sr$ as high as 0.7076, (2) trachytes, olivine tholeiites, and alkaline basalts of 7.5–5.0 Ma, and (3) basalts and evolved rocks of 3.5–1.3 Ma. Quaternary volcanoes that were active in the past 0.6 Ma represent a particular sequence started by high-Mg olivine melanephelinites and demonstrated focusing of magmatism along the Rukwa and Karonga basins (Rasskazov et al. 2003a).

In a hotspot frame, the African plate moved in the Cenozoic northeastwards with relatively high speed that decreased to 3 cm/year in a time range between 30 and 19 Ma and afterwards was as low as 2 cm/year (O'Connor et al. 1999). The speed decreasing could reflect collision between Africa and Eurasia. The collision produced also a tectonic stress affecting the African lithosphere. As a result, initial thermal erosion in the Western rift was concentrated in the Usangu and Albert basins, trending in accordance with the African plate motion. In the late Quaternary, the role of border faults in spatial control of volcanoes increased both in the northern and southern parts of the Western rift, although the active border faults had orthogonal orientation (northeast–southwestern in the north and northwest–southeastern in the south).

Unlike the northern part of the Western rift, initiated due to Africa–Asian convergence and conservatively developed along the northeast–southwestern extensional locus in the past 19 Ma, the southern portion of the Western rift showed temporal rotation of extensional loci. The initial convergence, caused by Early Miocene stress that provided volcanic activity of 19–17 Ma along the Usangu basin, changed to a neutral regime within a triple junction between the Usangu, Karonga, and Rukwa basins in a time interval of 9.2–1.3 Ma with subsequent transition of activity to an orthogonal (relative to convergence direction) extensional locus. This difference indicates that the northern part of the rift, located closer to the Africa–Asian convergence zone, remained strongly dependent on external stress from the Miocene through Quaternary, but the more remote southern part of the rift underwent orthogonal extension in the late Quaternary, probably due to a dominating effect of buoyant low-velocity material beneath axial rift basins.

13.3 Comparisons: Africa—Eurasia—North America

Comparative analysis of magmatic evolution in the three continents is focused on phenomena that generally resemble the deep nature of rift-related processes and deviations expected from different plate interactions.

On the one hand, magmatic territories of Asia and North America underwent uplifting, similar to Northeast Africa, where Ethiopian and East African plateaus arouse due to buoyancy of low-velocity mantle regions. East Asia is underlain by the 350–400 km deep Transbaikal low-velocity domain, whereas Central Asia by the 200 km deep Sayan–Mongolia low-velocity domain. The former was dynamically provided by late Phanerozoic subduction of oceanic plates beneath East Asia, whereas the latter was influenced by convergence between India and Asia. The sublithospheric low-velocity region beneath Africa was related to superplume activity.

On the other hand, lithospheric plates of the Pacific Ocean interacted with the eastern margin of Eurasia and western margin of North America. Unlike the latter, the former margin was affected by accretion of oceanic plates and terranes. Slab subduction and transform motions in both margins were accompanied by extension of the lithosphere and magmatism. Therefore, magmatic evolution in Asia and North America should demonstrate similarity due to principally the same simultaneously operated geodynamic mechanisms with some consequences resulting from the difference in plate interactions. Northeast Africa does not show any evidence on long-term subduction-related processes, but, nevertheless, magmatic evolution here might be affected by interaction between Africa and the European part of Eurasia similar to interaction between India and the Asian part of Eurasia. Major magmatic features should reflect collision- and convergence-caused consequences.

Representative tectonic and magmatic events, listed in Chaps. 12 and 13, are tabulated in Figs. 13.7 and 13.8. Tectonic features of continents, well pronounced in the late Tertiary, become obscure in the Late Cretaceous and early–middle Tertiary (Fig. 13.7). A lack of reliable information on tectonic events is partly compensated by available data on magmatic evolution (Fig. 13.8).

We have described Cenozoic magmatic dynamics in Asia in terms of spatial–temporal relations between high- and moderate-K basaltic magmatism. High- and ultra-K magmatism occurred in both North America and Africa. In the former continent, this type of magmatism manifested itself in early and advanced stages of landward migration of the magmatic arc (52–43 Ma—Montana; 35–27 Ma—Two Buttes, Great Plains; 28–19 Ma—Navajo, Colorado Plateau) and distributed sporadically afterwards (3.8–3.3 Ma—central Sierra Nevada; <1 Ma—Leucite Hills, Wyoming) (Van Kooten 1980; Vollmer et al. 1984; Leat et al. 1988; Gibson et al. 1993 etc.). In the latter continent, similar magmatism was characteristic only of the past $\sim$3 Ma activity in the Virunga and Toro-Ankole volcanic fields, located in the northern part of the Western rift (Logatchev 1977; Pasteels et al. 1989; Thompson et al. 1989; Rasskazov 1993; and references therein). Unlike East Asia and North America, where high-K magmatism was related to subduction–transform motions of plates, collisional interaction of continents (India and Eurasia, Africa and Eurasia) resulted in a temporal transition from low-moderate-K to high-ultra-K magmatism (Fig. 13.8).

Comparison of magmatic evolution in the three continents reveals time intervals of asynchronous and synchronous magmatic and tectonic processes. We connect the former periods with initial phases of structural reorganizations, when collisional or other geologic events induce initial instability, and the latter periods with advanced stages, when magmatic systems adapt to new geodynamic conditions. The collision or accretion might cause a specific magmatic phase, whereas the following magmatic propagation into adjacent areas demonstrates spatially growing instability in the lithosphere and sub-lithospheric mantle. When magmatic processes extremely satisfy new structural conditions, events take place synchronously in different areas.

Eurasia collided with India earlier than with Africa (i.e. 66–50 and $\sim$30 Ma, respectively). Initial collisional phases accompanied voluminous trap magmatism in the Deccan and Ethiopian provinces, respectively. As a key trigger, collision

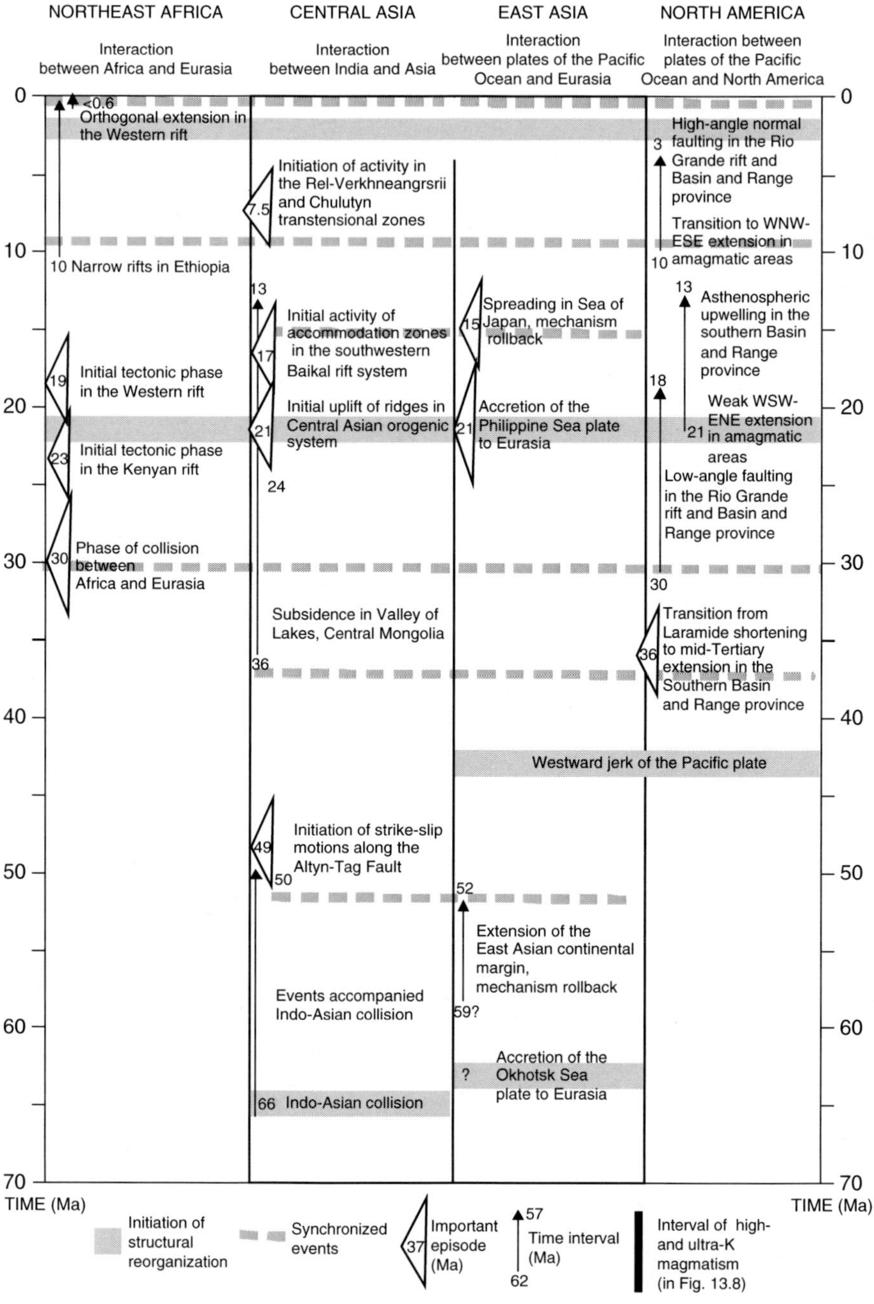

Fig. 13.7 Correlation of the main Cenozoic tectonic events in Asia, Africa, and North America (explanation in the text)

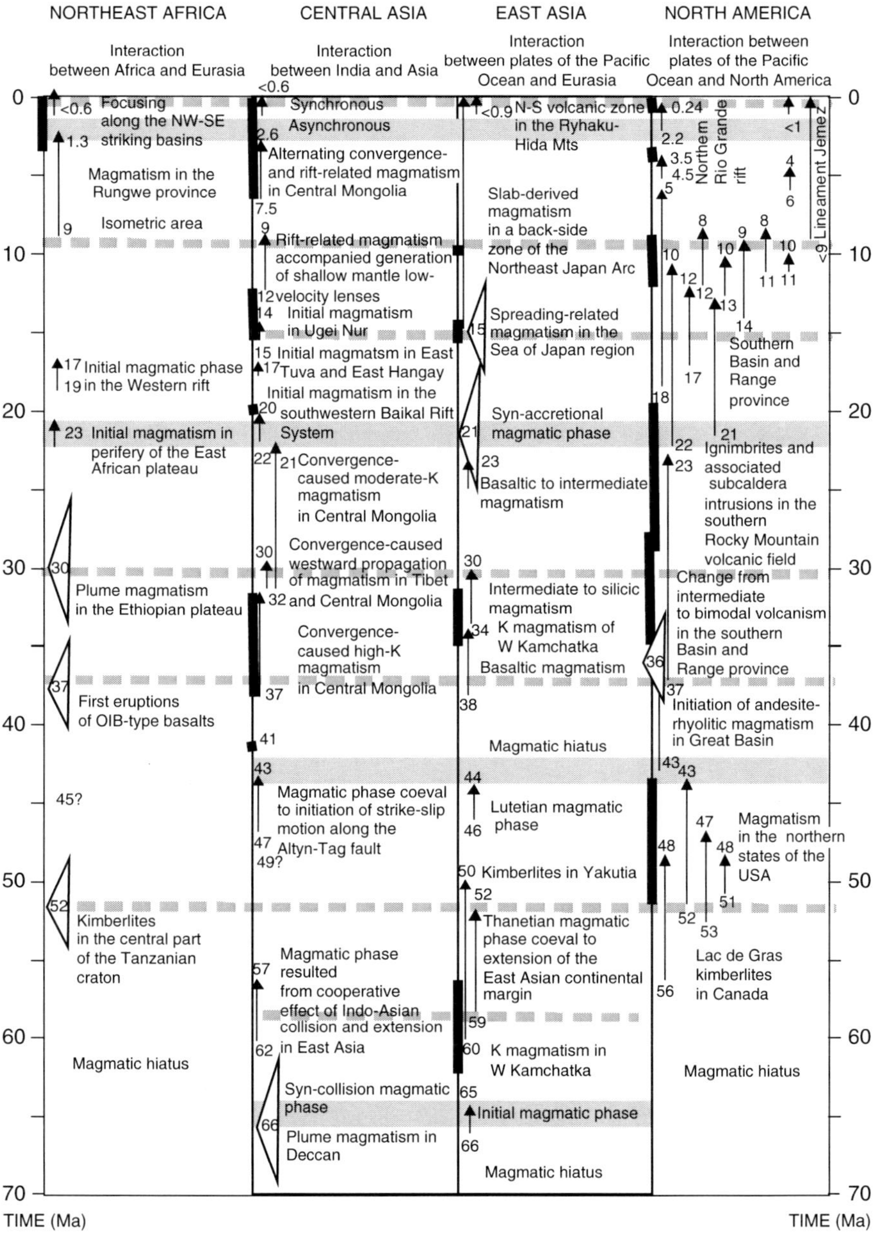

Fig. 13.8 Correlation of Cenozoic magmatic events in Asia, Africa, and North America. Legend as in Fig. 13.7

between the continents could provide additional increasing magma productivity within upwelling hot mantle regions.

Simultaneously with the first phase of Indo-Asian collision, short-term magmatic events occurred in Tien-Shan and Central Mongolia. Both regions were tectonically active in the Late Cretaceous. No specific coeval magmatic pulses took place in Africa or North America, however. The former continent was tectonically quiescent in that time; the latter was dominated by the Laramide tectonic events. Similarly, East Asia developed as a tectonically active region affected by intermittent slab subduction from the Pacific Ocean. The tectonic instability in Asia, induced by the Indo-Asian collision, was enforced by extension of the East Asian continental margin by a roll-back mechanism. The cooperative effect of collision and extension of the continental margins resulted in synchronization of magmatic pulses in different areas in time intervals of 62–50 and 49–43 Ma. Magmatism of these time intervals focused within the same areas (Chap. 12). Magmatic pulses of Africa and North America (including kimberlites) occurred chiefly between these intervals and coincided with the Eocene global flare-up of magmatic activity connected with a major reorganization of plate motion between 53.5 and 37.5 Ma (Rona and Richardson 1978).

Figure 13.9 demonstrates that voluminous trap magmatism did not affect temporal variations of $^{87}Sr/^{86}Sr$ in sea water. Periods of long-term minor variations and sufficient increasing of $^{87}Sr/^{86}Sr$ likely reflected plate interactions. We speculate that long-term minor variations of $^{87}Sr/^{86}Sr$ between 70 and 37 Ma resulted from the Indo-Asian collision and accompanied processes lasted until the Middle Eocene structural reorganization. Similar tectonic and magmatic processes might be responsible for minor variations of $^{87}Sr/^{86}Sr$ ratio between 140 and 90 Ma. Periods of sufficiently increasing $^{87}Sr/^{86}Sr$ ratio of 160–125 Ma, 90–65 Ma, and <40 Ma might be indicative for relatively monotonous plate interactions. The period of 90–65 Ma was marked, for instance, by long-term activity in the East Sikhote Alin volcano-plutonic belt (e.g. Rasskazov and Taniguchi 2006) and the period of the past 43 Ma by constant toroidal (transform/spin) and relative decreasing poloidal (spreading/subduction) plate motions (Lithgow-Bertelloni and Richards 1998).

The 52 Ma kimberlites of the Tanzanian craton were approximately contemporaneous to the latest kimberlites of the Siberian craton in Yakutia. Unlike the former that were precursors to the superplume activity in Northeast Africa, the latter likely had no relation to any plume. The kimberlites of Yakutia were triggered by subduction of the Kula-Izanagi plate beneath East Asia finalized by accretion of the Okhotsk Sea plate to Eurasia (Rasskazov and Taniguchi 2006).

The late-Laramide (70–55 Ma) magmatic quiescence in the Western USA was followed by renewed eruptions contemporaneous with the middle–late Tertiary magmatic events in Asia. The remarkable Asian magmatic episode of 46–44 Ma is accompanied by magmatic episodes in both North America (Fig. 13.2) and Northeast Africa (assumed episode of $\sim$45 Ma in the Turkana saddle).

While addressing mid-Tertiary magmatism in different regions of Asia, one can notice a "sliding" character of magmatic intervals: 48–26 Ma (Khilok basin), 38–22 Ma (Southwest Primorye), 38–25 Ma (Sakhalin island, central and northern parts of Eastern Sikhote Alin), 36–22 Ma (northeastern part of the Fen-Wei rift zone),

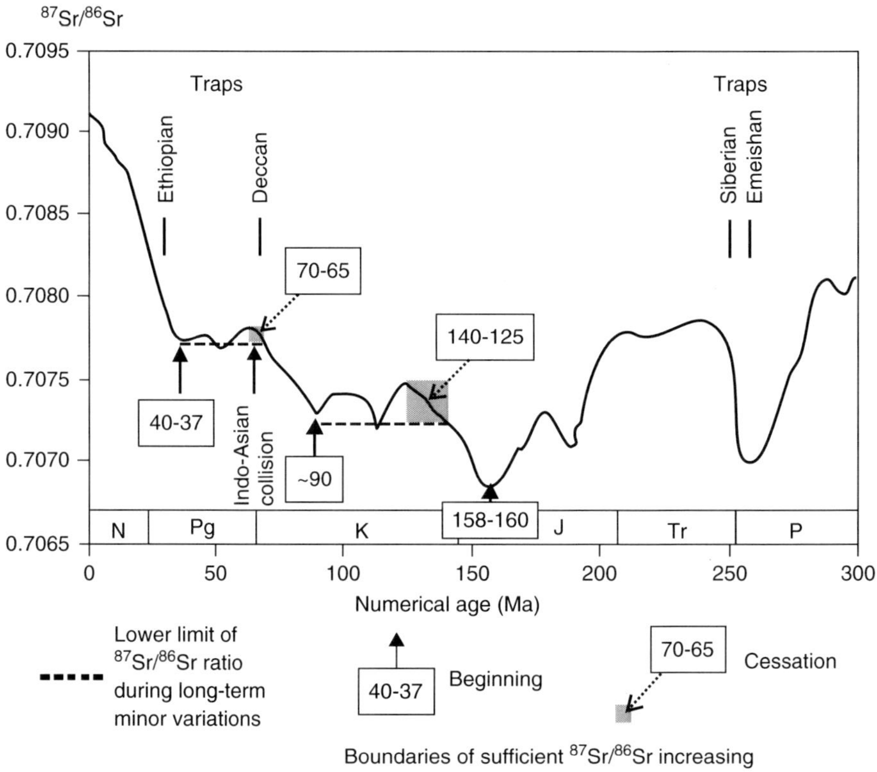

Fig. 13.9 Relation of the Indo-Asian collision and selected trap provinces to temporal variations of $^{87}Sr/^{86}Sr$ in sea water. Isotopically depleted traps from the Emeishan and Siberian provinces are not responsible for decreasing $^{87}Sr/^{86}Sr$ in the Permian and correspond to a minimum of this ratio at ~258 Ma and its following relative increase. Traps of the Ethiopian province have no reflection in a line of increasing $^{87}Sr/^{86}Sr$. Minor variations of $^{87}Sr/^{86}Sr$ ratio between 70 and 37 Ma correspond to a period that lasted from the beginning of the Indo-Asian collision and coeval emplacement of the Deccan traps to the Middle Eocene structural reorganization and accompanied change of magmatism. Similar tectonic and magmatic processes might be responsible for minor variations of the $^{87}Sr/^{86}Sr$ ratio between 140 and 90 Ma. The line of the marine $^{87}Sr/^{86}Sr$ record is adopted from McArthur et al. (2001)

32–21 Ma (Valley of Lakes). Similar "sliding" spatial–temporal distribution of magmatism occurred in the Western USA with a general trend to final events of 22–18 Ma (Fig. 13.2) and in Northeast Africa with propagation from the Ethiopian plateau to the East African one (Fig. 13.6).

In the Baikal Rift System, the Sayan-Khamar-Daban and Stanovoi uplifts are separated from each other by the Selenga saddle, which is comparable with the Turkana one that separates the Ethiopian and East African plateaus. The Selenga saddle spatially corresponds to the central part of the older (latest Cretaceous–Paleogene) Tunka-Eravna rift zone (Rasskazov 1994b; Rasskazov et al. 2007d). The late Tertiary rifts propagated from the South Baikal northeastwards and westwards (Rasskazov et al. 2000b, 2003a; Logatchev 2001, 2003). In the Baikal Rift System, voluminous magmatism of the southwestern part commenced in 22–21 Ma, although only minor eruptions occurred in the northeastern part (Udokan) ~14 Ma with transition to more voluminous eruptions ~9 Ma. This asynchronous magmatism is comparable with asynchronous magmatism in the East African Rift System that affected the Ethiopian rift ~30 Ma and shifted afterwards to the Kenyan and Western rifts ~23 and 19–17 Ma, respectively. The development of both the Baikal and East African rift systems reflected a mechanism of along-axis bilateral crack propagation.

The initial phases of ~30 and 22–20 Ma in the Ethiopian rift and southwestern part of Baikal Rift Systems are connected with triggering effects of external tectonic forces that were provided, in the former case, by collision between Africa and Eurasia and, in the latter case, by convergence between India and Asia, supplemented by structural reorganization due to accretion of the Philippine Sea plate. The delayed 23–17 Ma and ~14 Ma magmatic pulses marked respective phases of bilateral growing of these rift systems. The mechanism implies the rift development due to mantle instability induced, firstly, by continent–continent collision and then, enforced by upwelling of hot mantle material.

No evidence on bilateral crack propagation relative to principle rift systems can be traced in East Asia and North America, where one direction stepwise shift of magmatism prevailed. This structural style can be connected with subduction–transform geodynamics in these regions, when sub-lithospheric slab dynamics played a leading role.

In magmatic sequences of Central Asia, the 12.0–7.5 Ma asynchronous activity reflected instability finalized by the common magmatic pulse of 7.5–7.0 Ma. A specific reorganization of a magmatic system occurred in Central Mongolia between 11 and 7 Ma. A neutral regime (i.e. without influence of Indo-Asian convergence), resulted in moderate-K magmatism, was followed by synchronization of magmatic events ~7.5–7.0 Ma. A time interval of 12–9 Ma was characterized by voluminous eruptions in the Oka, Hubsugul, Tunka, and Vitim volcanic fields of the Baikal Rift System (Rasskazov et al. 2000b). From data on Central Mongolia (Chap. 12), we speculate that this period was favorable for development of melting processes in shallow mantle lenses responsible for upsetting of isostasy and basin subsidence in Central Asia. Such areas as Udokan and East Tuva, located, respectively, in the northeastern and western terminations of the Baikal Rift System, show transitional character in timing of volcanic evolution; i.e. no eruptions between 12 and 9 Ma but initiation of activity between 8.9 and 7.4 Ma, just before and contemporaneously with synchronization of events in Central Mongolia.

Simultaneously with initial subsidence of the north–south trending Hubsugul and Darkhat basins in the Baikal Rift System about 10 Ma, rifting was accelerated in East Africa, where narrow basins of the Ethiopian rift quickly subsided (Mohr and Zanettin 1988; Kieffer et al. 2004). Respectively, basins of the Rio Grande rift were bounded by high-angle faults between 10 and 3 Ma (Baldridge et al. 1984).

The recognized Quaternary transition from asynchronous to synchronous magmatism in Asia (Rasskazov et al. 2000b; Chap. 12) is comparable with a coeval change of magmatism that occurred in the Rungwe volcanic field of Africa, where the Western rift underwent structural reorganization that resulted in magmatic locus along the Rukwa and Karonga basins with initial eruptions of specific olivine melanephelinite lavas after a magmatic hiatus between 1.3 and 0.6 Ma (Fig. 13.4). The present-day magmatism within Asia and Africa likely characterizes processes dominated by the increased plate interaction of the past 0.6 Ma.

13.4 Summary: Two Geodynamic Patterns of Rift-Related Magmatic Evolution in Continents

In Chap. 12, we have emphasized the coeval character of magmatic evolution in Central and East Asia. From comparative analysis of magmatic evolution in Northeast Africa, Central and East Asia, and North America, we have distinguished here rifting processes developed in conditions of (1) continent–continent interaction with accompanied upwelling of hot mantle material and (2) subduction–transform relations between oceanic plates and continental margins with accompanied upwelling of hot mantle material.

In Northeast Africa, rifting was predated by diamond-bearing kimberlites emplaced in the central part of the Tanzanian craton 52 ± 2 Ma. The earliest ($\sim$37 Ma or probably $\sim$45 Ma) OIB-type magmatism occurred within the Turkana saddle located in the central part of the East African Rift System between the Ethiopian and East African plateaus. Erosion of the lithosphere, rifting, and magmatism propagated from the saddle towards both plateaus. The Ethiopian rift dissected the central part of the Ethiopian plateau, where the pulse of the most voluminous plume magmatism took place simultaneously with Africa–Eurasia collision $\sim$30 Ma. In the Kenyan and Western rifts, bounded by the East African plateau, magmatism initiated $\sim$23 Ma and 19–17 Ma, respectively. These data demonstrate the earlier processes of lithospheric erosion beneath the Ethiopian plateau and delayed beneath the East African one.

A geodynamic pattern of rifting in Northeast Africa and Central Asia corresponds to a model of along-axis bilateral crack propagation, provided by continent–continent interaction and accompanied upwelling of hot mantle material. Although rifted areas of Central Asia were, unlike Northeast Africa, tectonically active all through the late Phanerozoic, rift-related processes in these regions show resemble spatial–temporal evolution. The uplifted regions in the northeastern and southwestern parts of the Baikal Rift System are separated from each other by the Selenga structural saddle. Respectively, rifting began within the saddle and propagated bilaterally towards the uplifted regions.

Indo-Asian interaction provided coeval temporal variations of low-moderate-K and high-ultra-K magmatism in Tibet and Central Asia. Elevated potassium concentrations in liquids, erupted in the latter region, marked specific conditions of magma generating processes in the mantle depth interval of 120–200 km due to accommodation of Indo-Asian convergence between lithospheric keels of the Indian indenter and Siberian craton. We speculate that similar control of deep-seated processes by a craton lithospheric keel might cause magma generating processes with high potassium activity beneath the northern potion of the Western rift during the past 3 Ma.

In North America, rifting followed the Laramide orogeny, when subduction of the Farallon plate beneath the continental margin changed to transform motions of the Pacific plate relative to North America. Firstly, between 120 and 37 Ma, magmatism of intermediate–silicic composition migrated from a continental margin landwards, then, between 37 and 18 Ma, mostly intermediate–silicic liquids change to basaltic ones simultaneously with shift of magmatic processes along the strike of the continental margin with focusing in the southern Basin and Range province, and at last, in the past 15 Ma, basaltic magmatism predominated.

A geodynamic pattern of rifting in continental margins of North America and East Asia appears to have been controlled by long-term sub-lithospheric slab dynamics. In both North America and East Asia, magmatic evolution was governed by subduction–transform relations between oceanic plates and continental margins. Distribution of magmatism demonstrates one-direction propagation of processes from continental margins landwards and complicated spatial–temporal trends relative to principle rift structures. Intermittent low-moderate-K and high-ultra-K magmatism developed simultaneously.

Chapter 14
Separated Lead Isotopes

14.1 Dating Problems of Ore Mineralization

Radiogenic isotope dating of ore deposits is ambiguous. In many cases, different isotope systems yield inconsistent results. It remains unclear, which mineralization processes are marked in ore deposits by ages obtained and when ore substance was deposited relative to processes responsible for launch or reset of different isotope systems.

Direct radiogenic isotope dating of gold mineralization was performed by measurements of paleoatmospheric argon, which was detected in gold from the Paleozoic placer (unpublished experimental results of S.B. Brandt). The measured $^{40}Ar/^{36}Ar$ value was interpreted as corresponding to the captured Paleozoic atmosphere. At the beginning of the Paleozoic, $^{40}Ar/^{36}Ar$ was as low as 287.0, increasing afterward up to 295.5 in the modern atmosphere at sea level due to radioactive transformation $^{40}K \rightarrow {}^{40}Ar$ (Chap. 4).

Sulfide mineralization of gold or poly-metallic deposits is dated by the Re–Os isochron method on molybdenite (Stein et al. 1998, 2001). In long-lived ore systems, $^{40}Ar/^{39}Ar$ measurements on hydrothermal micas give underestimated dates, as compared to Re–Os ages of molybdenites, because of diffusion losses of ^{40}Ar during slow cooling of host rocks (Selby et al. 2002). $^{40}Ar/^{39}Ar$ spectra are informative for a post-ore stage, similar to a stage of relative decreasing temperatures after a peak of metamorphism. If an isotope system is not contaminated by excessive argon, K–Ar dates will yield minimal age for ore mineralization. In case of fast cooling of host rocks, ages in a K–Ar system will show dates consistent with results in Re–Os, Rb–Sr and/or Sm–Nd isotope systems.

Separated (ore) leads are informative in terms of both timing and origin of ore substance. Generally, their ages should be consistent with those obtained by other geochronological methods. A simple decision could be achieved for leads, passing from disseminated distribution to ore phase, due to a single event limited by magmatic or hydrothermal processes responsible for ore mineralization. More often, however, currently used models (e.g. Dickin 1997) show only approximate timing of ore lead separation. Inevitably, the question arises of which model parameters can be used for understanding isotope evolution of leads.

S.V. Rasskazov et al., *Radiogenic Isotopes in Geologic Processes*,
DOI 10.1007/978-90-481-2999-7_14,

14.2 Definition of Separated Leads

The isotope ratios $^{206}Pb/^{204}Pb$, $^{207}Pb/^{204}Pb$, $^{208}Pb/^{204}Pb$ vary sufficiently in natural ore leads. The isotopes ^{206}Pb, ^{207}Pb, ^{208}Pb are radiogenic, i.e. have originated due to radioactive decay of ^{238}U, ^{235}U, and ^{232}Th, respectively. Only specific samples show low isotope ratios. Such, in particular, are leads of the Canyon Diablo meteorite accepted as primeval (Faure 1989):

$$\frac{^{206}Pb}{^{204}Pb} = 9.307, \quad \frac{^{207}Pb}{^{204}Pb} = 10.294, \quad \frac{^{208}Pb}{^{204}Pb} = 29.476$$

The primeval leads have been never connected with any radioactive U–Th-bearing substance. Significant radiogenic additions assume that the leads used to belong to huge parent–daughter systems, but have been separated from them in the geologic past. These leads are frequently named as common leads, though their origin remains mysterious in many respects. We shall try to highlight the problem of separated leads in terms of evolving Th, U–Pb isotope system.

In the Earth, leads occur as: (1) disseminated in rocks and minerals, directly connected with parent uranium and thorium, (2) separated from uranium and thorium in accessory minerals of rocks, and (3) separated from uranium and thorium in ore deposits.

In geologic processes, Pb behaves as an extremely mobile chemical element. In the mantle, it concentrates in sulfides, as a siderophile element and in partial basaltic liquids, as an incompatible element. Due to transfer by water fluids, Pb enriches the continental crust generated in convergent zones (Miller et al. 1994). The undifferentiated mantle contains 0.185 ppm of lead. In the normal middle ocean ridge basalt (NMORB), the mean Pb content increases up to 0.3 ppm, in the enriched middle ocean ridge basalt (EMORB) up to 0.6 ppm, and in the ocean island basalt (OIB) up to 3.2 ppm (Sun and McDonough 1989). The continental crust shows 1–2 order higher Pb concentrations with the mean value 12.6 ppm (in the lower, middle, and upper crust 4.2, 15.3, and 20 ppm, respectively) (Rudnick and Fountain 1995).

While allocated within an ore phase (galena, pyrite, arsenopyrite etc.), Pb isotopes are separated from parent isotopes of U and Th kipping isotope ratios of the separation moment. At first approximation, an age of separated leads is estimated using the Holmes–Houtermans model. The closer data points to the Geochron ($t = 0$), the younger the age. In addition to exact parameters of the Th, U–Pb isotope system (constants of radioactive decay), the age calculation of separated leads requires assumptions on values of μ (^{238}U/^{204}Pb) and ω (^{232}Th/^{204}Pb) which existed for some periods of the geologic past. From measured values of isotope ratios in ore minerals from deposits, interpretation of linear groups yields results on two-stage (multistage) evolution of the Th, U–Pb isotope system in which the parameter μ changes with time. In fact, results of age calculations depend on additional arguments involved for substantiation of an accepted model (Doe and Stacey 1977; Brandt et al. 1978; Neymark 1988; Faure 1989; Rasskazov et al. 2005a).

Data points of conformable leads (appropriate to age of host rocks in sedimentary deposits and volcanic rocks from greenstone belts and island arcs) are distributed along a single-stage curve with $\mu = 9.0$. Those of anomalous leads, referred to as J-type or Joplin type, are plotted off this curve yielding a future model age in the Holmes–Houtermans model (Hamilton 1965; Doe 1970; Dickin 1997).

The two-stage model assumes that material formed at the initial and final moments of ore formation (t_1 and t_2) with relative increasing values from μ_1 to μ_2. The comparative analysis of lead isotope ratios of ore phases and leached feldspars from magmatic and metamorphic rocks of a selected area shows their spatial distribution along a single line interpreted as an isochron (Brandt et al. 1978; Neymark et al. 1993, 1995).

14.3 Diffusion Discordia

Data on the U–Pb isotope system are interpreted by the Concordia–Discordia model (Chap. 2) that characterizes open conditions and formally describes lead separation. Therefore, understanding isotope history of ore leads requires more detail consideration of the model characteristics.

Application of a rectilinear Discordia is acceptable, if a younger imposed process (metamorphic impact) is clearly traced. But frequently, data points are distributed approximately along a straight line with the upper and lower intersections between Concordia and Discordia without any visible geologic event that could approximate an imposed process. Such a situation, described by a diffusion model, assumes that radiogenic leads underwent continuous diffusion and that the crystalline structure, in which the diffusion occurred, had domains with different diffusion parameters. In a system with simultaneous radioactive accumulation and diffusion of daughter substances (i.e. lead), the following expressions are valid (Rasskazov et al. 2005a):

$$
\begin{aligned}
x &= \frac{^{206}Pb}{^{238}U} = \frac{8}{\pi^2}\sum_{\nu=0}^{\infty}\frac{1}{(2\nu+1)^2}\,\frac{e^{(1-\pi^2(2\nu+1)^2\frac{Fo}{\lambda_{.8}t_\nu})\lambda_{.8}t}-1}{1-\frac{\pi^2(2\nu+1)^2 Fo}{\lambda_{.8}\cdot t}} \\
y &= \frac{^{207}Pb}{^{235}U} = \frac{8}{\pi^2}\sum_{\nu=0}^{\infty}\frac{1}{(2\nu+1)^2}\,\frac{e^{(1-\pi^2(2\nu+1)^2\frac{Fo}{\lambda_{.5}t_\nu})\lambda_{.5}t}-1}{1-\frac{\pi^2(2\nu+1)^2 Fo}{\lambda_{.5}\cdot t}}
\end{aligned}
\qquad (14.1)
$$

where $Fo = Dt/h^2$ (Fo – Fourier number, D – diffusion coefficient, h – size of a mineral grain). This is an expression of infinite series that are almost rectilinear and start from the beginning of coordinates. In these calculations, the variable Fo ranges from 0.01 to 3.0.

As already noticed, strongly radiogenic ore leads are puzzling. Their radiogenic composition assumes that they used to enter into a Th, U–Pb isotope system and were separated from the latter. The Concordia–Discordia model formally describes a moment of this separation. As a rule, the model adequately explains processes that occurred in accessory minerals (zircons, apatites etc.). Released micro-quantities of leads should form micro-quantities of accessory lead-bearing minerals in host rocks.

In case of ore lead deposits, one has to assume that the Concordia–Discordia mechanism has operated on a huge scale. Somewhere nearby, there should be respective deposits of parent uranium and thorium.

Unlike the Wetherill model that assumes origin of the Discordia due to instant lead loss, the alternative (Wasserburg) model suggests continuous separation of leads by diffusion in the U–Pb system without identification of any separation moment. In the Wasserburg diffusion model of the Discordia, leads are accumulated according to expression:

$$^{208,207,206}Pb = {}^{238,235,232}P\frac{8}{\pi^2}\sum_{\nu=0}^{\infty}\frac{1}{(2\nu+1)^2}\frac{e^{(1-\pi^2(2\nu+1)^2\frac{Fo}{\lambda_{.8,5,2}t_\nu})\lambda_{.8,5,2}t}-1}{1-\frac{\pi^2(2\nu+1)^2 Fo}{\lambda_{.8,5,2}\cdot t}} \quad (14.2)$$

where $^{238,235,232}P$ – parent isotopes of uranium and thorium, $\lambda_{.8,5,2}$ – appropriate constants of radioactive decay. At $Fo = 0$, expression (14.2) becomes an ordinary formula of radioactive substance accumulation:

$$^{j}Pb = {}^{i}P(e^{\lambda t} - 1) \quad (14.3)$$

where j – masses of Pb isotopes, i – masses of U and Th isotopes.

Figure 14.1 shows an example of the diffusion Discordia calculated by formulas (14.1–14.2). The Discordia starts from the beginning of coordinates and crosses the Concordia at the unique point at $Fo = 0$.

Figures 14.2 and 14.3 present two series of similar calculated curves, which clearly demonstrate that only one measured point is necessary to plot the whole diffusion Discordia, and, consequently, to define an age of any event.

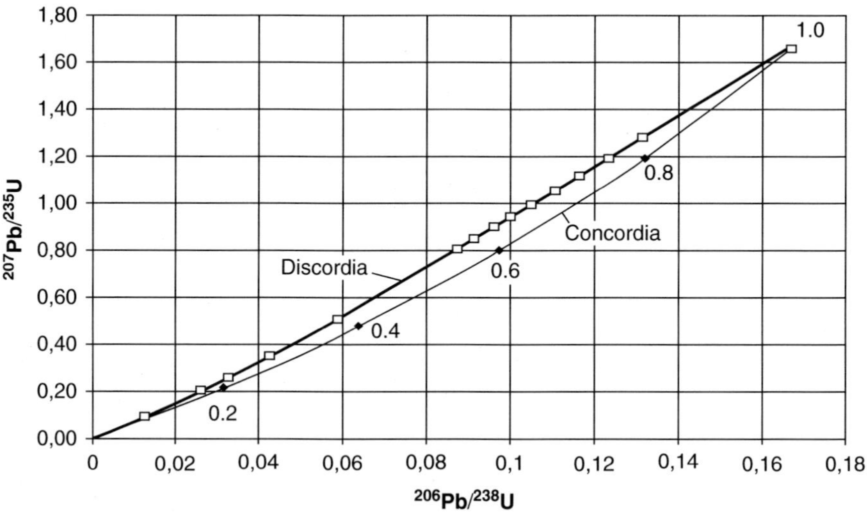

Fig. 14.1 The diffusion Discordia starts from the beginning of coordinates and with some curvature crosses the Concordia only at one point that approximates the nominal age of 1 Ga

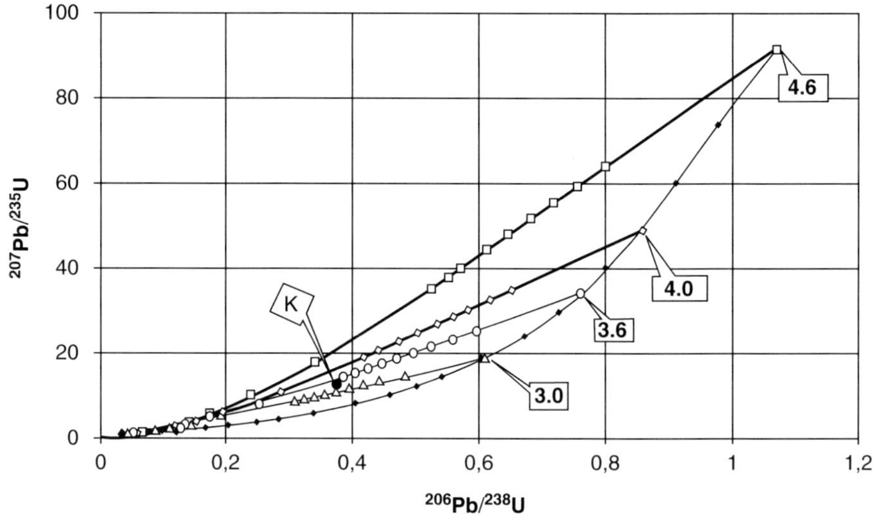

Fig. 14.2 A series of diffusion Discordia curves for a time interval of 4.6–3.0 Ga. A point *K* can belong only to the Discordia 3.3–3.5 Ga. The diffusion Discordia curves are calculated with precision of 300 members in a row

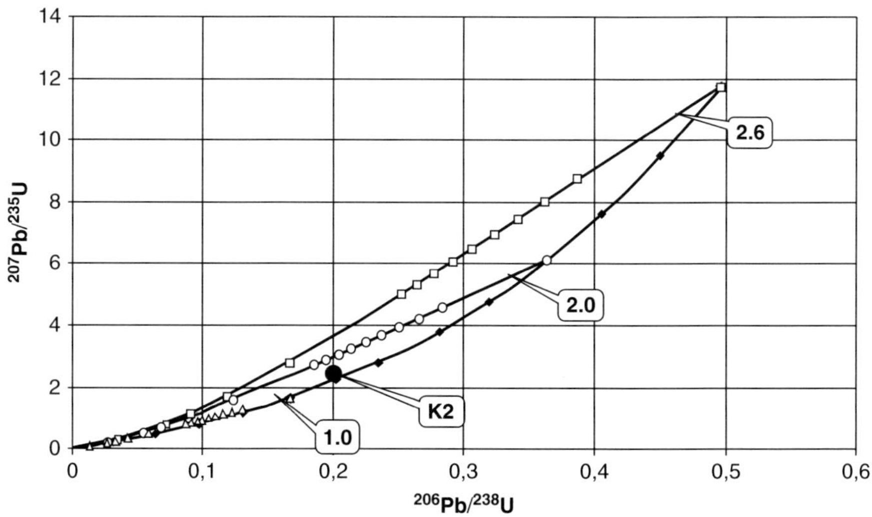

Fig. 14.3 A series of the diffusion Discordia curves for a time interval of 2.6–1.0 Ga. The discordant point *K*2 can belong only to the Discordia ~1.4 Ga. The diffusion Discordia lines are shown similar to those in Fig. 14.2

14.4 Electrical Charge of Pb Separated from U–Pb System

Pb is a four-valent chemical element. What charge will result in a grain of a mineral (zircon) at complete loss of radiogenic leads due to a metamorphic impact? How many particles of uranium will decay and what charge will they carry away at loss processes with transition from the Concordia to Discordia?

Let a zircon grain of the age 2 Ga have a diameter of 1 mm and be 2% made up of uranium. For the ball grain with a radius 0.05 cm, calculate volume $V = (4/3)\pi \times 1.25\,e^{-4} = 5\,e^{-4}\text{cm}^3$. From density of zircon 4.6 g/cm^3, obtain its mass $M = 2.5e^{-3}$ g = 2.5 mg. The grain contains $2.5e^{-3} \times 0.02\ 5e^{-5}$ g of uranium.

Calculate the quantity of uranium particles in this ball: 238 g of uranium = $6e^{23}$ of particles; $5e^{-5}$ g of uranium = x particles:

$$\frac{x}{6e^{23}} = \frac{5e^{-5}}{238}; \quad x = 6e^{23}\frac{5e^{-5}}{238} = 1.35e^{17}$$

Calculate lead particles, formed during a time interval of $2e^9$ years $N = 1.35e^{17}$ $(e^{9.5e^{-10}\times 2e^9} - 1) = 1.35e^{17} \times 5.685 = 7.67e^{17}$. Each atom of lead carries four elementary charges equal $1.6e^{-19}$ C. We receive $4 \times 1.6e^{-19} = 6.4e^{-19}$ C. Hence, the complete charge acquired by a particle for 2 Ga, makes $Q = 6.4e^{-19} \times 7.67e^{17} = 0.49$ C.

This charge is huge. The grain might have a high voltage resulting in a catastrophic electrical phenomena. But in reality, nothing happens. Hence, the lead migration operates by the mechanism called by Korzhinskii (1947) "bi-metasomatism" (exchange reaction), replacement of one particle by another with an equal electrical charge. In general, isomorphic elements can be used as indicators for lead loss.

14.5 Connected Points of Concordia and Discordia

Whichever mechanism of central elements replacement in a crystal lattice is used, points of the Concordia formally pass into points of the Discordia with an unequivocal conformity. How do they connect to each other?

The square root of the mass ratio for isotopes ^{235}U and ^{238}U $\sqrt{235/238} = 0.9936 \approx 1$ implies no differentiation of the isotopes during diffusion. An isotope composition A of a probe that corresponds to a Concordia age of 2.2 Ga will be changed along the direct line connecting the point A with the beginning of coordinates (Fig. 14.4). The point can move only along this trajectory and only up to intersection with the diffusion Discordia 2.6 Ga (up to a point B). Hence, the points A and B are mutually connected.

Now determine the mutually connected points graphically for abscissa and ordinate of the diagram in Fig. 14.4 and plot results in Fig. 14.5. Show the coordinate distance in millimeter. One millimeter for abscissa is equal to 0.002869 for

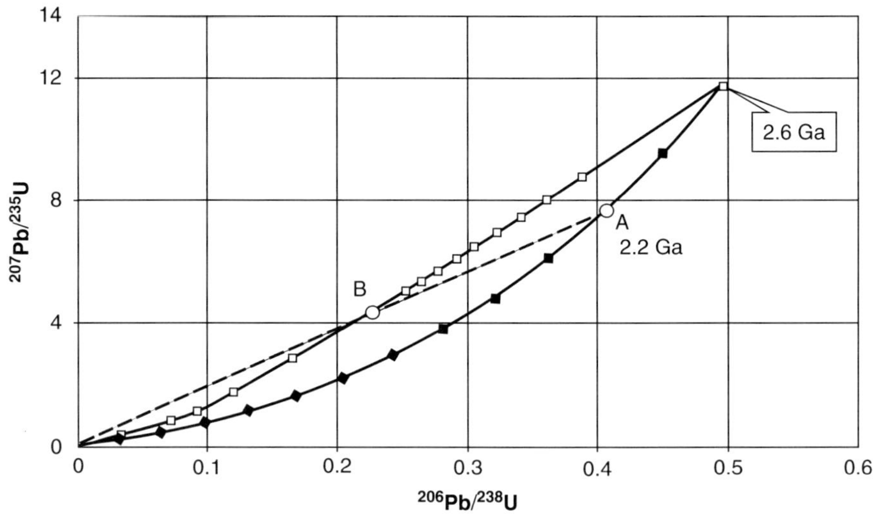

Fig. 14.4 Diffusion Discordia provides a move of a point *A* from the Concordia along a straight line to a point *B* at the Discordia. These points are mutually connected

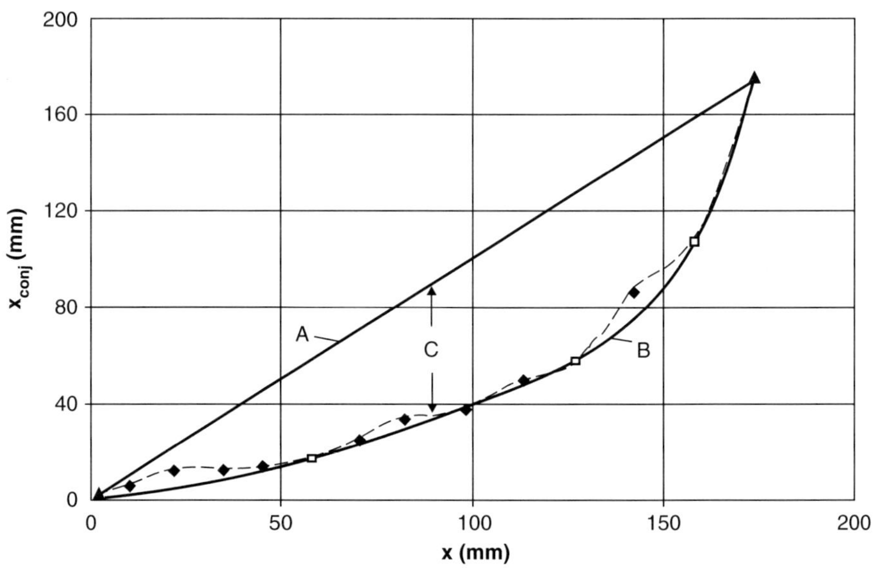

Fig. 14.5 Abscissa of the mutually connected points of Fig. 14.4 is a curve *B*. The separated lead is difference of curves *A* and *B* (i.e. a piece *C*). Dispersion of data points along a curve *B* is due to approximation done by a graphic-analytical method

^{206}Pb/^{238}U and 0.095012 for ^{207}Pb/^{235}U. The purpose of these diagrams is to define an ordinate difference for curves A and B equal to a piece C.

14.6 Leads Released by Continuous Diffusion

The piece C is expressed as:

$$\Delta\frac{^{206}Pb}{^{238}U};\quad \Delta\frac{^{207}Pb}{^{235}U}$$

For transition to $^{206}Pb/^{204}Pb$ and $^{207}Pb/^{204}Pb$ use coefficients $\mu = {}^{238}U/^{204}Pb$ and $\mu/137.88 = {}^{235}U/^{204}Pb$. We get:

$$\Delta\frac{^{206}Pb}{^{238}U}\times\mu = \Delta\frac{^{206}Pb}{^{204}Pb};\ \Delta\frac{^{207}Pb}{^{235}U}\times\frac{\mu}{137.88} = \Delta\frac{^{207}Pb}{^{204}Pb} \tag{14.4}$$

The value μ of the terrestrial rocks is equal to approximately 9/10 (Doe 1970). For formal calculations, first accept a high value $\mu = 50$. Figure 14.6 demonstrates that leads, separated by diffusion, are shifted to the right of the Geochron (i.e. these leads are anomalous). According to the Holmes–Houtermans model, their ages belong to "the future". Hereafter, we shall see that the separated leads are not all anomalous.

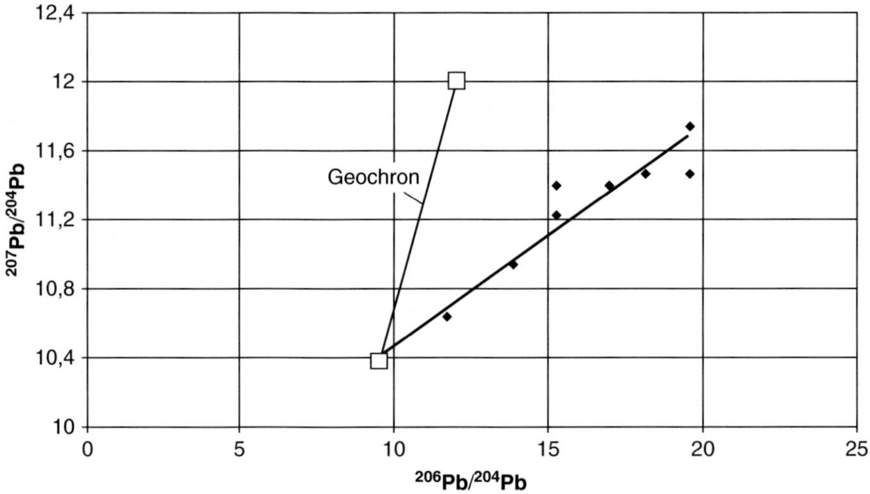

Fig. 14.6 Isotope compositions of leads separated from U–Pb isotope system characterized by the diffusion Discordia. The dispersion of data points is due to approximation done by graph-analytical methods. Accepted $\mu = 50$

14.7 Instant Release of Leads

Consider now a U–Pb isotope system, in which leads are separated from crystalline rocks instantly in different moments of time (Wetherill model). This is an usual interpretation of separated leads. A curve will be the Concordia itself with some changes of coordinate axes. Coordinates of the Concordia are calculated as:

$$x = \frac{^{206}Pb}{^{238}U} = e^{\lambda_8 t} - 1; \quad y = \frac{^{207}Pb}{^{235}U} = e^{\lambda_5 t} - 1$$

In Fig. 14.7, the Concordia for the age of the Earth $4.56 \cdot 10^9$ years is transformed to Pb–Pb coordinates:

$$x' = x \times \mu = x \times \frac{^{238}U}{^{204}Pb} = \frac{^{206}Pb}{^{204}Pb}; \quad y' = y \times (\mu/137.88) = \frac{1}{137.88}\frac{^{207}Pb}{^{204}Pb} \tag{14.5}$$

All leads, separated *today*, lie to the right of the Geochron (i.e. anomalous). If the separation took place t years ago, a composition of separated leads is obtained by subtraction of its amount, which would be formed during a time interval t.

Get expressions for a radiogenic addition in the Holmes–Houtermans model:

$$x = \frac{^{206}Pb}{^{204}Pb} = \mu(e^{\lambda_8 T} - e^{\lambda_8 t}); \quad y = \frac{^{207}Pb}{^{204}Pb} = \frac{\mu}{137.8}(e^{\lambda_5 T} - e^{\lambda_5 t}) \tag{14.6}$$

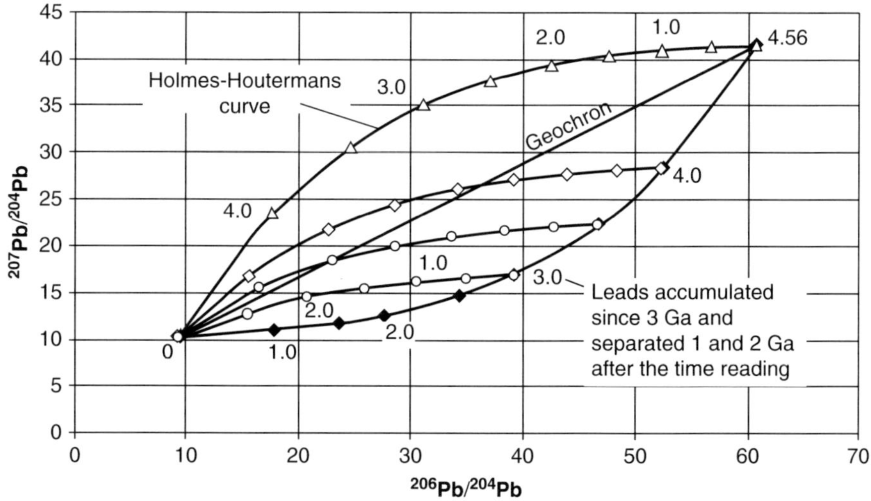

Fig. 14.7 Growth curves for separated leads of: (1) the closed Holmes–Houtermans model (*open triangles*), (2) the open Concordia model (*filled diamonds*), and (3) conjugated Holmes–Houtermans—Concordia–Discordia model (*open circles* and *diamonds*). Accepted $\mu = 50$

We present the Holmes–Houtermans model itself:

$$x = \frac{^{206}Pb}{^{204}Pb} = \mu(e^{\lambda_8 T} - e^{\lambda_8 t}) + 9.307$$
$$y = \frac{^{207}Pb}{^{204}Pb} = \frac{\mu}{137.88}(e^{\lambda_5 T} - e^{\lambda_5 t}) + 10.294 \qquad (14.7)$$

According to the model, leads started accumulating at the moment (T) of the Earth formation (4.56 Ga) and continued to the moment (t) of their separation from uranium-bearing phases. In Fig. 14.7 this process is exhibited by the upper curve with normal leads occurring to the left of the Geochron.

Assume now that the moment (T) of a U–Pb isotope system is less than 4.56 Ga and corresponds to 3.0 Ga. All released leads are anomalous. If the U–Pb isotope system begins releasing leads 4.0 Ga with separating 2 Ga later on, they will be anomalous, but the earlier separated leads will be normal. In case of release since 3.5 Ga, a piece of a growth curve with normal leads will be shorter and that with anomalous leads longer.

It is noteworthy that anomalous leads, separated today, will stay anomalous in all subsequent geologic time. To substantiate this statement, consider the Geochron that is a straight line between the point of primeval leads and the point of the present (4.56 Ga). One can show that its slope increases in time and aspires to ∞. In coordinates of lead isotopes, the Geochron is characterized by parameters:

$$x = \mu(e^{\lambda_8 T} - 1) + 9.307 \quad \text{and} \quad y = \frac{\mu}{137.88}(e^{\lambda_5 T} - 1) + 10.294 \qquad (14.8)$$

A variable is μ. Exclude this parameter from the equations with transition to the Geochron equation:

$$\frac{^{207}Pb}{^{204}Pb} = \frac{^{206}Pb}{^{204}Pb}\frac{e^{\lambda_5 T} - 1}{137.88(e^{\lambda_8 T} - 1)} - \frac{9.307}{137.88}\frac{(e^{\lambda_5 T} - 1)}{(e^{\lambda_8 T} - 1)} + 10.294 \qquad (14.9)$$

The Geochron slope $k = y/x$ is a growing function of T. With increasing T the slope aspires to ∞. Therefore, leads, located to the right of the Geochron today, will stay to the right later on. Alternatively, leads that are normal today may become anomalous in the geologic future. In this respect, the terms "normal" and "anomalous" leads lose their original geochemical sense. These categories have no special properties, though normal leads in general indicate the earlier closure of the U–Pb isotope system than anomalous leads of J-type.

14.8 Tuning of T, t, and μ

Anomalous leads require interpretations by two or multi-stage models. A choice of the correct model is a matter of skills and intuition of the experimenter. Recollect that the term "anomalous" has arisen, because "the negative age" (anomalous leads of J-type) put experimenters in impasse in interpretation by the Holmes–Houtermans model. Then anomalous leads of J-type were explained in terms of the two-stage model using a Geiss parallelogram with $T = 4.56$ Ga. But the parameter T in a model of separated leads should be less than 4.56 Ga.

Gravitational accretion of the Earth was accompanied by release of considerable amounts of energy due to core separation, bombardment of the Earth by space bodies, radioactive transformation of short- and long-living nuclides etc. As a result, the Earth melted (Chap. 10). The extraordinary event—a collision of the proto-Earth with a Mars-sized impactor Theia—resulted in generation of the Moon from the silicate portion of the latter. Pb isotopic compositions of the Bulk Silicate Earth (BSE) are shifted to the right of the Geochron. That shift indicates continuous core formation with exponentially decreasing planetary growth of the Earth less than about 40 Ma. The same effect is provided by the two collided (previously evolved) planets—the proto-Earth and Theia (Halliday 2000).

Separated leads were derived from a silicate portion of the Earth, no composition of which corresponding to the Geochron assumes respective correction of the Holmes–Houtermans model. The magmatic ocean of the early Earth appeared to cool irregularly in space and time. Some of its parts could solidify relatively quickly, other retained convective melting domains with open evolution of the U–Pb isotope system. A lack of crystalline rocks within the primary melting domains could not provide a mechanism of lead separation. Alternatively, extinction of the melting domains resulted in transition from open to closed isotopic evolution of the U–Pb system. The integrated correction of the T value is provided by conjugation of the Holmes–Houtermans and Concordia–Discordia models as suggested here.

We propose interpretation of separated leads by choice of moments for closure of the U–Pb isotope system (T) and lead separation (t) with a common μ. Figure 14.8 demonstrates curves calculated at $\mu = 50$, 40, and 8. As μ reduces, the curves

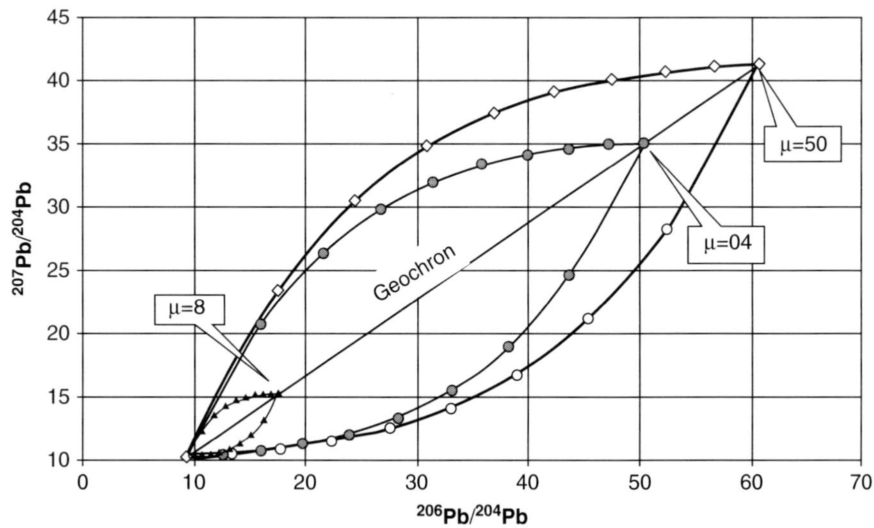

Fig. 14.8 Growth curves of ore leads in the closed (Holmes–Houtermans) and open (Concordia) models. Calculations are done for $\mu = 8$, 40, and 50

are shifting towards the beginning of coordinates with simultaneous change of the growth curve slopes. Every linear data point in these coordinates, therefore, can be approximated by tuning μ and T.

14.9 Interpreting Separated Leads from Ore Deposits in Asia

In ore deposits of East Siberia and Far East of Russia with adjacent areas of Mongolia and China, separated leads have been studied by using different models (Chernyshev and Shpikerman 2001; Rasskazov et al. 2002b; Tveritinov et al. 2006; and references therein). As a rule, researches are anticipated to determine two stages of lead evolution, i.e. to estimate values t_1, t_2, μ_1, and μ_2. Some models were developed for deposits of large regions, other for local ore fields and single deposits.

East Asia represents a collage of terranes accreted to the North Asian continent in the late Precambrian and Phanerozoic (Sengör and Natal'in 1996; Parfenov et al. 2003). From this point of view, separated leads have been analyzed in ore deposits from different geologic settings (Tveritinov et al. 2006). To demonstrate application of the conjugated Holmes–Houtermans—Concordia–Discordia model, we focus only on estimates of the T–t–μ parameters obtained for separated leads from the Precambrian terranes: the southern part of the Siberian craton and Gargan block of the Tuva-Mongolian massif (Fig. 14.9).

14.9.1 The Southern Part of the Siberian Craton

The Siberian craton was assembled from microcontinents at 1.9–1.8 Ga. The geologic structure of the craton basement is ambiguous. Its southern portion was suggested to exhibit the Onot granite-greenstone equi-dimensional region (Khiltova et al. 2003), polimetamorphic volcano-sedimentary structures distributed along the margin of the Tunguska granite-greenstone and tonalite province (Smelov et al. 2003) or different terranes comprised of the Tunguska and Anabar provinces (e.g. Rosen et al. 2006). Study of separated leads of ore deposits is critical for evaluation of the suggested geologic schemes. We emphasize the particular origin of the southern part of the Siberian craton, which, unlike its other parts, is characterized by J-type leads.

The deposits with these leads are distributed in the northwestern coast of Lake Baikal (along the craton margin) in distance over 300 km both in metamorphic rocks of the craton basement and sedimentary cover. Isotopic compositions of galena leads were firstly reported by Brandt et al. (1978). The rectilinear arrangement of data points was interpreted using a parallelogram of Geiss (Hamilton 1965) with parameters: $t_1 = 1.7$ Ga, $t_2 = 0.46$ Ga, $\mu_1 = 8.93$, and $\mu_2 = 16$–40. Lead compositions were measured in this range also for galenas from kimberlite-like rocks of the Ingashi pipe, which is located at the southwestern suture zone of the craton (data of Fefelov and Kostrovitskii).

Lead isotope compositions, presented by Neymark et al. (1993), are comparable to those determined by Brandt et al. (1978). In lead isotope coordinates, galenas

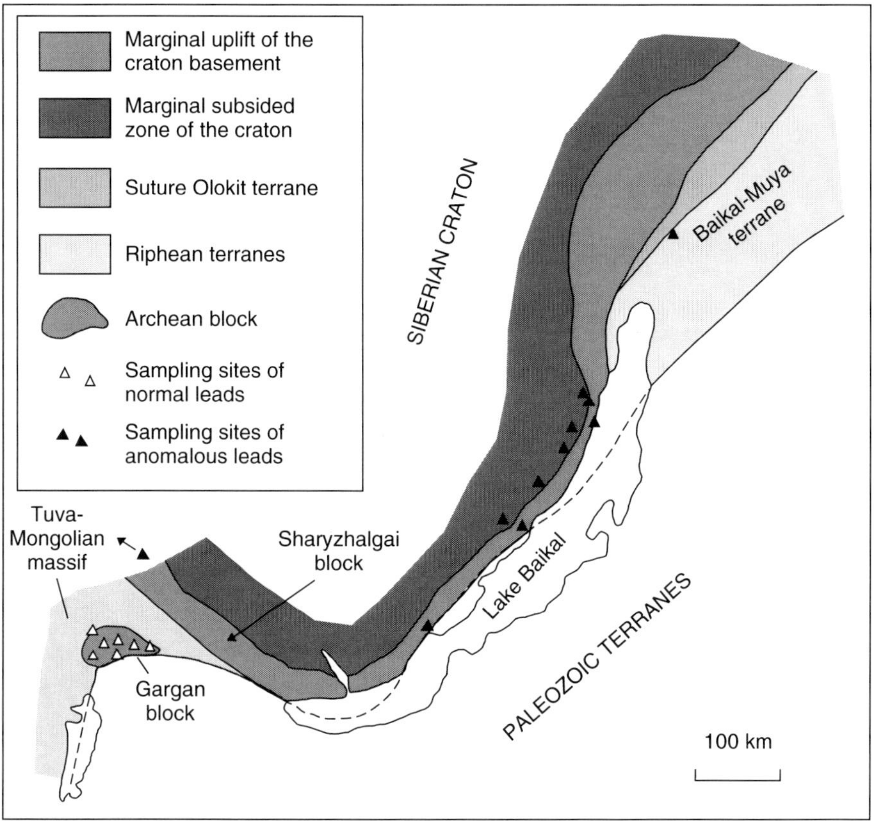

Fig. 14.9 Spatial relations between Paleozoic and Riphean terranes with the southern margin of the Siberian craton. Modified after Belichenko et al. (2006). A point of anomalous leads with array indicates location of the Ingashi pipe of kimberlite-like rocks in the southwestern suture zone of the Siberian craton

and leached feldspars from magmatic and metasomatic rocks belong to a single line interpreted as an isochron. Neymark et al. have inferred the initial and imposed process as corresponding to t_1 =1860 Ma and t_2 =320 Ma with $\mu = 11.8$–13.0 and $\omega = 43.6$–48.4 in the Stacey and Kramers (1975) model.

In terms of the conjugated Holmes–Houtermans—Concordia–Discordia model, the lead compositions appear to result from relatively late closure of the U–Pb isotope system ($T = 3.82$ Ga) with elevated μ (20.1) (Fig. 14.10).

Accepting $T = 3.82$ Ga implies that all preceded open evolution of the isotope system was determined by shift from the Geochron along the Concordia with $\mu = 20.1$ (this line is not shown). Afterwards, leads were separated from crystalline rocks in time intervals of 1.8–1.50 Ga, 1.2–0.9 Ga, 0.6–0.5 Ga, and about 0.25 Ga. This timing is consistent with magmatic and metamorphic events that affected the rocks of the southern craton margin and adjacent Riphean and Paleozoic terranes.

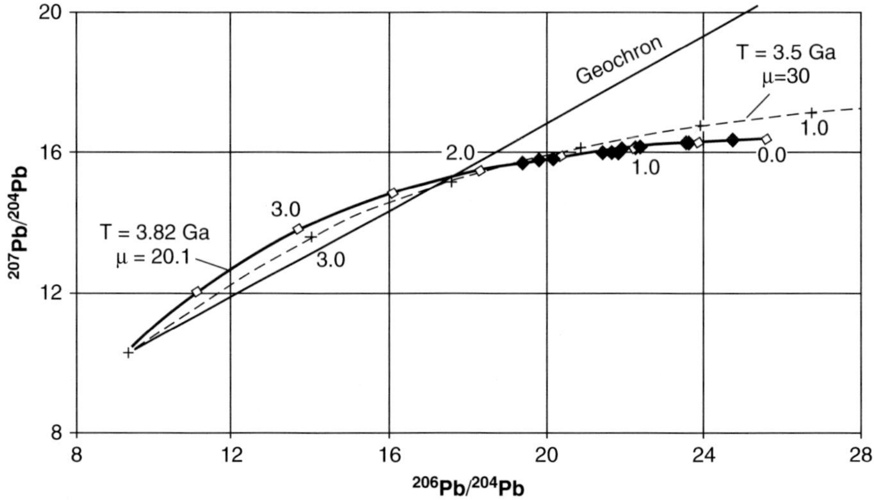

Fig. 14.10 Interpretation of separated leads from ore deposits in the southern part of the Siberian craton. A Discordia curve of the conjugated Holmes–Houtermans—Concordia–Discordia model approximates lead compositions of galenas (*black diamonds*) with parameters: $T = 3.82$ Ga and $\mu = 20.1$. For comparison, we show a Discordia growth curve $T = 3.5$ Ga and $\mu = 30$ (*broken line*) that crosses the accepted trend at a point of the oldest separated leads and goes above the trend. A Discordia growth curve with $T > 3.82$ Ga and $\mu < 20.1$ (not shown), which can be accommodated by iteration at the same intersection of the trend, will go below the latter

The earliest and latest phases of lead separation were characterized, respectively, by Rb–Sr isochrones of granites and $^{40}Ar/^{39}Ar$ ages of mica reset that was coeval to the final low-temperature calcite and quartz-carbonate veins (Brandt et al. 1978, 2006; Chap. 5).

We infer that general assembling of the Siberian craton accompanied by granite magmatism, dated mostly at ca. 1.86 Ga (Aftalion et al. 1991; Neymark et al. 1993; Didenko et al. 2003), was followed by long-term processes providing intermittent ore mineralization in its southern part consistent with timing of geologic processes.

14.9.2 The Gargan Block

Lead isotopes of galenas from gold deposits, concentrated within the Gargan block, were reported by Neymark et al. (1995), Tveritinov et al. (2006), and Airiyants et al. (2007) (Fig. 14.11). In the first publication, data points were interpreted by a growth curve of Pb with $\mu_2 = 10.55$. A source of leads was referred to a continental crust with an age of ca. 2.75 Ga and ore-generating processes to granite magmatism affected the Gargan block at ca. 450 Ma. In the second, data set was interpreted using a parallelogram of Geiss with parameters: $T = 4.56$ Ma, $t_1 = 1.8$ Ma, $t_2 = 0.5$ Ma, $\mu_1 = 8.02$, and $\mu_2 = 9.12$. At last, in the third, variations of lead isotope ratios in galenas from the Vodorazdel'noe deposit (northern margin of the Gargan

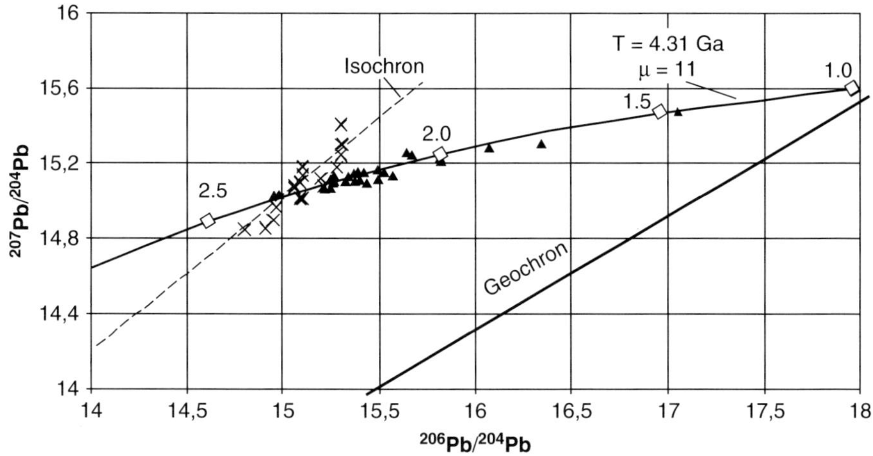

Fig. 14.11 Interpretation of separated leads from ore deposits in the Gargan block. A Discordia curve of the conjugated Holmes–Houtermans—Concordia–Discordia model approximates lead compositions of galenas (*black triangles*) with parameters: $T = 4.31$ Ga and $\mu = 11$. A broken line shows the isochron for leads from the Vodorazdel'noe deposit (*oblique crosses*). The isochron intersects the accepted Discordia growth curve at ~2.35 Ga

block) were explained by mixing of material derived from the mantle and lower crust sources in terms of plumbotectonics (Doe et al. 1979).

We approximate data points of separated leads from the Gargan block by a Discordia growth curve with parameters $T = 4.31$ Ga and $\mu = 11$. In this curve, the data points belong largely to a piece of 2.25–2.15 Ga. Five points fall within an interval of 2.1–1.7 Ga and one point shows the latest event of 1.5–1.4 Ga. Two deposits in the central part of the Gargan block (Taysaevskii and Sumsu-Nur Lake) represent the earliest event ~2.35 Ga.

Data points of the Vodorazdel'noe deposit show notable differentiation of the U–Pb isotope system (distribution along an isochron line) and separation of leads simultaneously with the earliest event. Variable μ of the protolith for the Vodorazdel'noe deposit reflects its specific geochemical signature. Unlike homogeneous crustal material of the Gargan block, its northern portion, hosted by the Vodorazdel'noe deposit, appears to include material provided both the increasing and decreasing U/Pb ratio relative to the mean value of the Gargan block.

Recent geochronological studies showed that the Gargan block was affected by magmatism in a time interval from 800 to 154 Ma. The Gargan block was obducted by ophiolites with U–Pb zircon age of 1020 Ma (Kuz'michev 2000; Rasskazov et al. 2001; Khain et al. 2002). The conjugated Holmes–Houtermans—Concordia–Discordia model implies that separated leads of the Gargan block originated in the Early Precambrian from 4.31 to 1.4 Ga and have no connection with the Late Precambrian and Phanerozoic geologic events. These leads are older than host rocks and are referred to as B-type or Bleiberg type (Doe 1970).

14.9.3 Geologic Implication

Taking into account the common parameter $T = 3.82$ Ga for leads from the southern part of the Siberian craton, we suggest that this part has exhibited a single tectonic unit since the Early Precambrian. If the Pb anomaly resulted from a primary accretional and liquid dynamics of the early Earth, it marked a primary center, where crust was enriched by U relative to Pb (and probably by other incompatible elements). We speculate that a lithospheric weakness of the early Earth was likely inherited by the angular shape of the southern Siberian craton edge and was formed at a triple junction of the center providing boundaries of a craton segment with an angle of 120°. One third of the whole circular projection has been preserved in this segment; the other two thirds have been removed due to break-up and spreading of the former continental crust. The removed parts might be found somewhere on the Earth, so occurrence of anomalous separated leads could be indicative for geologic connection of the southern Siberian craton with other old tectonic units. Searching for possible links between cratons is beyond the major purposes of this study and will be done elsewhere. Here we notice only that the parameters similar to those in the southern margin of the Siberian craton (i.e. $T = 3.82$ Ga and $\mu = 20.1$) can be attributed to separated J-type leads from ore deposits of other regions, for instance, Yukon (Sei-Li Kuo and Folinsby 1977).

Geologic relations between the Gargan cratonic block and Siberian craton are disputable. It has been suggested that the Gargan and other early Precambrian blocks amalgamated with formation of the Riphean Tuva-Mongolian massif as a unit, which had no connection with the Siberian continent in the Precambrian (Belichenko et al. 1994). According to other authors (Il'in 1971; Mossakovskii et al. 1993; Didenko et al. 1994), the Gargan and some other blocks of Eastern Sayan did not belong to the Tuva-Mongolian massif and originated due to tectonic separation from the Siberian proto-continent.

Arguments in favor of the former hypothesis were provided by study of lead isotopes in Late Tertiary lavas (Rasskazov et al. 2002d). Within the Tuva-Mongolian massif, the lavas showed elevated Δ208/204Pb in a range of the enriched mantle anomaly DUPAL of the southern hemisphere, whereas those that erupted within the adjacent East Tuva and Dzhida zones of the Early Caledonides revealed no anomaly. From paleomagnetic reconstructions (Didenko et al. 1994), it was inferred that the Tuva-Mongolian massif was a fragment of the East Gondwana that drifted across the paleo-Asian Ocean in the late Precambrian and accreted to the Siberian continent by the middle Cambrian. Respectively, the DUPAL anomaly was likely brought by the lithosphere of this drifted unit. In Fig. 14.9 the Gargan block is shown as the early Precambrian fragment within the Riphean Tuva-Mongolian massif.

The late accretion of the Gargan block to the Siberian craton margin is emphasized by contrasting the geochemical history of separated leads. Unlike strict control of lead separation in ore deposits at the southern part of the craton by geologic processes of 1.80–0.25 Ga, lead separation in ore deposits of the Gargan block had no relation to magmatic or accompanied events of 1.02–0.15 Ga. Instead of the direct connection, interpretation by the conjugated Holmes–Houtermans—Concordia–Discordia model demonstrates lead separation in a time interval from 2.35 to 1.40 Ma, when the Gargan block was likely a part of East Gondwana.

14.10 Summary: Application of Separated Leads for Understanding Timing of Ore-Forming Processes

For interpretation of separated leads, the Holmes–Houtermans and Concordia–Discordia models have been mutually conjugated. The combined model is based on assumption that radiogenic leads are accumulated in addition to the primeval lead of the meteorite Canyon Diablo firstly along a Concordia and then along a Discordia. These two stages are characterized by a common μ value; the Concordia stage is attributed to accretion and melting of the early Earth and the Discordia stage to lead separation by diffusion from U–Th-bearing mineral phases. A closure of the U–Pb isotope system in the early Earth (T) and a time interval of subsequent lead separation (t_1–t_2) are defined.

The theoretically substantiated model has been applied for interpretation of separated leads from ore deposits in Archean terranes of Central Asia: the Gargan block and southern margin of the Siberian craton. Leads of B-type from the Gargan block are interpreted as separated from a protolith of 4.31 Ga in a time interval of 2.35–1.40 Ga and leads of J-type from the southern margin of the Siberian craton as separated from another protolith of 3.82 Ga in a time interval of 1.80–0.25 Ga. The former protolith ($\mu = 11.0$) is referred to the end of the initial accretionary history of the early Earth and the latter one ($\mu = 20.1$) to the end of the late meteoritic bombardment about 3.9 Ga (Chap. 10).

Ore-forming processes of the Gargan block developed before 1.4 Ga, when this block was likely a part of East Gondwana, and had no subsequent manifestation, when it was tectonically separated from East Gondwana and included into the Tuva-Mongolian massif. Ore-forming processes of the southern Siberian craton started developing after general assembling of the whole craton at 1.9–1.8 Ga and reflected subsequent intermittent interaction and accretion of the younger terranes.

We speculate that the southern margin of the Siberian craton existed as a single tectonic unit since the Early Precambrian. The Pb isotopic anomaly marked a primary center of the early Earth, where crust was enriched by U relative to Pb (and probably by other incompatible elements). This primary crustal weakness was likely inherited by angular shape of the southern margin of the Siberian craton. A triple junction of the center provided boundaries of a craton segment with an angle of 120°. One third of the whole circular projection has been preserved in the craton segment; the other two thirds have been removed due to break-up and spreading of the former continental crust. These fragments might be found somewhere on the Earth due to occurrence of anomalous separated leads. Possible geologic links between the southern margin of the Siberian craton and an unknown tectonic unit, separated due to the 3.9–3.8 Ma break-up of continental crust, will be considered elsewhere.

References

Abbott D, Mooney W (1995) The structural and geochemical evolution of the continental crust: support for the oceanic plateau model of continental growth. Rev Geophys 33(Suppl):231–242

Aftalion M, Bibikova EV, Bowes DR et al (1991) Timing of Early Proterozoic collisional and extensional events in the granulite–gneiss–charnokite–granite complex, Lake Baikal, USSR: a U–Pb, Rb–Sr, and Sm–Nd isotopic study. J Geol 99:851–861

Ageeva LI, Genshaft US, Saltykovskii AY (1988) The new data on absolute ages of Cenozoic basalts from Mongolia. Dokl Acad Nauk USSR 300(1):166–168 (in Russian)

Airiyants EV, Zhmodik SM, Mironov AG et al (2007) Gold mineralization in siliceous-carbonate rocks of southeastern East Sayan. Russ Geol Geophys 48(5):497–510 (in Russian)

Akinin VV, Hourigan JK (2002) $^{40}Ar/^{39}Ar$ geochronology of the Arman and Maltano–Ol'sk volcanic fields: a call for a revised chronostratigraphy of the Okhotsk–Chukotka volcanic belt. 4th symposium of IGCP, ITIG FEB RUS. Khabarovsk, pp 23–24 (in Russian)

Al-Bataina B, Jänecke J (1987) Half-lives of long-lived alpha emitters. Radiochim Acta 42:159–164

Albertão GA, Grassi AA, Marini F et al (2004) The K–T boundary in Brazilian marginal sedimentary basins and related spherules. Geochem J 38:121–128

Aldrich MJ, Deither DP (1990) Stratigraphic and tectonic evolution of the northern Espanola basin, Rio Grande rift, New Mexico. Geol Soc Am Bull 102: 1695–1705

Aldrich MJ, Laughlin AW (1984) A model for the tectonic development of the southeastern Colorado Plateau boundary. J Geophys Res 89:10207–10218

Alibert C, Michard A, Albarède F (1986) Isotope and trace element geochemistry of Colorado Plateau volcanics. Geochim Cosmochim Acta 50:2735–2750

Alibert C, Norman MD, McCulloch MT (1994) An ancient Sm–Nd age for a ferroan noritic anorthosite clast from lunar breccia 67015. Geochim Cosmochim Acta 58:2921–2926

Allan D (1956) Accumulation of radiogenic product in a system with diffusion. Am Math Mon 5:315

Allègre ÑJ, Manhès G, Göpel C (1995) The age of the earth. Geochim Cosmochim Acta 59(8):1445–1456

Alvarez LW, Alwarez W, Asaro F et al (1980) Extraterrestrial cause for the Cretaceous–Tertiary extinction. Science 208:1095–1108

Amirkhanov HI, Brandt SB, Bartnitsky EN (1960) Radiogenic argon in minerals and rocks. Dagestan Publisher, Makhachkala (in Russian)

Amirkhanov HI, Brandt SB, Bartnitsky EN (1961) Radiogenic argon in minerals and its migration. Ann N Y Acad Sci 8:235–275 (in Russian)

Amthor JE, Grotzinger JP, Schroder S et al (2003) Extinction of Cloudina and Namacalathus at the Precambrian–Cambrian boundary in Oman. Geology 31(5):431–434

Arakawa Y, Li X, Ebihara M et al (2003) Element profiles and Ir concentration of Cretaceous–Tertiary (K–T) boundary layers at Medetli, Gölpazari, northwestern Turkey. Geochem J 37:681–693

Arakelyants MM, Akhmet'ev MA, Filimonova LG (1982) Age of volcanics from South Sikhote-Alin. Dokl Acad Nauk 262:946–948 (in Russian)

Arnaboldi C, Brofferio C, Cremonesi O et al (2003) Bolometric bounds on the Antineutrino mass. Phys Rev Lett 91(16):161802

Arnaud NO, Vidal Ph, Tapponnier P et al (1992) The high K_2O volcanism of northwestern Tibet: geochemistry and tectonic implications. Earth Planet Sci Lett 111:351–367

Asmeron Y, Jacobsen SB, Wernicke BP (1994) Variations in magma source regions during large-scale continental extension, Death Valley region, western United States. Earth Planet Sci Lett 125:235–254

Atwater T (1970) Implications of plate tectonics for the Cenozoic tectonic evolution of Western North America. Geol Soc Am Bull 81:3513–3536

Audi G, Bersillon O, Blachot J et al (1997) The NUBASE evaluation of nuclear and decay properties. Nucl Phys 624:1–124

Axen GJ, Taylor WJ, Bartley JM (1993) Space–time patterns and tectonic controls of Tertiary extension and magmatism in the Great Basin of the western United States. Geol Soc Am Bull 105:56–76

Bagdasarov YA, Voronovskii SN, Ovchinnikova LV et al (1980) The new data on K–Ar age of ultrabasic alkaline Carbonatite massifs in Eastern Sayan and some problems of their formation. Dokl Acad Nauk 254(1):171–179 (in Russian)

Bahcall JN (2000) How the sun shines. Focus on science. http://www.nobel.se/physics/articles/fusion/index.html

Baker JA, Snee L, Menzies MA (1996) A brief Oligocene period of flood volcanism in Yemen: implications for duration and rate of continental flood volcanism at the Afro-Arabian triple junction. Earth Planet Sci Lett 138:39–55

Baksi AK (2007) A quantitative tool for detecting alteration in undisturbed rocks and minerals—II: application to argon ages related to hotspots. In: Foulger GR,

Jurdy DM (eds) Plates, plumes, and planetary processes. Geological Society of America Special Paper 430:305–333. doi:10.1130/2007.2430(16)

Baksi AK, Archibald DA, Farrar E (1996) Intercalibration of $^{40}Ar/^{39}Ar$ dating standards. Chem Geol 129:307–324

Baldridge WS, Damon PE, Shafiqullah M, Bridwell RJ (1980) Evolution of the central Rio Grande rift, New Mexico: new K–Ar ages. Earth Planet Sci Lett 51:309–321

Baldridge WS, Olsen K, Callender JF (1984) Rio Grande rift: problems and perspectives. New Mexico Geological Society Guidbook, 35th field conference. Rio Grande rift: Northern New Mexico, pp 1–12

Barruol G, Deschamps A, Dévershèr J et al (2008) Upper mantle flow beneath and around the Hangay dome, Central Mongolia. Earth Planet Sci Lett 274(1–2):221–233

Barry TL, Ivanov AV, Rasskazov SV et al (2007) Helium isotopes provide no evidence for deep mantle involvement in widespread Cenozoic volcanism across Central Asia. Lithos 95:415–424

Barry TL, Kampunzu AB, Rasskazov S et al (2000) Contrasting styles of volcanism and rifting. Proceedings of the 31th geological congress. Rio de Janeiro

Barry TL, Saunders AD, Kempton PD et al (2003) Petrogenesis of Cenozoic basalts from Mongolia: evidence for the role of asthenospheric versus metasomatized lithospheric mantle sources. J Petrol 44(1):55–91

Basu AR, Poreda RJ, Renne PR et al (1995) High-^{3}He plume origin and temporal–spatial evolution of the Siberian flood basalts. Science 269:822–825

Bayanova TB (2003) Baddeleyite—perspective geochronometer for alkaline and basic magmatism. Isotope geochronology in solution of geodynamic and ore genesis problems. 2nd Russian conference on isotope geology. St. Petersburg, pp 50–54 (in Russian)

Bayasgalan A, Jackson J, McKenzie D (2005) Lithosphere rheology and active tectonics in Mongolia: relations between earthquake source parameters, gravity and GPS measurements. Geophys J Int 163:1151–1179

Beck RA, Burbank DW, Sercombe WJ et al (1995) Stratigraphic evidence for an early collision between northwest India and Asia. Nature 373:55–58

Beckinsale RD, Gale NH (1969) A reappraisal of the decay constants and branching ratio of ^{40}K. Earth Planet Sci Lett 6:289–294

Begemann F, Ludwig KR, Lugmair GW et al (2001) Call for improved set of decay constants for geochronological use. Geochim Cosmochim Acta 65:111–121

Belichenko VG, Geletii NK, Barash IG (2006) Barguzin microcontinent (Baikal mountain area): the problem of outlining. Russian Geol Geophys 47(10):1035–1045

Berger GW (1975) $^{40}Ar/^{39}Ar$ step heating of thermally overprinted biotite, hornblende and potassium feldspar from Eldora, Colorado. Earth Planet Sci Lett 26:387–408

Berger M, Braun I (1997) Pb–Pb dating by a stepwise dissolution technique. Chem Geol 142:23–40

Berger WH, Meyer LA (1987) Cenozoic paleoceanography 1986: an introduction. Paleoceanography 2:613–623

Best MG, Christiansen EH, Deino AL et al (1989) Eocene trough Miocene volcanism in the Great Basin of the Western United States. In: Chapin CE, Zidek J (eds) Field excursions to volcanic terranes in the western United States, vol 2: cascades and intermountain west. New Mexico Bureau of Mines & Mineral Resources, Socorro, pp 91–133

Bibikova EV, Klaessen S, Bogdanova SV (2003) The oldest granitoids of the Dnestr–Bug domain (western Ukrainian shield): U–Pb ages of zircons measured on a ion microprobe NORDSIM. Isotope geochronology in solution of geodynamic and ore genesis problems. 2nd Russian conference on isotope geology. St. Petersburg, pp 65–68 (in Russian)

Bijwaard H, Spakman W, Engdahl ER (1998) Closing the gap between regional and global travel time tomography. J Geophys Res 103:30055–30078

Bird P (2002) Stress direction history of the western United States and Mexico since 85 Ma. Tectonics 21(3):1014. doi:10.1029/2001TC001319

Black LP, Williams IS, Compston W (1986) Four zircon ages from one rock: the history of a 3930 Ma-old granulite from Mount Sones, Enderby Land, Antarctica. Contrib Mineral Petrol 94:427–437

Bohannon RG, Parsons T (1995) Tectonic implications of post-30-Ma Pacific and North American relative plate motions. Geol Soc Am Bull 107(8):937–959

Bonneville A, Capolsini P (1999) THERMIC: a 2-D finite-element tool to solve conductive and advective heat transfer problems in Earth Sciences. Comput Geosci 25:1137–1148

Borukaev CB (1996) Tectonics of lithospheric plates in the late Archean. Geol geophys 37(1):33–41 (in Russian)

Boven A (1992) The applicability of the K–Ar and $^{40}Ar/^{39}Ar$ techniques on Quaternary volcanics: a fundamental and experimental investigation. Vrije Universiteit Brussel

Boven A, Pasteels P, Kelley SP et al (2001) $^{40}Ar/^{39}Ar$ study of plagioclases from the Rogaland anorthosite complex (SW Norway): an attempt to understand argon ages in plutonic plagioclase. Chem Geol 176:105–135

Bowring SA, Housh T (1999) The Earth's early evolution. Science 269:1535–1540

Bowring SA, Williams IS (1999) Priscoan (4.00–4.03 Ga) orthogneisses from northwestern Canada. Contrib Mineral Petrol 134:3–16

Bowring SA, Ramezani J, Grotzinger JP (2003) INTERNET

Bowring SA, Grotzinger JP, Isachsen CE et al (1993) Cambrian rates of Early Cambrian evolution. Science 261:1293–1298

Bowring SA, Erwin DH, Jin YG et al (1998) U/Pb zircon geochronology of the end-Permian mass extinction. Science 280:1039–1045

Boztug D, Jonckheere R, Wagner GA et al (2004) Slow Senonian and fast Paleocene–Early Eocene uplift of the granitoids in the Central Eastern Pontides, Turkey: apatite fission–track results. Tectonophysics 382:213–228

Bradshaw TK, Hawkesworth CJ, Gallagher K (1993) Basaltic volcanism in the southern basin and range: no role for a mantle plume. Earth Planet Sci Lett 116:45–62

Bragin VY, Reutskii VN, Litasov KD et al (1999) Late Mesozoic episode of intraplate magmatism in the Severo-Minusinskii basin according to paleomagnetic and geochronological data. Russ Geol Geophys 40:576–582 (in Russian)

Brakhfogel FF (1995) The age division of the kimberlitic and related magmatites in the NE of the Siberian platform (methods and results). In: Sobolev NV (ed) 6th international Kimberlite conference, extended abstracts, Novosibirsk. pp 60–62

Brakhfogel FF, Zaitsev AI, Shamshina EA (1997) Age of kimberlitic magmatites—basis for prospecting of diamond-bearing areas. Otechestvennaya Geologia (9):20–24 (in Russian)

Brandt IS, Brandt SB, Rasskazov SV et al (2000) Radiogenic argon in a cooling dike. Vestnik GeoIGU (2):173–183 (in Russian)

Brandt IS, Rasskazov SV, Brandt SB et al (2002a) Thermal-field propagation in an exocontact zone of a magmatic body and its impact on radiogenic isotope concentrations in minerals. Isotopes Environ Health Stud 38(1):47–58

Brandt IS, Rasskazov SV, Ivanov AV et al (2002b) Influence of radiogenic argon distribution inside a mineral on definition of diffusion parameters. Geochem Int (1):63–69 (in Russian)

Brandt IS, Rasskazov SV, Ivanov AV et al (2006) Radiogenic argon distribution within a mineral grain: implications for dating of hydrothermal mineral-forming event in Sludyanka complex, Siberia, Russia. Isotopes Environ Health Stud 42(2):189–201

Brandt IS, Rasskazov SV, Popov VK et al (2009) Potassic specificity of basalts of the Sineutesovskaya depression: geochemical correlations and problems of K–Ar dating. Geol Pac Ocean 28(4):75–89 (in Russian)

Brandt SB (1965) Methodological aspects of argon geochronometry. Doctor dissertation, Institute of Geochemistry SB RAS, Irkutsk (in Russian)

Brandt SB (1973) A new approach to the determination of temperatures of intrusives from radiogenic argon loss in contact aureoles. In: Hofmann AW, Giletti BJ, Yoder HS Jr, Yund RA (eds) Geochemical transport and kinetics. Carnegie Institution of Washington Publication 634:295–298

Brandt SB, Voronovskii SN (1967) Dehydration and diffusion of radiogenic argon in micas. Int Geol Rev 9:1504–1507

Brandt SB, Brandt IS, Rasskazov SV et al (2003) Formal considerations on argon–argon diagram in $^{40}Ar/^{39}Ar$ geochronometry. Wissenschaftliche Mitteilungen Institut für Geologie 24:19–26

Brandt SB, Bukharov AA, Perminov AV (1978) About absolute age of magmatic complexes in the Baikal and Primorskii complexes. Soviet Geol (3):142–148 (in Russian)

Brandt S, Rasskazov S, Brandt I et al (1996) Betrachtungen zur Anspeicherung radiogenen Argons in Mineralen bei gleichzeitiger Diffusion. Freiberger Isotopenkolloquium Proceedings, pp 35–46 (in German)

Brandt S, Rasskazov S, Brandt I et al (1997) The inverse problem of argon diffusion from minerals: Determination of kinetic parameters from stepwise-heating experiments. Isotopes Environ Health Stud 33:399–409

Brandt SB, Rasskazov SV, Brandt IS et al (2004) Formal analysis of isochron construction in $^{40}Ar/^{39}Ar$ geochronology. Geochem Int 42(8):728–735 (in Russian)

Brick JL (1986) Precision K–Rb–Sr isotopic analysis: application to Rb–Sr chronology. Chem Geol 56:73–83

Bromham L, Rambaut A, Fortey R et al (1998) Testing the Cambrian explosion hypothesis by using a molecular dating technique. Proc Natl Acad Sci USA 95:12386–12389

Brooks C, Hart SR, Wendt I (1972) Realistic use of two-error regression treatment as applied to rubidium–strontium data. Rev Geophysand Space Phys 10(2): 551–577

Burchfield JD (1975) Lord Kelvin and the age of the Earth. University of Chicago Press, Chicago

Burk K, Wilson JT (1972) Is the African plate stationary? Nature 239:387–390

Butler R (1995) When did India hit Asia? Nature 373:20

Calais E, Vergnolle M, San'kov V et al (2003) GPS measurements of crustal deformation in the Baikal–Mongolia area (1994–2002): implications for current kinematics of Asia. J Geophys Res 108(B10):2501. doi:10.1029/2002JB002373

Campbell IH, Czamanske GK, Fedorenko VA et al (1992) Synchronism of the Siberian traps and the Permian–Triassic boundary. Science 258:1760–1763

Canaghan PJ, Shaw SE, Veevers JJ (1994) Sedimentary evidence of the Permian–Triassic global crisis induced by the Siberian hotspot. Canad Soc Petrol Geol 17:785–795

Cather SM (1990) Stress and volcanism in the northern Mogollon–Datil volcanic field, New Mexico: effects of the post-Laramide tectonic transition. Geol Soc Am Bull 102:1447–1458

Chapin CE (1979) Evolution of the Rio Grande rift—a summary. In: Riecker RE (ed) Rio Grande rift: tectonics and magmatism. American Geophysical Union, Washington, pp 1–5

Chashchin AA, Martynov YA, Rasskazov SV et al (2007) Isotopic and geochemical characteristics of the Late Miocene subalkali and alkali basalts of the southern part of the Russian Far East and the role of continental lithosphere in their genesis. Petrology 15(6):575–598 (in Russian)

Chen JH, Moore JG (1982) Uranium–lead isotopic ages from the Sierra Nevada batholith, California. J Geophys Res 87(B6):4761–4784

Chen X-B (2007) Present-day horizontal deformation status of continental China and its driving mechanism. Sci China Ser D Earth Sci 50(11):1663–1673

Chernyshev IV, Shpikerman VI (2001) Isotope composition of ore lead as reflection of block structure in the central part of Northeast of Russia. Dokl Acad Nauk 377(4):530–533 (in Russian)

Chernyshev IV, Lebedev VA, Goltsman YV et al (2000) Isotope systematics of ore-bearing scarns and related granites in the W–Mo Tyrnyauz deposit (Northern

Caucasus): K–Ar, Rb–Sr, and Sm–Nd data. Isotope dating of geologic processes: new methods and results. In: Abstracts of the 1st Russian conferences on isotope geochronology. GEOS, Moscow, 387–398 (in Russian)

Chernyshova EV, Sandimirova GP, Pakhol'chenko YA et al (1992) Rb–Sr age and some features of genesis of Bol'shetagninkii carbonatite complex (East Sayan). Dokl Acad Nauk 323(5):942–948 (in Russian)

Choi B-G, Ouyang X, Wasson JT (1995) Classification and origin of IAB and IIICD iron meteorites. Geochim Cosmochim Acta 59(3):593–612

Christiansen RL, Lipman PW (1972) Cenozoic volcanism and plate-tectonic evolution of the western United States. II. Late Cenozoic. Philosophical transactions of the Royal society of London A271:249–284

Chumakov NM, Zharkov MA (2002) A climate in the time of Permian–Triassic biospheric reorganizations. 1.A climate in Early Permian. Stratigr Geol Correl 10(6):62–81 (in Russian)

Chuvashova IS, Rasskazov SV (2008) Collision-derived Late Cenozoic dynamics of a melting anomaly beneath Central Mongolia: magmatic and tectonic evidence. Geophys Res Abstr 10:00301 (SRef–ID:1607–7962/gra/EGU2008–A–00301)

Chuvashova IS, Rasskazov SV, Yasnygina TA et al (2007) Holocene volcanism in Central Mongolia and Northeast China: asynchronous decompressional and fluid melting of the mantle. J Volcanol Seismol 1(6):19–45 (in Russian)

Claoue-Long JC, Zichao Z, Guogan M et al (1991) The age of the Permian–Triassic boundary. Earth Planet Sci Lett 105:182–190

Clube SVM, Napier WM (1982) The role of episodic bombardment in geophysics. Earth Planet Sci Lett 57:251–262

Coccioni R, Galeotti S (1994) K–T boundary extinction: geologically instantaneous or gradual even? Evidence from deep-sea benthic foraminifera. Geology 22:779–782

Collerson KD, Campbell LM, Weaver BL et al (1991) Evidence for extreme mantle fractionation in early Archean ultramafic rocks from northern Labrador. Nature 349:209–214

Compston W (1994) Time points within the Vendian by ion microprobe. In: Abstracts of the 8th international conference on geochronology US geological survey circular 1107:65

Compston W, Froude DO, Ireland TR et al (1985) The age of (a tiny part of) the Australian continent. Nature 317:559–560

Compston W, Sambridge MS, Reinfrank RF (1995) Numerical ages of the volcanic rocks and the earliest faunal zone within the Late Precambrian of east Poland. J Geol Soc London 152(5):599–611

Compston W, Williams IS, Kirschvink JL (1992) Zircon U–Pb ages relevant to the Early Cambrian time-scale. J Geol Soc London 149(1):171–184

Condie KC (1998) Episodic continental growth and supercontinents: a mantle avalanche connection? Earth Planet Sci Lett 163:97–108

Coney PJ, Reynolds SJ (1977) Cordilleras Benioff zones. Nature 270:404–406

Copeland P, Harrison TM, Yun P et al (1995) Thermal evolution of the Gangdes batholith, southern Tibet: a history of episodic unroofing. Tectonics 14(2): 223–236

Corsetti FA, Hagadorn JW (2000) Relation of the Canyon Wood Formation and Stelling quartzites to the Precambrian–Cambrian boundary. Geology 28(4): 299–302

Courant R, Hilbert D (1951) Methoden der mathematischen Physik, Bd 2. GITTL, Moscow–Leningrad (in Russian, translated from German)

Courtillot V (1994) Mass extinctions in the last 300 million years: one impact and seven flood basalts. Israel J Earth Sci 43:255–266

Courtillot V, Féraud G, Maluski H (1988) Deccan flood basalts and the Cretaceous/Tertiary boundary. Nature 833:843–846

Courtillot V, Gallet Y, Rocchia R et al (2000) Cosmic markers, $^{40}Ar/^{39}Ar$ dating and paleomagnetism of the K–T sections in the Anjar area of the Deccan large igneous province. Earth Planet Sci Lett 182:137–156

Creaser RA, Papanastassiou DA, Wasserburg GJ (1991) Negative thermal ion mass spectrometry of osmium, rhenium and iridium. Geochim Cosmochim Acta 55:397–401

Dalrymple GB (1969) $^{40}Ar/^{36}Ar$ analyses of historic lava flows. Earth Planet Sci Lett 6:47–55

Dalrymple GB, Lanphere LA (1969) Potassium–Argon dating principles, techniques, and application to geochronology. WH Freeman, San Francisco, p 258

Dalrymple GB, Czamanske GK, Fedorenko ON et al (1995) A reconnaissance $^{40}Ar/^{39}Ar$ study of ore-bearing and related rocks, Siberian Russia. Geochim Cosmochim Acta 59:2071–2083

Dalrymple GB, Izzet GA, Snee LW et al (1983) $^{40}Ar/^{39}Ar$ age spectra and total-fusion ages of tektites from Cretaceous–Tertiary boundary sedimentary rocks in the Beloc Formation, Haiti. US Geol Surv Bull (B2065):395–447

Damask A, Dins J (1966) Dot defects in metals. Mir Publisher, Moscow (in Russian, translated from English)

Dao-Yi X, Shu-Lan M, Zhi-Fang C et al (1985) Abundance variation of iridium and trace elements at the Permian–Triassic boundary at Shangsi in China. Nature 314:154–156

Davis DW, Gray J, Cumming GL (1977) Determination of the ^{87}Rb decay constant. Geochim Cosmochim Acta 41:1745–1749

Davis JM, Hawkesworth CJ (1995) Geochemical and tectonic transitions in the evolution of the Mogollon–Datil Volcanic Field, New Mexico, USA. Chem Geol 119:31–53

Deng W, Sun H, Zhang Y (2000) K–Ar age of the Cenozoic volcanic rocks in the Nangqen basin, Qinghai Province and its geological significance. Chin Sci Bull 45(11):1015–1019

De Paolo DJ (1981a) A neodymium and strontium isotopic study of the Mesozoic calc-alkaline granitic batholiths of the Sierra Nevada and Peninsular ranges, California. J Geophys Res 86(11):10470–10488

De Paolo DJ (1981b) Neodymium isotope in the Colorado Front Range and crust–mantle evolution in the Proterozoic Nature 291:193–196

Deschamps A, Lallemand S (2002) The West Philippine basin: an Eocene to early Oligocene back arc basin opened between two opposed subduction zones. J Geophys Res 107:2322. doi:10.1029/2001JB001706

Devirts AL, Rasskazov SV, Polyakov AI et al (1981) Radiocarbon age of young volcanoes in the Udokan Range (Northeastern Pribaikal). Geochem Int (8):1250–1253 (in Russian)

Devyatkin EV (1981) The Cenozoic of inner Asia. Nauka, Moscow (in Russian)

Devyatkin EV (2004) Geochronology of Cenozoic basalts from Mongolia and their connection with structures of the recent stage. Stratigr Geol Correl 12(2):102–114 (in Russian)

De Wit MJ, Ghosh JG, De Villiers S et al (2002) Multiple organic carbon isotope reversals across the Permo–Triassic boundary of terrestrial Gondwana sequences: clues to extinction patterns and delayed ecosystem recovery. J Geol 110:227–240

Dickens GR (2003) Rethinking the global carbon cycle with a large, dynamic and microbially mediated gas hydrate capacitor. Earth Planet Sci Lett 213:169–183

Dickin AP (1997) Radiogenic isotope geology. University Press, Cambridge

Dickinson WR (2002) The Basin and Range Province as a composite extensional domain. Int Geol Rev 44:1–38

Didenko AN, Kozakov IK, Bibikova EV et al (2003) Paleomagnetism of Early Proterozoic granitoids in Sharyzhalgai uplift of the basement of the Siberian craton and geodynamic consequences. Dokl Acad Nauk 390(3):368–373 (in Russian)

Didenko AN, Mossakovskii AA, Pecherskii DM et al (1994) Geodynamics of Paleozoic oceans of Central Asia. Geol Geophys (7-8):59-75 (in Russian)

Ding L, Kapp P, Zhong D, Deng W (2003) Cenozoic volcanism in Tibet: evidence for a transition from oceanic to continental subduction. J Petrol 44(10): 1833–1865

Ding L, Zhou Y, Zhang J et al (2000) Geologic relationships and geochronology of the Cenozoic volcanoes and interbedded weathering mantles of Yulinshan in Qiangtang, North Tibet. Chin Sci Bull 45(24):2214–2220

Doblas M, Lopez-Ruiz J, Cebria J-M (2002) Mantle insulation in West African craton during the Precambrian–Cambrian transition. Geol Soc Am Bull 30(9): 839–842

Dobretsov NL (1997a) Permian–Triassic magmatism and sedimentation in Eurasia as reflection of superplume. Dokl Acad Nauk 354(2):220–223 (in Russian)

Dobretsov NL (1997b) Mantle superplumes as the cause of the major periodicity and global reorganizations. Dokl Acad Nauk 357(6):797–800 (in Russian)

Dodson MH (1973) Closure temperature in cooling geochronological and petrological systems. Contrib Mineral Petrol 40:259–274

Doe BR (1970) Lead isotopes. Springer-Verlag, Berlin

Doe B, Stacey J (1977) Outline of studies on application of lead isotopes for solution of problems of ore origin and perspectives for prospecting of ore deposits. In: Skinner BJ (ed) Economic Geology. Stable isotopes as applied to problems of ore deposits 69(6):12–57 (in Russian, translated from English)

Doe B, Zartman RE, Stacey J (1979) General theory of plumbotectonics. Nauka, Moscow (in Russian)

Domyshev VG (1974) Pyroclastic strata, trap volcanism, and tectonics of the southeast of the Tunguska basin. Nauka, Novosibirsk (in Russian)

Dugarmaa T, Schlupp A (2003) One century of seismicity in Mongolia (1900–2000). Research Center of Astronomy and Geophysics Mongolian Academy of Sciences, Department Analyse Surveillance Environnement France

Duncan RA, Pyle DG (1988) Rapid eruption of the Deccan flood basalts at the Cretaceous–Tertiary boundary. Nature 333(6176):841–843

Dungan MA, Lindstrom MM, McMillan NJ et al (1986) Open system magmatic evolution of the Taos Plateau volcanic field, Northern New Mexico, vol 1. The petrology and geochemistry of the servilleta basalt. J Geophys Res 91B: 5999–6028

Dungan MA, Thompson RA, Stormer JS et al (1989) Rio Grande rift volcanism: northeastern Jemez zone, New Mexico. In: Chapin CE, Zidek J (eds) Field excursions to volcanic terranes in the western United States, vol 1: southern Rocky Mountain region. New Mexico Bureau of Mines & Mineral Resources, Socorro, pp 435–486

Dunlap WJ, Fossen H (1998) Early Paleozoic orogenic collaps, tectonic stability, and late Paleozoic continental rifting revealed through thermochronology of K-feldspars, southern Norway. Tectonics 17(4):604–620

Eaton GP (1979) A plate-tectonic model for late Cenozoic crustal spreading in the Western United States, Rio Grande rift: tectonics and magmatism. Geophys Union Spec Pub, pp 7–30

Ebinger CJ, Deino AL, Drake RE et al (1989) Chronology of volcanism and rift basin propagation: Rungwe volcanic province, East Africa. J Geophys Res 94(B11):15785–15803

Ebinger CJ, Sleep NH (1998) Cenozoic magmatism throughout east Africa resulting from impact of a single plume. Nature 395:788–791

Ebinger CJ, Yemane T, Woldegabriel G (1993) Late Eocene–Recent volcanism and faulting in the southern main Ethiopian rift. J Geol Soc London 150: 99–108

Endt PM, Van der Leun C (1973) Energy levels of A = 21–44 nuclei (V). Nucl Phys A 214:1–625

England P and Molnar P (1997) Active deformation of Asia: from kinematics to dynamics. Science 278: 647–650

Enkhtuvshin H (1995) A petrological study on the Late Mesozoic and Cenozoic volcanic rocks of the Mongolian Plateau. Master thesis. Shimane University, Shimane

Ermakov VA, Garagash IA (2003) Petrologic-dynamic model of the early Earth. News of section of Earth sciences RAEN 10:38–46 (in Russian)

Ernst WG, Tsujimory T, Zhang R et al (2007) Permo–Triassic collision, subduction-zone metamorphism, and tectonic exgumation along the East Asian continental margin. Annu Rev Earth Planet Sci 35:73–110. doi:10.1146/annurev.earth.35.031306.140146

Erwin DH (1994) The Permo–Triassic extinction. Nature 367:231–236

Erwin DH (1998) The end and the beginning: recoveries from mass extinctions. Trends Ecol Evol 13:344–349

Esin SV, Parkhomenko VS, Travin AV et al (1994) Petrogenesis of tholeiitic–alkali-basaltic plateau in East Sikhote-Alin: (K–Ar ages, petrochemical and REE characteristics). Geol Geophys 35(9):21–33 (in Russian)

Evans NJ, Gregoire DC, Goodwellow WD et al (1993) Ru/Ir ratios at the Cretaceous–Tertiary boundary: implications for PGE source and fractionation within the eject cloud. Geochim Cosmochim Acta 57:3149–3158

Farmer GL, Glazner AF, Manley CR (2002) Did lithospheric delaminating trigger late Cenozoic potassic volcanism in the southern Sierra Nevada, California? Geol Soc Am Bull 114:754–768

Farmer GL, Glazner AF, Wilshire HG et al (1995) Origin of late Cenozoic basalts at the Sima volcanic field, Mojave desert, California. J Geophys Res 100B: 8399–8415

Faure G (1989) Principles of isotope geology. Mir Publisher, Moscow (in Russian, translated from English)

Faure G (2001) Origin of igneous rocks: the isotopic evidence. Springer-Verlag, Berlin

Faure G, Powel JL (1972) Strontium isotope geology. Springer-Verlag, Berlin

Fedoseev GS, Sotnikov VI (2003) Petrology and isotope geochronology of dike associations in the Tomsk magmatic area (Western Siberia). Isotope geochronology in solution of geodynamic and ore genesis problems. 2nd Russian conference on isotope geology. St. Petersburg, pp 521–525 (in Russian)

Fefelov NN, Sharov VN, Yablonovskii BV (2000) Pb–Pb isochron age of metamorphosed carbonaceous sediments of the Imnyakh suit (Bodaibo basin). Geol Geophys. 41(1):86–89 (in Russian)

Feoktistov GD (1979) Petrology and conditions of formations of trap sills. Nauka, Novosibirsk (in Russian)

Feuerbach DL, Smith EI, Walker JD et al (1993) The role of the mantle during crustal extension: Constraints from geochemistry of volcanic rocks in the Lake Mead area, Nevada and Arisona. Geol Soc Am Bull 105:1561–1575

Florensov NA (1948) Geomorphology and recent tectonics in Transbaikal. Izvestiya academy of sciences of USSSR. Ser Geol (2):3–16 (in Russian)

Foley SF, Venturelly G, Green DH et al (1987) The ultrapotassic rocks: characteristics, classification, and constraints for petrogenetic models. Earth Planet Sci Lett 24:81–134

Forsyth D, Uyeda S (1975) On the relative importance of the driving forces of plate motion. Geophys J Int 43:163–200

Frank F, Mises R (1937) Differential and integrated equations of mathematical physics. Part 2. GTTI, Moscow (in Russian)

Frei R, Villa IM, Nägler ThF et al (1997) Single mineral dating by the Pb–Pb step-leaching method: assessing the mechanisms. Geochim Cosmochim Acta 61(2):393–414

Fukuyama H (1985) Heat fusion of basaltic magma. Earth Planet Sci Lett 73:407–414

Furman T, Bryge JG, Karson J et al (2004) East African Rift System (EARS) plume structure: insight from Quaternary mafic lavas of Turkana, Kenya. J Petrol 45(5):1069–1088

Furman T, Graham D (1999) Erosion of lithospheric mantle beneath the East African Rift system: geochemical evidence from the Kivu volcanic province. Lithos 48:237–262

Galeazzi M, Fontanelli F, Gatti F et al (2001) End-point energy and half-life of the ^{187}Re β decay. Phys Rev C 63:014302

Galer SJG (1999) Optimal double and triple spiking for high precision lead isotopic measurement. Chem Geol 157:255–274

Gao W, Grand SP, Baldridge WS et al (2004) Upper mantle convection beneath the central Rio Grande rift imaged by P and S wave tomography. J Geophys Res 109:B03305. doi:10.1029/2003JB002743

Garagash IA, Ermakov VA (2004) Probable geodynamical model of early Earth. Dokl Acad Nauk 394(2):247–256

Genshaft YS, Saltykovskii AY (1990) The catalogue of inclusions of deep-seated rocks and minerals in Mongolian basalts. Nauka, Moscow (in Russian)

Genshaft YS, Klimenko GV, Saltykovskii AY et al (1990) The new data on compositions and ages of Cenozoic basalts from Mongolia. Dokl Acad Nauk USSR 311(2):420–424 (in Russian)

Gentner W, Kley W (1957) Argonbestimmungen an kaliummineralen. Geochim Cosmochim Acta 12:323–337

Geyh MA, Schleicher H (1990) Absolute age determination. Springer-Verlag, Berlin

Gibson SA, Thompson RN, Leat PT et al (1993) Ultrapotassic magmas along the flanks of the Oligo-Miocene Rio Grande rift, USA: monitors of the zone of lithospheric mantle extension and thinning beneath a continental rift. J Petrol 34:187–228

Gillot PY, Cornette Y (1986) The Cassignol technique for K–Ar dating, precision and accuracy: examples from the late Pleistocene to recent volcanics from southern Italy. Chem Geol 59:205–222

Gladenkov YB, Sal'nikov BA, Barinov KB (1999) Ecosystems of the Cenozoic in the Okhotsk Sea region. A basic section of the Paleogene and Neogene in the Northern Sakhalin (Schmidt peninsula): stratigraphy, paleogeography, and geologic events. GEOS, Moscow (in Russian)

Glazner AF, Supplee JA (1982) Migration of Tertiary volcanism in the southwestern United States and subduction of the Mendocino fracture zone. Earth Planet Sci Lett 60:429–436

Golashvili TB, Chechev VP, Lbov AA (2002) Nuclide guide-2. Central Scientific Atom Inform, Moscow (in Russian)

Göpel G, Manhès G, Allègre CJ (1985) U–Pb systematics in iron meteorites: Uniformity of primordial lead. Geochim Cosmochim Acta 49:1681–1696

Gordienko IV, Bayanov VD, Klimuk VS et al (1999) Composition and age (^{40}Ar/^{39}Ar) of volcanogenic rocks of the Chikoi–Khilok riftogenic basin in Transbaikal. Russ Geol Geophys 40:583–591

Gorokhov IM (1985) Rubidium–strontium method of isotope geochronology. Energoatomizdat, Moscow (in Russian)

Gouchi N, Omata S, Katoh K et al (1992) K–Ar ages of mica fractions from the Susunai metamorphic rocks in Sakhalin, Far East Russia. J Fac Sci Hokkaido Univ Ser 4(23):281–286

Govorov GI (2002) Phanerozoic magmatic belts and origin of the Okhotsk Sea geoblock structure. Dalnauka, Vladivostok (in Russian)

Gradstein FM, Ogg JG, Smith AG et al (2004) Geologic time scale. Cambridge University Press, Cambridge

Grossman L, Ebel DS, Simon SB et al (2000) Major element chemical and isotopic compositions of refractory inclusions in C3 chondrites: the separate roles of condensation and evaporation. Geochim Cosmochim Acta 64: 2879–2894

Grotzinger JP, Bowring SA, Saylor BZ et al (1995) Biostratigraphic and geochronologic constraints on early animal evolution. Science 270:598–604

Grudinin MI, Gerasimov NS, Rasskazov SV (2007) Heterogeneous gabbro-syenite complex of Pribaikal. In: Scherbak NP, Kogarko LN (eds) Alkaline magmatism of the Earth and its ore formation. Institute of Geochemistry, Mineralogy and Ore Formation, Kiev, pp 56–57 (in Russian)

Grunt TA (2005) Global and East-European stratigraphic scales of the Permian system: probable applications within off-tropical zones of sedimentation. Stratigr Geol Correl 13(1):41–55 (in Russian)

Guillou H, Carracedo JC, Torrado FP et al (1996) K–Ar ages and magnetic stratigraphy of a hotspot-induced, fast grown oceanic island: El Hierro, Canary Island. J Volcanol Geotherm Res 73:141–155

Guillou H, Garcia MO, Turpin L (1997) Unspiked K–Ar dating of young volcanic rocks from Loihi and Pitcairn hot spot seamounts. J Volcanol Geotherm Res 78:239–249

Guo F, Fan W, Wang Y, Li C (2004) When did the Emeishan mantle plume activity start? Geological and geochemical evidence from ultramafic–mafic dikes in Southwestern China. Int Geol Rev 46:226–234

Guo Z, Wilson M, Liu J (2007) Post-collisional adakites in southern Tibet: products of partial melting of subduction-modified lower crust. Lithos 96: 205–224

Haase R (1967) Thermodynamik der irreversiblen prozesse. Dr. Dietrich Steinkopff Verlag, Darmstadt (in German)

Habfast K (1998) Fraction correction and multiple collectors in thermal ionization isotope ratio mass spectrometry. Int J Mass Spectrom 176:133–148

Halim N, Cogne J-P, Chen Y et al (1998) New Cretaceous and Early Tertiary paleomagnetic results from Xining-Lanzhou basin, Kunlun and Qiangtang blocks, China: implications on the geodynamic evolution of Asia. J Geophys Res 103(B9):21025–21045

Hallem A (1985) Great geological disputes. Mir Publisher, Moscow (in Russian, translated from English)

Halliday AN (2000) Terrestrial accretion rates and the origin of the Moon. Earth Planet Sci Lett 176:17–30

Hamilton EI (1965) Applied geochronology. Academic Press, London

Hamilton WB (2003) An alternative Earth. GSA Today 13:4–12

Hanson GN, Gast PW (1967) Kinetic studies in contact metamorphic zones. Geochim Cosmochim Acta 31:1119–1153

Harkin DA (1960) The Rungwe volcanics at the northern end of lake Nyasa. Geol Surv Tanganyika Mem 2:1–172

Harris N (1997) Radiogenic isotopes and the interpretation of granitic rocks. Episodes 19:107–112

Harrison TM, Bé K (1983) $^{40}Ar/^{39}Ar$ age spectrum analysis of detritial microclines from the southern San Joaquin Basin, California: an approach to the thermal evolution of sedimentary basins. Earth Planet Sci Lett 64:244–256

Harrison TM, Lovera OM, Heizler T (1991) $^{40}Ar/^{39}Ar$ results for alkali feldspars containing diffusion domains with differing activation energy. Geochim Cosmochim Acta 55:1435–1448

Hart SR (1964) The petrology and isotopic-mineral age relations of a contact zone in the Front Range, Colorado. J Geol 72:493–525

Hart SR (1981) Diffusion compensation in natural silicates. Geochim Cosmochim Acta 45:279–291

Hawkesworth C, Turner S, Gallagher K et al (1995) Calc-alkaline magmatism, lithospheric thinning and extension in the basin and range. J Geophys Res 100(B7):10271–10286

Hawkins DP, Bowring SA (1999) U–Pb monazite, xenotime and titanite geochronological constraints on the prograde to post-peak metamorphic thermal history of Paleoproterozoic migmatites from the Grand Canyon, Arizona. Contrib Miner Petrol 134:150–169

Hay DE, Wendlandt RF, Keller GR (1995) Origin of Kenia Rift plateau-type flood phonolites: integrated petrologic and geophysical constraints on the evolution of the crust and upper mantle beneath the Kenya Rift. J Geophys Res 100(B7):10549–10557

Heaman LM, LeCheminant AN (1993) Paragenesis and U–Pb systematics of baddeleyite (ZrO_2). Chem Geol 110:95–126

Heaman LM, Kjars BA, Creaser RA (2003) The timing of kimberlitic magmatism in North America: implications for global kimberlite genesis and diamond exploration. Lithos 71:153–184

Heizler MT, Harrison TM (1988) Multiple trapped isotope components revealed by $^{40}Ar/^{39}Ar$ isochron analysis. Geochim Cosmochim Acta 52:1295–1303

Hildebrand AR, Penfield GT, Kring DA et al (1991) Chicxulub crater: a possible Cretaceous/Tertiary boundary impact crater on the Yucatan Peninsula, Mexico. Geology 19:867–871

Höck V, Daxner-Höck G, Schmid HP et al (1999) Oligocene–Miocene sediments, fossils and basalts from the Valley of Lakes (Central Mongolia)—an integrated study. Mitt Österr Geol Ges 90:83–125

Hofmann AW (1997) Early evolution of continents. Science 275:498–499

Hofmann C, Courtillot V, Feraud G et al (1997) Timing of the Ethiopian flood basalt event and implications for plume birth and global change. Nature 389:838–841

Hofmann C, Feraud G, Courtillot V (2000) $^{40}Ar/^{39}Ar$ dating of mineral separates and whole rocks from the Western Ghats lava pile: further constraints on duration and age of Deccan traps. Earth Planet Sci Lett 180:13–27

Holser WT, Schonlaub HP, Jr MA et al (1989) A unique geochemical record at the Permian–Triassic boundary. Nature 337(5):39–44

Horan MF, Smoliar MI, Walker RJ (1998) ^{182}W and ^{187}Re–^{187}Os systematics of iron meteorites: chronology for melting, differentiation, and crystallization in asteroids. Geochim Cosmochim Acta 62:545–554

Horn I, Rudnick RL, McDonough WF (2000) Precise elemental and isotope ratio determination by simultaneous solution nebulization and laser ablation–ICP–MS: application to U–Pb geochronology. Chem Geol 164:281–301

Horwitz EP, Chiarizia R, Dietz ML (1992) A novel strontium-selective extraction chromatographic resin. Solvent Extr Ion exch 10(2):313–336

Hou Z-Q, Gao Y-F, Qu X-M et al (2004) Origin of adakitic intrusives generated during mid-Miocene east–west extension in southern Tibet. Earth Planet Sci Lett 220:139–155

Huang S-J, Quing H-R, Huang P-P et al (2008) Evolution of strontium isotope composition of seawater from Late Permian to Early Triassic based on study of marine carbonates, Zhongliang mountain, Chongqing, China. Sci China Ser D–Earth Sci 51(4):528–539

Hudson TL, Magoon LB (2002) Tectonic controls on greenhouse gas flux to the Paleogene atmosphere from the Gulf of Alaska accretionary prism. Geology 30(6):547–550

Iizumi S, Imaoka T, Kagami H (2000) Sr–Nd isotope ratios of gabbroic and dioritic rocks in a Cretaceous–Paleogene granite terrain, Southwest Japan. Isl Arc 9:113–127

Il'in AV (1971) On the Tuva–Mongolian massif. In: Moralev VM (ed) Materials on regional geology of Africa and foreign Asia. Nedra, Moscow, pp 67–73 (in Russian)

Imaoka T, Nakashima K, Itaya T et al (2001) Timing and duration of hydrothermal activity in the Oligocene Hamada Cauldron, SW Japan: evidence from K–Ar ages of sericites. Resour Geol 51:55–62

Ingersoll RV (1988) Tectonics of sedimentary basins. Geol Soc Am Bull 100: 1704–1719

Ingersoll RV, Cavarza W, Baldridge WS et al (1990) Cenozoic sedimentation and paleotectonics of north–central New Mexico: implication for imitation and evolution of the Rio Grande rift. Geol Soc Am Bull 102(9):1280–1296

Isachsen CE, Bowring SA, Landing E et al (1994) New constraint on the division of Cambrian time. Geology 22(6):496–498

Ishiwatari A, Tsujimory T (2003) Paleozoic ophiolites and blueschists in Japan and Russian Primorye in the framework of East Asia: a synthesis. Isl Arc 12:190–206

Isozaki Y (1997) Permo–Triassic boundary superanoxia and stratified superocean: records from lost deep sea. Science 276:235–238

Ivanenko VV, Karpenko MI, Litzarev MA (1990) Age of phlogopite deposits in Sludyanka (data of ^{40}Ar–^{39}Ar method). Izvestiya Academii Nauk. Ser Geol (5):92–98 (in Russian)

Ivanov AV, Rasskazov SV, Boven A et al (1998) Late Cenozoic alkali-ultrabasic and alkali-basaltic magmatism in Rungwe Province, Tanzania. Petrology 6(3):228–250 (in Russian)

Ivanov AV, Rasskazov SV, Feoktistov GD et al (2005) $^{40}Ar/^{39}Ar$ dating of Usol'skii sill in the southeastern Siberian Traps Large Igneous Province: evidence for long-lived magmatism. Terra Nova 17:203–208

Ivanov VG, Yarmolyuk VV, Smirnov VN (1995) New data on ages of volcanism manifestation in the western Transbaikal Late Mesozoic–Cenozoic volcanic region. Dokl Acad Nauk 345:648–652 (in Russian)

Ivanov VS, Tararin IA, Ignat'ev AV et al (1989) Geochemical features of alumina granitoids of the Aniva massif, Sakhalin island. New data on magmatism and metallogeny of Far East. Dalnauka, Vladivostok, pp 22–32 (in Russian)

Izett GA, Cobban WA, Dalrymple GB et al (1998) $^{40}Ar/^{39}Ar$ age of the Manson impact structure, Iowa, and correlative impact ejecta in the Crow Creek Member of the Pierre Shale (Upper Cretaceous), South Dakota and Nebraska. GSA Bull 110:361–376

Izett GA, Cobban WA, Obradovich JD et al (1993) The manson impact structure: $^{40}Ar/^{39}Ar$ age and its distal impact ejecta in the Pierre shale in southeastern South Dakota. Science 262:729–732

Jackson ED, Shaw HR, Bargar KE (1975) Calculated geochronology and stress field orientations along the Hawaiian chain. EPSL 26:145–155

Jaffey AH, Flynn KF, Glendenin LF et al (1971) Precision measurement of half-lives and specific activities of ^{235}U and ^{238}U. Phys Rev C4:1889–1906

Johnson GL, Rich JE (1986) A 30 Million year cycle in arctic volcanism? J Geodyn 6(1–4):111–116

Jolivet L, Tamaki K (1992) Neogene kinematics in the Japan Sea region and volcanic activity of the Northeast Japan Arc. Proc Ocean Drill Program Sci Results 127–128:1311–1331

Jolivet L, Brunel M, Seward D et al (2003) Neogene extension and volcanism in the Kunlun Fault Zone, northern Tibet: new constraints on the age of the Kunlun fault. Tectonics 22(5):1052. doi:10.1029/2002TC001428

Jolivet L, Tamaki K, Fournier M (1994) Japan Sea, opening history and mechanism: a synthesis. J Geophys Res 99(B11):22237–22259

Jolley DW, Clarke B, Kelley S (2002) Paleogene time scale miscalibration: evidence from the dating of North Atlantic igneous province. Geology 30(1): 7–10

Kaiho K, Kajiwara Y, Nakano T et al (2001) End-Permian catastrophe by a bolide impact: Evidence of a gigantic release of sulfur from the mantle. Geology 29(9): 815–818

Kamo SL, Czamanske GK, Amelin Y et al (2000) U–Pb zircon and baddeleyite and U–Th–Pb perovskite ages for Siberian flood volcanism, Maimecha-Kotuy area, Siberia. J Goldshmit Conf Abs 5(2):569

Kamo SL, Czamanske GK, Krogh TE (1996) A minimum U–Pb age for Siberian flood-basalt volcanism. Geochim Cosmochim Acta 60(18):3505–3511

Kampunzu AB, Caron J-P, Lubala RT (1986) The East African rift, magma genesis and asthenospheric dynamics. Episodes 9:211–216

Kaneoka L (1980) Rare gas isotopes and mass fractionation: An indicator of gas transport into or from magma. Earth Planet Sci Lett 48:284–292

Kashik SA, Lomonosova TK (2006) Cenozoic sediments of underwater Akademicheskii Ridge in Lake Baikal. Lithol Miner Resour (4):303–316

Keller G, Li L, MacLeod N (1995) The Cretaceous/Tertiary boundary stratotype section at El Kef, Tunisia: how catastrophic was the mass extinction? Palaeogeogr, Palaeoclymatol, Palaeoecol 119:221–254

Kemkin IV (2006) Geodynamical evolution of Sikhote Alin and Sea of Japan region. Nauka, Moscow (in Russian)

Kennett JP, McBirney AR, Thunell RC (1977) Episodes of Cenozoic volcanism in the circum-Pacific region. J Volcanol Geotherm Res 2(2):145–163

Kepezhinskas VV (1979) Cenozoic alkaline basaltoids of Mongolia and their deep inclusions. Nauka, Moscow (in Russian)

Kerrich R, Polat A, Wyman D et al (1999) Trace element systematics of Mg, to Fe-tholeiitic basalt suites of the Superior Province: implications for Archean mantle reservoirs and greenstone belt genesis. Lithos 46:163–187

Khain EV, Bibikova EV, Kröner A et al (2002) The most ancient ophiolite of the Central Asian fold belt: U–Pb and Pb–Pb zircon ages for the Dunzhugur Complex, Eastern Sayan, Siberia, and geodynamic implications. Earth Planet Sci Lett 199:311–325

Khanchuk AI (2000) Paleogeodynamic analysis of ore deposit formation in the Far East of Russia. Ore deposits of continental margins. Dalnauka, Vladivostok, pp 5–34 (in Russian)

Khanchuk AI (ed) (2006) Geodynamics, magmatism and metallogeny of the Russian East. Book 1. Dalnauka, Vladivostok (in Russian)

Khiltova VY, Berkovskii AN, Kozakov IK (2003) Major elements in structure of the Siberian platform basement: geological-geophysical, geochronological isotopic-geochemical data. In: Karyakin YuV (ed) Tectonics and geodynamics of continental lithosphere, vol 2. GEOS, Moscow, pp 276–279 (in Russian)

Khomentovskii VV, Karlova GA (2005) The foot of the Tommotian—the lower boundary of the Cambrian in Siberia. Stratigr Geol Correl 13(1):26–40 (in Russian)

Kieffer B, Arndt N, Lapierre H et al (2004) Flood and shield basalts from Ethiopia: magmas from African Superswell. J Petrol 45(4):793–834

Kinoshita N, Yokoyama A, Nakanishi T (2003) Half-life of samarium–147. J Nucl Radiochem Sci 4:5–7

Kirillova GL (2000) Cretaceous environmental changes of east Russia: sedimentation, geodynamics, biodiversity, climate. Dal'nauka, Vladivostok (in Russian)

Kirillova GL (2002) Structure of Jurassic accretion prism in Priamurye: aspects of nonlinear geodynamics. Dokl Acad Nauk 380(4):515–518 (in Russian)

Kleine T, Münker C, Mezger K et al (2002) Rapid accretion and early core formation on asteroids and the terrestrial planets from Hf–W chronometry. Nature 418:952–955

Knight KB, Renne PR, Halkett A et al (2003) $^{40}Ar/^{39}Ar$ dating of the Radjahmundry Traps, Eastern India and their relationships to the Deccan Traps. Earth Planet Sci Lett 208:85–99

Kober B (1987) Single-zircon evaporation combined with Pb-emitter-bedding for $^{207}Pb/^{206}Pb$-age investigations using thermal ion mass spectrometry, and implications to zirconology. Contrib Mineral Petrol 96:63–71

Koeberl C, Gilmour I, Reimold WU et al (2002) End-Permian catastrophe by a bolide impact: Evidence of a gigantic release of sulfur from the mantle: comment and reply. Geology 30:855–856

Koeberl C, Shirey SB (1997) Re–Os isotope systematics as a diagnostic tool for the study of impact craters and distal ejecta. Paleogeogr, Paleoclimatol, Paleoecol 132:25–46

Kojima S, Kemkin IV, Kametaka M et al (2000) A correlation of accretionary complexes of southern Sikhote-Alin of Russia and Inner Zone of Southwest Japan. Chin Sci J 4:175–185

Konev AA, Chernenko AI, Fefelov NN et al (1975) Potassium–argon age of nepheline rocks in Pribaikal. Geol Geophys (4):141–146 (in Russian)

Konstantinovskaya EA (2002) Continental crust accretion mechanism: West Kamchatka case history. Geotectonics 5:59–78 (in Russian)

Kontorovich AE, Varlamov AI, Grazhdankin DV et al (2008) A section of the Vendian in the east of west Siberian plate. Geol Geophys 49(12):1238–1247 (in Russian)

Kopylova MG, O'Reilly SY, Genshaft YS (1995) Thermal state of the lithosphere beneath Central Mongolia: evidence from deep-seated xenoliths from Shavaryn-Saram volcanic centre in the Taryat depression, Hangai, Mongolia. Lithos 36:243–255

Korina NA (1982) Upland of Hangay. Geomorphology of the Mongolian national republic. In: Florensov NA, Korzhuev SS (eds) Geomorphology of Mongolian people republic. Nauka, Moscow, pp 87–108 (in Russian)

Korzhinskii DS (1947) Archean bi-metasomatic phlogopite and lazurite deposits in Pribaikal. Publisher of Academy of Sciences of the USSSR, Moscow

Kostitsyn YA (2002) Origin of rare-metal granites: isotope-geochemical approach. Abstract of doctor dissertation. Institute of Geochemistry RAS, Moscow (in Russian)

Kostitsyn YA, Kremenetsky AA (1995) Age of final magmatic stage of the Eldjurtinskii granite. Geochem Int 7:925–932

Kostitsyn YA, Zhuravlev DZ (1987) Analysis of errors and optimization of the isotope dilution method. Geochemistry (7):1024–1035 (in Russian)

Kostoyanov AI, Pushkarev YD (1996) Mass spectrometric determination of Os isotope composition by means of registration of electronegative ions OsO^{3-}. Fact Lab 64:24–28 (in Russian)

Kotlyar IN, Zhulanova IL, Rusakova TB, Gagieva AM (2001) Isotope systems in magmatic and metamorphic rock assemblages of Northeastern Russia. NEISRI FEB RAS, Magadan

Kotlyar IN, Zhulanova IL, Rusakova TB et al (2001) Isotope systems in magmatic and metamorphic rock assemblages of North-Eastern Russia. NEISRI FEB RAS, Magadan (in Russian)

Kotov AB (2003) Boundary conditions in geodynamic models of the Aldan shield crust formation. Doctor of sciences dissertation. St Petersburg (in Russian)

Kotov AB, Kozakov IK, Bibikova EV et al (1995) Stages of continental crust formation in the central part of the Aldan granulite–gneiss region: U–Pb and Sm–Nd isotopic data on granitoids. Petrology 3(6):622–631 (in Russian)

Kotov AB, Sal'nikova EB, Reznitskii LZ (1997) About the age of metamorphism in the Sludyanka complex (Southern Pribaikalye): results of U–Pb geochronological research of granitoids. Petrology 5(4):380–393 (in Russian)

Kovalenko DV, Yarmolyuk VV, Solovyev AV (1997) Parameters of spatial migration of volcanic centers in South Hangay hotspot based on paleomagnetic data. Geotectonics 3:66–73 (in Russian)

Kravchinsky VA, Konstantinov KM, Courtillot V et al (2002) Paleomagnetism of East Siberian traps and kimberlites: two new poles and paleogeographic reconstractions at about 360 and 250 Ma. Geophys J Int 148:1–33

Krough TE (1973) A low-contamination method for hydrothermal decomposition of zircon and extraction of U and Pb for isotopic age determination. Geochim Cosmochim Acta 37:485–494

Krough TE (1982) Improved accuracy of U–Pb zircon ages by the creation of more concordant systems using an air abrasion technique. Geochim. Cosmochim Acta 46:637–649

Krummenacher D (1970) Isotopic composition of argon in modern surface volcanic rocks. Earth Planet Sci Lett 8:109–117

Kuski TM, Li J-H, Tucker RD (2001) The archean dongwanzi ophiolite complex, North China craton: 2.505-billion-year-old oceanic crust and mantle. Science 292:1142–1145

Kuz'michev AB (2000) Tectonic history of the Tuva–Mongolian massif: early Baikalian, late Baikalian, and early Caledonian stages. Probel–2000, Moscow (in Russian)

Landing E (1994) Precambrian–Cambrian boundary global stratotype ratified and a new perspective of Cambrian time. Geology 22(2):179–182

Larin AM, Kotov AB, Sal'nikova EB et al (2003) Tectonic evolution of the central part of the Dzhugdzhur–Stanovoi fold region: results of U–Pb geochronological and isotope-geochemical (Nd, Sr, Pb) studies. Isotope geochronology in solution of geodynamic and ore genesis problems. 2nd Russian conference on isotope geology. St. Petersburg, pp 253–257 (in Russian)

Laughlin AW, Aldrich MJ, Shafiqullah M et al (1996) Tectonic implications of the age, composition and orientation of lamprophyre dikes, Navajo volcanic field, Arizona. Earth Planet Sci Lett 76:361–374

Laverov NP (ed) (2005) Modern and Holocene volcanism in Russia. Nauka, Moscow

Leat PT, Thompson RN, Morrison MA et al (1988) Compositionally-diverse Miocene–recent rift-related magmatism in northwest Colorado: partial melting and mixing of mafic magmas from 3 different asthenospheric and lithospheric mantle sources. J Petrol, Special Lithospheric Issue:351–377

Le Bas MJ (1989) Nephelinitic and basanitic rocks. J Petrol 30:1299–1312

Le Bas MJ, Streckeisen AL (1991) The IUGS systematics of igneous rocks. J Geol Soc London 148:825–833

Lee DC, Halliday AN (1995) Hafnium–tungsten chronometry and the timing of terrestrial core formation. Nature 378:771–774

Lee T-Y, Lawver LA (1995) Cenozoic plate reconstruction of Southeast Asia. Tectonophysics 251:85–138

Lelikov EP (2001) Geology of phosphorites on the sea of Japan bottom. Dalnauka, Vladivostok (in Russian)

Leven EY (2004) Fusulinides and the Permian scale of Tethys. Stratigr Geol Correl 12(2):33–47 (in Russian)

Libby WF, Anderson EC, Arnold JR (1949) Age determination by Radiocarbon content: world-wide assay of natural radiocarbon. Science 109:227–228

Lightfoot PC, Hawkesworth CJ, Sethna SF (1987) Petrogenesis of rhyolites and trachytes from the Deccan trap: Sr, Nd and Pb isotope and trace element evidence. Contrib Mineral Petrol 95:44–54

Lindner M, Leich DA, Russ GP et al (1989) Direct determination of the half-life of ^{187}Re. Geochim Cosmochim Acta 53:1597–1606

Lipman PW (1969) Alkalic and tholeiitic basaltic volcanism related to the Rio Grande depression. Geol Soc Am Bull 80(7):1343–1354

Lipman PW (2007) Incremental assembly and prolonged consolidation of Cordilleran magma chambers: evidence from the Southern Rocky Mountain volcanic field. Geosphere 3:42–70

Lipman PW, Mehnert HH (1975) Late Cenozoic basaltic volcanism and development of thc Rio Grande depression in the southern Rocky Mountains. Geol Soc Am Mem 144:119–154

Lipman PW, Doe BR, Hedge CE et al (1978) Petrologic evolution of the San Juan volcanic field, southwestern Colorado: Pb and Sr isotope evidence. Geol Soc Am Bull 89:59–82

Lipman PW, Logatchev NA, Zorin YA et al (1989) Intracontinental rift comparisons: Baikal and Rio Grande Rift systems. Am Geophys Union EOS 70(19):578–579, 586–588

Lister GS, Baldwin SL (1996) Modeling the effect of arbitrary P-T-t histories on argon diffusion in minerals using the MacArgon program for Apple Macintosh. Tectonophysics 253: 83–109

Lithgow-Bertelloni C, Guynn JH (2004) Origin of the lithospheric stress field. J Geophys Res 109:B01408. doi:10.129/2003JB002467

Lithgow-Bertelloni C, Richards MA (1998) The dynamics of Cenozoic and Mesozoic plate motions. Rev Geophys 36:27–78

Liu J, Taniguchi H (2001) Active volcanoes in China. Northeast Asian Stud 6:173–189

Lo C-H, Chung S-L, Lee T-Y et al (2002) Age of the Emeishan flood magmatism and relations to Permian–Triassic boundary events. Earth Planet Sci Lett 198:449–458

Lo C-H, Lee JKW, Onstott TC (2000) Argon release mechanisms of biotite in vacuo and the role of short-circuit diffusion and recoil. Chem Geol 165:135–166

Logatchev NA (1977) Volcanogenic and sedimentary formations in rift zones of East Africa. Nauka, Moscow (in Russian)

Logatchev NA (2001) About a historical core of the Baikal rift zone. Dokl Acad Nauk 376(4):510–513 (in Russian)

Logatchev NA (2003) History and geodynamics of the Baikal rift. Russ Geol Geophys 44(5):391–406

Logatchev NA, Brandt IS, Rasskazov SV et al (2002) K–Ar dating of Paleocene weathering crust in Pribaikalye. Dokl Acad Nauk 385:797–799 (in Russian)

Lovera OM, Richter FM, Harrison TM (1989) The $^{40}Ar/^{39}Ar$ thermochemistry for slowly cooled samples having a distribution of diffusion domain sizes. J Geophys Res 94:17917–17935

Lovera OM, Richter FM, Harrison TM (1991) Diffusion domains determined by argon 39 released during step heating. J Geophys Res 96:2057–2069

Luchitskaya MV, Hourigan J, Bondarenko GE et al (2003) New data of SHRIMP U–Pb study of zircons from granites of the Pribrezhno-Taigonos and East-Taigonos belts, southern part of the Taigonos peninsula. Dokl Acad Nauk 389:786–789 (in Russian)

Ludwig KR (1990) Isoplot: a plotting and regression program for radiogenic—isotope data for IBM–PC compatible computers, version 2.01. USGS open-file report, pp 88–557

Ludwig KR (1999) User's manual for ISOPLOT/Ex version 2.05. A geochronological toolkit for Microsoft Excel. Berkley Geochronology Center Spec Publ 1a:1–48

Ludwig KR (2000) Decay constant errors in U–Pb Concordia intercept age. Chem Geol 166:315–318

Luedke RG, Smith RL (1983) Map showing distribution and composition of late Cenozoic volcanic centers in the Western United States. US Geological Survey

Lugmair GW, Galer SJG (1992) Age and isotopic relationships among the angrites Lewis Cliff 86010 and Angra dos Reis. Geochim Cosmochim Acta 56:1673–1694

Lugmair GW, Marti K (1978) Lunar initial $^{143}Nd/^{144}Nd$: Differential evolution of the lunar crust and mantle. Earth Planet Sci Lett 39:349–357

Lugmair GW, Shukolyukov A (1998) Early solar system timescales according to 53Mn–53Cr systematics. Geochim Cosmochim Acta 62:2863–2886

Lykov AV (1964) The theory of heat conductivity. Higher school, Moscow (in Russian)

Lynch MA (1999) Linear ridge groups: evidence for tensional cracking in the Pacific plate. J Geophys Res 104:29321–29333

Maeda J (1990) Opening of the Kuril basin deduced from the magmatic history of Central Hokkaido, North Japan. Tectonophysics 174:235–255

Madelung E (1961) Die mathematischen hilfsmittel des physikers. State Publisher on physico-mathematical literature, Moscow (in Russian, translated from German)

Mahoney JJ, Duncan RA, Khan W et al (2002) Cretaceous volcanic rocks of the South Tethyan suture zone, Pakistan: implications for the Réunion hotspot and Deccan Traps. Earth Planet Sci Lett 203:295–310

Makagon VM, Lepin VS, Brandt SB (2000) Rubidium–strontium dating of rare metal pegmatites in the Vishnyakovskii deposit (East Sayan). Geol Geophys 41(12):1783–1789 (in Russian)

Maksimov SO, Sakhno VG (2008) Geochronology of basaltic volcanism in the Shufan plateau (Primorye). Dokl Acad Nauk 422(3):359–264 (in Russian)

Malkovets VG, Litasov YuD, Travin AV et al (2003) Volcanic pipes as clues to upper mantle petrogenesis: Mesozoic Ar/Ar dating of the Minusinsk basalts, South Siberia. Int Geol Rev 45:133–142

Markey R, Stein H, Morgan J (1998) Highly precise Re–Os dating for molybdenite using alkaline fusion and NTIMS. Talanta 45:935–946

Martin H (1994) The Archean gray gneisses and genesis of continental crust. In: Condie K (ed) Archean crustal evolution. Elsevier, Amsterdam, 205–259

Martin H (1999) Adakitic magmas: modern analogues of Archean granitoids. Lithos 46:411–429

Martins JB, Terranova ML, Moreira Correa M (1992) Half-life for alpha-decay of ^{147}Sm. II Nuovo Cimento 105:1621–1627

Martynov YA (1999) Geochemistry of basalts of continental margins and mature island arcs (case study of Northwestern Pacific). Dalnauka, Vladivostok (in Russian)

Martynov YA, Chashchin AA, Rasskazov SV et al (2002) Late Miocene–Pliocene basaltic volcanism in the south of Far East of Russia as indicator of heterogeneity of lithospheric mantle in an ocean–continent transitional zone. Petrology 10(2):189–209 (in Russian)

Martynov YA, Kovalenko SV, Rasskazov SV et al (2001) Geochemistry and problems of metallogeny of Cenozoic post-subduction calc-alkaline volcanics from southwestern Primorye. Ore deposits in continental margins. In: Khanchuk AI (ed) Dalnauka, Vladivostok 2(1):5–21 (in Russian)

Maruyama S, Liou JG (1998) Initiation of ultrahigh-pressure metamorphism and its significance on the Proterozoic–Paleozoic boundary. Isl Arc 7:6–35

Maruyama S, Seno T (1986) Orogeny and relative plate motions: example of the Japanese islands. Tectonophysics 127:305–329

Maruyama S, Isozaki Y, Mimura G et al (1997) Paleogeographic maps of the Japanese islands: plate tectonic synthesis from 750 Ma to the present. Isl Arc 6:121–142

Matheney RK, Brookins DG, Wallin ET et al (1990) Incompletely reset Rb–Sr systems from a Cambrian red-rock granophyre terrane, Florida Mountains, New Mexico Chem Geol 86:29–47

Matsumoto A, Kobayashi T (1995) K–Ar age determination of late Quaternary volcanic rocks using the "mass fractionation correction procedure": application to the Younger Ontake Volcano, central Japan. Chem Geol 125:123–135

Matsuura H (1989) Radiometric ages of Late Cretaceous to Paleogene igneous rocks in central San'in region, Southwest Japan. Bull Geol Surv Japan 40:479–495

McArthur JM, Howarth RJ, Bailey TR (2001) Strontium isotope stratigraphy: LOWESS version 3: best fit to the marine Sr-isotope curve for 0–509 Ma and accompanying look-up table for deriving numerical age. J Geol 109:155–170

McCormack FG, Hogg AG, Blackwell PG (2004) SHCal04 Southern Hemisphere Calibration 0–11 cal Kyr BP. Radiocarbon 46(3):1087–1092

McDougall I, Harrison MT (1988) Geochronology and termochronology by ^{40}Ar/^{39}Ar method. New York–Oxford: Oxford University Press

McDougall I, Harrison TM (1999) Geochronology and thermochronology by the $^{40}Ar/^{39}Ar$ method, 2nd edn. Oxford University Press, New York

Menzies M, Gallagher K, Yelland A et al (1997) Volcanic and nonvolcanic rifted margins of the Red Sea and Gulf of Aden: crustal cooling and margin evolution in Yemen. Geochim Cosmochim Acta 61(12):2511–2527

Menzies MA, Kyle PR, Jones M et al (1991) Enriched and depleted source components for tholeiitic and alkaline lavas from Zuni-Bandera, New Mexico: inferences about interplate processes and stratified lithosphere. J Geophys Res 96B:13645–13671

Merrihue CM, Turner G (1966) Potassium–argon dating by activation with fast neutrons. J Geophys Res 71:2852–2857

Milanovskii EE (1995) Pulsations of the Earth. Geotectonics (5):3–24 (in Russian)

Miller DM, Goldstein SL, Langmuir CH (1994) Cerium/lead and lead isotope ratios in arc magmas and the enrichment of lead in the continents. Nature 368:514–519

Min K, Mundil R, Renne PR et al (2000) A test for systematic errors in $^{40}Ar/^{39}Ar$ geochronology through comparison with U/Pb analysis of a 1.1-Ga rhyolite. Geochim Cosmochim Acta 64:73–98

Minster J-F, Birk J-L, Allègre CJ (1982) Absolute age of formation of chondrites studied by the $^{87}Rb–^{87}Sr$ method. Nature 300:414–419

Mitchel JG (1968) The argon–40/argon–39 method for potassium–argon age determination. Geochim Cosmochim Acta 32:781–790

Mo X, Hou Z, Niu Y et al (2007) Mantle contribution to crustal thickening during continental collision: evidence from Cenozoic igneous rocks in southern Tibet. Lithos 96:225–242

Mohr P, Zanettin B (1988) The Ethiopian flood basalt province. In: MacDougal JD (ed) Continental flood basalts. Kluwer Academic, Dohert, pp 63–110

Molnar P (2005) Mio-Pliocene growth of the Tibetan Plateau and evolution of East Asian climate. Paleontolo Electronica 8(1, 2A):23

Molnar P, Tapponnier P (1975) Cenozoic tectonics of Asia: effects of a continental collision. Science 189(4201):419–426

Montero P, Bea F (1998) Accurate determination of $^{87}Rb/^{86}Sr$ and $^{147}Sm/^{144}Nd$ ratios by inductively-coupled plasma mass spectrometry in isotope geoscience: an alternative to isotope dilution analysis. Anal Chem Acta 358:227–233

Moorbath S, Whitehouse MJ, Kamber BS (1997) Extreme Nd-isotope heterogeneity in the early Archean: fact or fiction? Case histories from northern Canada and West Greenland. Chem eol 197:213–231

Mordvinova VV, Deshamp A, Dugarmaa T et al (2007) Research of velocity structure of the lithosphere on Mongolia–Baikal transect 2003 on exchange SV-waves. Phys Earth (2):21–32 (in Russian)

Morgun EG, Kovda IV, Ryskov YG et al (2008) Prospects and problems of using in the method of geochemistry of stable carbon isotopes in soil studies. Soil Stud (3):299–310 (in Russian)

Mossakovskii AA, Ruzhentsev SV, Samygin SG et al (1993) Central Asian fold belt: geodynamic evolution and history of formation. Geotectonics 6:3–33 (in Russian)

Mueller PA, Wooden JL, Nutman AP (1992) 3.96 Ga zircons from an Archean quartzite, Beartooth Mountains, Montana. Geology 20:327–330

Mueller PA, Wooden JL, Nutman AP et al (1998) Early Archean crust in the northern Wyoming province: evidence from U–Pb age of detrital zircons. Precambrian Res 91:295–307

Müller PH, Neumann P, Storm R (1979) Tafeln der mathematischen Statistik. VEB Fachbuchverlag Leipzig

Nakamuro K (1977) Volcanoes as possible indicators of tectonic stress orientation-principle and proposal. J Volcanol Geotherm Res 2(1):1–16

Narbonne GM, Kaufman AJ, Knoll AH (1994) Integrated biostratigraphy and chemostratigraphy of the upper Windermere Supergroup, MacKenzie Mountains, northwestern Canada: implications for Neoproterozoic correlations and the early evolution of animals. Geol Soc Am Bull 106:1281–1292

Natal'in B (1993) History and modes of Mesozoic accretion in southeastern Russia. Isl Arc 2:15–34

Natal'in BA, Faure G, Mone P et al (1994) The Anui metamorphic dome (Sikhote-Alin) and its significance for Mesozoic geodynamic evolution of East Asia. Geol Pac Ocean (6):2–25 (in Russian)

Neumann W, Huster E (1976) Discussion of the ^{87}Rb half-life determined by absolute counting. Earth Planet Sci Lett 33:277–288

Neymark LA (1988) Isotopes of ore lead and some questions of genesis of deposits. In: Shukolyukov YA (ed) Isotope geochemistry of processes of ore formation. Nauka, Moscow, pp 99–116 (in Russian)

Neymark LA, Larin AM, Rytzk EYu et al (1993) U-Pb-geochronologic and Pb-isotope evidence on the Hercynian stage of ore mineralization within the Proterozoic framing of the south of the Siberian platform (northeastern Pribaikal). Dokl Acad Nauk 333(6):765–767 (in Russian)

Neymark LA, Rytzk EY, Ovchinnikova GV et al (1995) Lead isotopes in the gold-ore deposits of East Sayan (Russia). Geol Ore Depos 37(3):237–249 (in Russian)

Nikishin AM, Ziegler PA, Abbott D et al (2002) Permo–Triassic intraplate magmatism and rifting in Eurasia: implications for mantle plumes and mantle dynamics. Tectonophysics 351:3–39

Nittler LR (2003) Presolar stardust in meteorites: recent advances and scientific frontiers. Earth Planet Sci Lett 209:259–273

Nivin VA, Ikorskii SV, Kamenskii IL (2001) Isotope-gas (He, Ar) indicators for sources of substance in Paleozoic alkaline complexes of the Kola province and related ore deposits. In: Vladykin NV (ed) Alkaline magmatism and problems of mantle sources. Irkutsk Technical University, Irkutsk, pp 129–142 (in Russian)

Nutman AP, McGregor VR, Friend CRL et al (1996) The Itsaq Gneiss Complex of southern West Greenland: the world' most extensive record of early crustal evolution (3900–3600 Ma). Precambrian Res 78:1–39

Nutman AP, Mojzsis SJ, Friend CRL (1997) Recognition of $\geq$ 3850 Ma water-lain sediments in west Greenland and their significance for the early Archean Earth. Geochim Cosmochim Acta 61:2475–2484

O'Connor JM, Stoffers P, van den Bogaard P et al (1999) First seamount age evidence for significantly slower African plate motion since 19 to 30 Ma. Earth Planet Sci Lett 171(4):575–589

Oding S (1994) Echelle des temps geologiques. C R Acad Sci Paris 318(2):59–71

Ohki J, Shuto K, Kagami H (1994a) Middle Miocene bimodal magmatism by asthenospheric upwelling: Sr and Nd isotopic evidence from the back-arc region of the Northeast Japan Arc. Geochem J 28(6):473–487

Ohki J, Watanabe N, Shuto K et al (1994b) Shifting of the volcanic fronts during Early to Late Miocene in the Northeast Japan Arc. Isl Arc 2:87–93

Okamoto K, Shinjoe H, Katayama I et al (2004) SHRIMP U–Pb zircon dating of quartz-bearing eclogite from the Sanbagawa Belt, southwest Japan: implications for metamorphic evolution of subducted protolith. Terra Nova 16: 75–80

Okamura S, Martynov YA, Furuyama K et al (1998) K–Ar ages of the basaltic rocks from Far East Russia: constraints on the tectono-magmatism associated with the Japan Sea opening. Isl Arc 7:271–282

Omar GI, Steckler MS (1995) Fission track evidence of the initial rifting of the Red Sea: two pulses, no propagation. Science 270:1341–1344

Otofuji Y-I, Itaya T, Wang HC et al (1995a) Paleomagnetism and K–Ar dating of Pleistocene volcanic rocks along the Altyn Tagh fault, northern border of Tibet. Geophys J Int 120:367–374

Otofuji Y-I, Kambara A, Matsuda T et al (1994) Counterclockwise rotation of Northeast Japan: paleomagnetic evidence for regional extent and timing of rotation. Earth Planet Sci Letters 121:503–518

Otofuji Y-I, Matsuda T, Enami R et al (2003) Late Cretaceous paleomagnetic results from Sikhote Alin, far eastern Russia: tectonic implications for the eastern of the Mongolia Block. Geophys J Int 152:202–214

Otofuji Y-I, Matsuda T, Itaya T et al (1995b) Late Cretaceous to early Paleogene paleomagnetic results from Sikhote-Alin, Far Eastern of Russia: implications for deformation of East Asia. Earth Planet Sci Lett 130:95–108

Ovchinnikova GV, Vasil'eva IM, Semikhatov MA et al (2000a) Opportunities of Pb–Pb dating of carbonaceous rocks with open U–Pb systems: the Minyar suit of the stratotype for the upper Riphean, Southern Ural. Stratigr Geol Correl 8(6):3–19 (in Russian)

Ovchinnikova GV, Vasil'eva IM, Semikhatov MA et al (2000b) U–Pb systematics of Proterozoic carbonaceous rocks: the Inzer suit of the stratotype for the upper Riphean (Southern Ural). Stratigr Geol Correl 6(4):20–31 (in Russian)

Ozima M, Podosek FA (1999) Formation age of Earth from $^{129}I/^{127}I$ and $^{244}Pu/^{238}U$ systematics and the missing Xe. J Geophys Res 104:25493–25499

Ozima M, Podosek FA (2002) Noble gas geochemistry, 2nd edn. Cambridge University Press, Cambridge

Palandzhyan SA (2002) West-Koryakian belt of dike and hipabyssal rocks as indicator of extension and destruction of the forearc of the Okhotsk–Chukotka volcanic belt in the Late Senonian–Paleocene. Dokl Acad Nauk 385:800–804 (in Russian)

Papanastassiou DA, Wasserburg GJ (1969) Initial strontium isotopic abundances and the resolution of small time differences in the formation of planetary objects. Earth Planet Sci Lett 5:361–376

Parfenov LM, Berzin NA, Khanchuk AI et al (2003) A model for formation of orogenic belts in Central and Northeast Asia. Geol Pac Ocean 22(6):7–41 (in Russian)

Parfenov LM, Kuzmin MI (ed) (2001) Tectonics, geodynamics and metallogeny of the Sakha Republic (Yakutia). Maik "Nauka/Interperiodica", Moscow, p 571

Pasteels P, Villeneuve M, De Paepe P et al (1989) Timing of the volcanism of the southern Kivu province: implications for the evolution of the western branch of the East African rift system. Earth Planet Sci Lett 94(3/4):353–363

Patchett PJ, Chase CG (2002) Role of transform continental margins in major crustal growth episodes. Geology 30(1):39–42

Patriat P, Achache J (1984) India–Eurasia collision chronology has implications for crustal shortening and driving mechanism of plates. Nature 311:615–617

Patterson CC (1956) Age of meteorites and the Earth. Geochim Cosmochim Acta 10:230–237

Patterson C, Smith AB (1987) Is the periodicity of extinctions a taxonomic artefact? Nature 330:248–251

Pavlenkova GA, Priestley K, Cipar J (2002) 2D model of the crust and uppermost mantle along rift profile, Siberian craton. Tectonophysics 355:171–186

Penfield GT, Camargo AZ (1981) Definition of a major igneous zone in the central Yucatán platform with aeromagnetics and gravity. Society of Exploration Geophysicists, 51st annual international meeting. Society of Exploration Geophysicists, Los Angeles, p 37

Pentel'kov VG, Voronovskii SN (1977) Absolute ages of carbonatites in Mbalizi, Tanzania and their correlation with ages of other carbonatites from the Rukwa–Malawi rift zone. Dokl Acad Nauk USSSR 235(5):1136–1139 (in Russian)

Perchuk LL, Safonov OG, Yapaskurt VO et al (2002) Crystal–melt equilibrium involving potassium-bearing clinopyroxene as indicator of mantle-derived ultrahigh-potassic liquids: an analytical review. Lithos 60:89–111

Perepelov AB, Ivanov AV, Macintosh WS et al (2003) Potassium alkaline magmatism of Western Kamchatka: Late Eocene–Early Oligocene episode of inversion of geodynamic regimes in evolution of an arc system. In: Kozakov IK, Kotov AB (ed) Isotopic geochronology in solutions of problems in geodynamics and ore genesis. II-nd Russian conference on isotopic geochronology, St Petersburg, pp 396–400 (in Russian)

Petterson MG, Windley BF (1991) Changing source regions of magmas and crustal growth in the Trans-Himalayas: evidence from the Chalt volcanics and Kohistan batholith, Kohistan, northern Pakistan. Earth Planet Sci Lett 102: 326–341

Pidgeon RT, Wilde SA (1998) The interpretation of complex zircon U–Pb systems in Archean granitoids and gneisses from the Jack Hills, Narryer Gneiss Terrane, Western Australia. Precambrian Res 91:309–332

Pin C, Bassin C (1992) Evaluation of a strontium-specific extraction chromatographic method for isotopic analysis in geological materials. Anal Chim Acta 269: 249–255

Pin C, Briot D, Bassin C et al (1994) Concomitant separation of strontium and samarium–neodymium for isotopic analysis in silicate samples, based on specific extraction chromatography. Anal Chim Acta 298:209–217

Pin C, Zalduegui JFS (1997) Sequential separation of light rare-earth elements, thorium and uranium by miniaturized extraction chromatography: application to isotopic analyses of silicate rocks. Anal Chim Acta 339:79–89

Pissarsky BI, Nambar B, Ariyadagva B (2003) Map of mineral waters in Mongolia. Scale 1:2500000. Ulaanbaatar

Pollat A, Kerrich R (2001) Magnesian andesites, Nb-enriched basalt-andesites, and adakites from late-Archean 2.7 Ga Wawa greenstone belts, Superior Province, Canada: implications for late Archean subduction zone petrogenesis processes. Contrib Mineral Petrol 141:36–52

Popov VK, Grebennikov AV (2001) New data on age of effusives of the Bogopolye Formation in Primorye. Geol Pac Ocean 20(3):47–54 (in Russian)

Popov VK, Grebennikov AV, Kovalenko SV, Kutub-Zade TK (2001) Petrological and geochemical characteristics of Cenozoic volcanic complexes in the Kraskino basin, Primorye. In: Mesozoic and Cenozoic magmatic and metamorphic formations of Far East. Proceedings of the 5th far east regional petrographic meeting. Khabarovsk-Geology, Khabarovsk, p 124 (in Russian)

Popov VK, Rasskazov SV, Chekryzhov IY et al (2005) Potassium–argon dates and geochemical signatures of Cenozoic trachybasalts and trachyandesites of Primorye. Geochemistry of magmatic rocks, Institute of Geochemistry RAS, Moscow. http://alkaline(2005).narod.ru/abstracts/Popov1.html (in Russian)

Pouclet A, Lee J-S, Vidal P et al (1995) Cretaceous to Cenozoic volcanism in South Korea and in the Sea of Japan: magmatic constraints on the opening of the back-arc basin. In: Smellie JL (ed) Volcanism associated with extension at consuming plate margins. Geol Soc Spec Publication 81:169–191

Qin C, Papazachos C, Papadimitriou E (2002) Velocity field for crustal deformation in China derived from seismic moment tensor summation of earthquakes. Tectonophysics 359:29–46

Quitté G, Birk JL, Allègre CJ (2000) ^{182}Hf–^{182}W systematics in eucrites: the puzzle of iron segregation in the early solar system. Earth Planet Sci Lett 184:83–94

Raaben ME (ed) (1969) The Tommotian and the problems of the lower Cambrian boundary. Nauka, Moscow (in Russian)

Rapp RP, Shimizu N, Norman MD (2003) Growth of early continental crust by partial melting of eclogite. Nature 425:605–609

Rasskazov SV (1982) Simultaneous changes of Cenozoic basalts compositions and stress field in the Udokan area. In: 10th conference of young researches on geology and geophysics of East Siberia abstracts, Institute of the Earth's Crust, Irkutsk, pp 19–21 (in Russian)

Rasskazov SV (1985) Basaltoids of Udokan (Baikal rift zone). Nauka , Siberian Branch, Novosibirsk (in Russian)

Rasskazov SV (1991) Volcanism of a hot spot and structure of the western part of the Baikal rift system. Geol Geophys 9:72–81 (in Russian)

Rasskazov SV (1993) Magmatism of the Baikal Rift System. Nauka, Siberian Publishing Firm, Novosibirsk (in Russian)

Rasskazov SV (1994a) Comparisons of volcanism and recent structures in the Yellowstone and Eastern Sayan hotspots. Geol Geophys 10:67–75 (in Russian)

Rasskazov SV (1994b) Magmatism related to the East Siberia rift system and the geodynamics. Bull Centres Rech Explor-Prod Elf Aquitaine 18(2):437–452

Rasskazov SV (1996) Volcanism and structure of the northeastern flank of the Baikal Rift System. Russ Geol Geophys 37(4):60–70 (in Russian)

Rasskazov SV (1999) A Middle Holocene stress change in the Udokan range volcanic zone, Eastern Siberia. Volcanol Seismol 21:261–267 (in Russian)

Rasskazov SV (2002) Rhythms of Late Cenozoic volcanism in southeastern part of the Eurasian plate. Atlas of temporal variations in natural, anthropogenic and social processes, Janus, Moscow 3:184–189 (in Russian)

Rasskazov SV, Chuvashova IS (2007) Structural indicators for sin-kinematical evolution of middle–late Cenozoic volcanism in Central Mongolia. Problems of present-day seismogeology and geodynamics in Central and East Asia. Institute of the Earth's crust, Irkutsk 2:90–97 (in Russian)

Rasskazov S, Taniguchi H (2006) Magmatic response to the Late Phanerozoic plate subduction beneath East Asia. Monograph Series No. 21. Tohoku University, Sendai

Rasskazov SV, Boven A, Andre L et al (1997) Evolution of magmatism in the northeastern Baikal Rift System. Petrology 5:115–136

Rasskazov SV, Boven A, Ivanov AI et al (2000a) A Middle Quaternary volcanic impulse in Olekma-Stanovoi mobile system: $^{40}Ar/^{39}Ar$ dating of volcanics from Tokinskii Stanovik. Geol Pac Ocean 19(4):19–28 (in Russian)

Rasskazov SV, Bowring SA, Harris N et al (1999a) Lithospheric domains in East Sayan: Pb, Sr, and Nd isotope constraints on sources of the Late Cenozoic magmatism in the northern segment of the Riphean Tuva–Mongolian massif. In: Logatchev NA (ed) Rifting in intracontinental setting: Baikal Rift System and other Continental Rifts, Irkutsk, pp 163–166

Rasskazov SV, Bowring SA, Housh T et al (2002a) Isotope systematics of Pb, Nd, and Sr in heterogeneous continental lithosphere above the convective mantle domain. Dokl Acad Nauk 387(4):519–523

Rasskazov SV, Brandt SB, Brandt IS et al (2005a) Radiogenic isotope geology in problems and examples. Academic Publishing House "GEO", Novosibirsk (in Russian)

Rasskazov S, Chebykin E, Chuvashova I et al (2008a) Isotope disequilibrium of ^{238}U-series in latest Pleistocene through Holocene volcanic rocks from Hangay, Central Mongolia and Eastern Sayan, Siberia. Geophys Res Abstr 10:04872 (SRef–ID:1607–7962/gra/EGU2008–A–04872)

Rasskazov SV, Chuvashova IS, Brandt IS et al (in press) Evolution of high- and moderate-K Cenozoic basaltic magmatism in Central Mongolia: relation to interaction between India and Asia and to rifting. Int J Earth Sci

Rasskazov SV, Chuvashova IS, Mordvinova VV et al (2007a) Role of the cratonic Lehman discontinuity in Cenozoic dynamics of the upper mantle in Central Asia: interpretation of models for seismic waive velocities in light of spatial–temporal evolution of volcanism. Regions of active tectonics in recent and ancient history of the Earth. In: Karyakin YV (ed) Proceedings of XL Tectonic Meeting. Moscow State University, Moscow, (2):126–129 (in Russian)

Rasskazov SV, Chuvashova IS, Yasnygina TA et al (2008b) Slab-derived and above-slab Late Cenozoic melts from convergent boundary zones of Asia and East Hangay, Central Mongolia. Izvestia of the Irkutsk State University. Ser Earth Sci 1:129–149 (in Russian)

Rasskazov SV, Demonterova EI, Ivanov AV et al (2007b) Evolution of late Cenozoic magmatism at a boundary of the Tuva–Mongolian massif (East Tuva). Institute of the Earth's crust SB RAS, Irkutsk (in Russian)

Rasskazov SV, Il'yasova AM, Konev AA et al (2007c) Geochemical evolution of the Zadoi alkaline ultrabasic massif of Cis-Sayan area, Southern Siberia. Geochem Int 45(1)3–18

Rasskazov SV, Ivanov VV, Khanchuk AI et al (2002b) Isotopic heterogeneity of lead in galena from ore deposits in the Primor'e region. Dokl Earth Sci 387A:1083–1087

Rasskazov SV, Kunk MJ, Luhr JF et al (1996) Episodes of eruptions and compositional variations of the Quaternary lavas in the Baikal Rift System (Ar–Ar and K–Ar dating of volcanism in the Dzhida River area). Geol Geophys 37:3–15 (in Russian)

Rasskazov SV, Logatchev NA, Brandt IS et al (2000b) Geochronology and geodynamics in the Late Cenozoic (South Siberia–South and East Asia). Nauka, Novosibirsk (in Russian)

Rasskazov SV, Logatchev NA, Brandt IS et al (2000c) Uplift of the Baikal rift system and change of vegetation in its flanks as inferred from variations of spores, pollen, and diatoms in sediments. Terra Nostra 9:148–163

Rasskazov SV, Logatchev NA, Ivanov AV et al (2003a) A magmatic episode in the Western Rift of East Africa (19–17 Ma). Russian Geol Geophys 44: 317–324

Rasskazov SV, Logatchev NA, Ivanov AV (1998) Correlation of Late Cenozoic tectonic and magmatic events between the Baikal rift system and the southeastern Eurasian plate. Geotectonics 32(4):25–40

Rasskazov SV, Logatchev NA, Kozhevnikov VM et al (2003b) Multistage dynamics of the upper mantle in East Asia: relationships between wandering volcanism and low-velocity anomalies. Dokl Earth Sci 390(4):492–496

Rasskazov SV, Luhr JF, Bowring SA et al (2003c) Late Cenozoic volcanism in the Baikal Rift System: evidence for formation of the Baikal and Khubsugul basins due to thermal impacts on the lithosphere and collision-derived tectonic stress. Proceedings of the international symposium speciation in ancient lakes, SIAL III Berliner paläobiologische abhandlungen 1(4), Freie Universität, Berlin, pp 33–48

Rasskazov SV, Lyamina NA, Chernyaeva GP et al (2007d) Cenozoic stratigraphy of the Vitim Plateau: Phenomenon of long rifting in the south of East Siberia, Academic Publishing House "GEO", Novosibirsk (in Russian)

Rasskazov SV, Maslovskaya MN, Saranina EV et al (1999b) Strontium isotope study of Late Cenozoic evolved volcanic rocks from the Rungwe province, Tanzania: Isochron ages of 16.1 and 463 Myr for phonolite eruptions and mantle–crust differentiation event. In: Logatchev NA (ed) Rifting in intracontinental setting: Baikal Rift System and other Continental Rifts, Irkutsk, pp 170–172

Rasskazov SV, Maslovskaya MN, Skopintsev VG et al (2003d) Late-Paleozoic subalkaline and rare-metallic granitoids in the south-eastern part of East Sayan (Geochemical characteristics and Rb–Sr isotope systematics). Geol Geophys 44(11):1133–1144 (in Russian)

Rasskazov SV, Mel'nikov OA, Rybin AV et al (2005b) Spatial change of deep sources of Cenozoic volcanic rocks in western coast of South Sakhalin. Geol Pac Ocean 24(2):10–32 (in Russian)

Rasskazov SV, Prikhodko VS, Ivanov AV et al (2003e) Dating of volcanic rocks of the Kizi Formation from Eastern Sikhote Alin in the stratotype area by the $^{40}Ar/^{39}Ar$ method: identification of anomalously low $^{40}Ar/^{36}Ar$ ratio in a contaminating fluid component. In: Kozakov IK, Kotov AB (eds) Isotope geochronology in solution of geodynamic and ore genesis problems. 2nd Russian conference on isotope geology. St. Petersburg, pp 396–400 (in Russian)

Rasskazov SV, Prikhodko VS, Saranina EV et al (2003f) Space-and-time variations of mantle and crustal components in the Late Cenozoic volcanic rocks from the Middle Amur basin, Southeastern Russia. Geol Pac Ocean 22(3):3–27 (in Russian)

Rasskazov SV, Saranina EV, Demonterova EI et al (2002c) Mantle components in late Cenozoic volcanics, East Sayan (from Pb, Sr, and Nd isotopes). Russ Geol Geophys 43(12):1014–1028

Rasskazov SV, Saranina EV, Logatchev NA et al (2002d) The DUPAL mantle anomaly in the Tuva–Mongolian massif and its paleogeodynamic implication. Dokl Earth Sci 382(1):44–48

Rasskazov SV, Saranina EV, Martynov YA et al (2003g) Evolution of Late Cenozoic magmatism at the active continental margin of Southern Primorye. Geol Pac Ocean 22(1):92–109 (in Russian)

Rasskazov SV, Simanenko VP, Malinovskii AI et al (2007e) Geochemical evolution of late Eocene–Oligocene magmatism of the Schmidt peninsula (northern Sakhalin). Russ Geol Geophys 48:247–256

Rasskazov SV, Skopintsev VG, Maslovskaya MN et al (2001) Rb–Sr isotopic systematics of granitoids from the Gargan and Oka zones in Eastern Sayan. In: Sizykh AI (ed) Geodynamic regimes of formation of Central-Asian fold belt. Intermit Engineering, Moscow, pp 106–136 (in Russian)

Rasskazov S, Taniguchi H, Goto A et al (2004a) Magmatic expression of plate subduction beneath East Asia in the Mesozoic through Cenozoic. Northeast Asian Stud 9:179–219

Rasskazov SV, Yasnygina TA, Fefelov NN et al (1999c) Geodynamics of the continental lithosphere: comparisons of magmatism in Rocky mountains, USA and East Sayan, Russia. In: Logatchev NA (ed) Rifting in intracontinental setting: Baikal Rift System and other Continental Rifts, Irkutsk, pp 173–179

Rasskazov SV, Yasnygina TA, Fefelov NN et al (2010) Geochemical evolution of the middle–late Cenozoic magmatism in the northern part of the Rio Grande Rift. Geol Pac Ocean 29(1):15–43 (in Russian)

Rasskazov SV, Yasnygina TA, Saranina EV et al (2004b) Middle–Late Cenozoic magmatism at a continental margin of the Sea of Japan basin, Southwestern Primorye: impulse melting of the mantle and crust. GeolPac Ocean 23:3–31 (in Russian)

Reddy SM, Potts GJ (1999) Constraining absolute deformation ages: the relationship between deformation mechanisms and isotope systematics. J Struct Geol 21: 1255–1265

Rees PM (2002) Land-plant diversity and the end-Permian mass extinction. Geology 30:827–830

Reichow MK, Saunders AD, White RV et al (2002) $^{40}Ar/^{39}Ar$ dates from the West Siberian Basin: Siberian flood basalt province doubled. Science 296: 1846–1849

Reichow MK, Pringle MS, Al'mukhamedov AI et al (2009) The timing and extent of the eruptions of the Siberian traps large igneous province: implications for the end-Permian environmental crisis. Earth Planet Sci Lett 277:9–20. doi:10.1016/j.epsl.2008.09.030

Reid DL, Cooper AF, Rex DC et al (1990) Timing of post-Karroo alkaline volcanism in southern Namibia. Geol Mag 127(5):427–433

Reimer PJ, Baillie MGL, Bard E (2004) IntCal04 terrestrial radiocarbon age calibration, 0–26 Cal Kyr BP. Radiocarbon 46(3):1029–1058

Renne PR (1995) Excess ^{40}Ar in biotite and hornblende from the Norilsk 1 intrusion, Siberia: implication for the age of Siberian Traps. Earth Planet Sci Lett 131:165–176

Renne PR, Basu AR (1991) Rapid eruption of the Siberian Traps flood basalts at the Permo–Triassic boundary. Science 253:176–179

Renne PR, Swisher CC, Deino AL et al (1998) Intercalibration of standards, absolute ages and uncertainties in $^{40}Ar/^{39}Ar$ dating. Chem Geol 145:117–152

Renne PR, Zichao Z, Richards MA et al (1995) Synchrony and causal relations between Permian–Triassic boundary crises and Siberian flood volcanism. Science 269:1413–1416

Repina LN, Rozanov AYu (eds) (1992) Cambrian of Siberia. Nauka, Novosibirsk (in Russian)

Richter FM, Lovera OM, Harrison T et al (1991) Tibetan tectonics from argon 40/argon 39 analysis of a single potassium-feldspar sample. Earth Planet Sci Lett 105:266–278

Ritsema J, van Heijst H (2000) New seismic model of the upper mantle beneath Africa. Geology 28(1):63–66

Roddick JC (1978) The application of isochron diagrams in ^{40}Ar–^{39}Ar dating: a discussion. Earth Planet Sci Lett 41:233–244

Roddick JC, Cliff RA, Rex DC (1980) The evolution of excess argon in Alpine biotites. Earth Planet Sci Lett 48:185–208

Roger F, Tapponnier P, Arnaud N et al (2000) An Eocene magmatic belt across central Tibet: mantle subduction triggered by the Indian collision? Terra Nova 12:102–108

Rogers N, Macdonald R, Fitton JG et al (2000) Two mantle plumes beneath the East African rift system: Sr, Nd and Pb isotope evidence from Kenya Rift basalts. Earth Planet Sci Lett 176:387–400

Rogozhina VA, Kozhevnikov VM (1979) Region of abnormal mantle under Baikal rift. Nauka, Novosibirsk (in Russian)

Rona PA, Richardson ES (1978) Early Cenozoic global plate reorganization. Earth Planet Sci Lett 40:1–11

Rosen OM, Turkina OM (2007) The oldest rock assemblages of the Siberian craton. In: van Kranendonk MJ, Smithies RH, Bennet VC (eds) Earth's oldest rocks, development in Precambrian geology (Condie KC, Series ed). 6:495–541. doi:10.1016/S0166–2635(07)15064–7

Rosen OM, Manakov AV, Zinchuk NN (2006) Siberian craton: origin and the diamond control. Scientific World, Moscow (in Russian)

Rozanov AY, Semikhatov MA, Sokolov BS et al (1997) The decision on a choice of stratotype for the Precambrian–Cambrian boundary: breakthrough or mistake? Stratigr Geol Correl 5(1):21–31 (in Russian)

Rozhdestvenskii VS (1988) Geological structure and tectonic development of the Schmidt peninsula. Geol Pac Ocean 3:62–71 (in Russian)

Rublev AG (1998) On isotope dating of sedimentary rocks. Stratigr Geol Correl 6(3):107–112 (in Russian)

Rudnick RL, Fountain DM (1995) Nature and composition of the continental crust: a lower crustal perspective. Rev Geophys 33(3):267–309

Rudnick VA, Sobotovich AV (1984) An early history of the Earth. Nedra, Moscow (in Russian)

Rumshinskii LZ (1976) Elements of theory of probability. Moscow, Nauka (in Russian)

Ryabov VV, Shevko AY, Gora MP (2001) Magmatic formations of the Norilsk area, vol 1. Petrology of traps. Nonaprel, Novosibirsk (in Russian)

Safonov OG, Perchuk LL, Litvin YA (2005) Equilibrium of potassium-bearing clinopyroxene with liquid as a model for barometry of deep associations. Russ Geol Geophys 46(12):1318–1334

Sakhno VG (2001) Late Mesozoic and Cenozoic continental volcanism of East Asia. Vladivostok, Dalnauka (in Russian)

Sakhno VG, Maksimov SO, Popov VK et al (2004) Leucite basanites and potassium shonkinites from the Uglovka basin (South Primorye). Dokl Acad Nauk 399:818–824 (in Russian)

Sakhno VG, Moiseenko VG (2001) Magnesian andesites in rift structures of Amur craton. Dokl Acad Nauk 380(1):103–108 (in Russian)Saltykovskii AY, Genshaft

US, Arakelyants MM (1984) Potassium basalts in the Cenozoic of Mongolia. Volcanol Seismol 5:60–73 (in Russian)

Sal'nikov BA (ed) (2004) State geological map of Russian Federation of scale 1: 200 000, 2nd edn. (Sakhalin series. Sheet N–54–XVII). St. Petersburg map factory VSEGEI (in Russian)

Sal'nikova EB, Kotov AB, Levitskii VI et al (2007) Age constraints of high-temperature metamorphic events in crystalline complexes of the Irkut block, the Sharyzhalgai ledge of the Siberian platform basement: results of the U–Pb single grain zircon dating. Stratigr Geol Correl 15(4):3–19.

Sambridge MS, Compston W (1994) Mixture modeling of multi-component data sets with application to ion probe zircon ages. Earth Planet Sci Lett 128(3):373–390

Samson SD, Alexander EC (1987) Calibration of the interlaboratory ^{40}Ar–^{39}Ar dating standard MMhb–1. Chem Geol 66:27–34

Samsonov AV, Chernyshev IV, Nutman AP et al (1996) Evolution of the Archean Aulian Gneiss Complex, Middle Dnepr gneiss–greenstone terrain, Ukrainian Shield: SHRIMP U–Pb zircon evidence. Precambrian Res 78:65–78

San'kov VA, Lukhnev AV, Radziminovich NA et al (2005) Quantitative estimation of present-day deformations of the earth's crust in the Mongolian block as inferred from data on GPS-geodesy and seismotectonics. Dokl Acad Nauk 403(5):685–688 (in Russian)

Sarda P, Staudacher T, Allègre CJ (1985) $^{40}Ar/^{36}Ar$ in MORB glasses: constraints on atmosphere and mantle evolution. Earth Planet Sci Lett 72(4):357–375

Sasada T, Hayagon H, Bell K et al (1997) Mantle-derived noble gases in carbonatites. Geochim Cosmochim Acta 61(19):4219–4228

Sato H (1994) The relationship between late Cenozoic tectonic events and stress field and basin development in northeast Japan. J Geophys Res 99(B11):22261–22274

Saylor BZ, Kaufman AJ, Grotzinger JP et al (1998) A composite reference section for terminal Proterozoic strata of southern Namibia. J Sediment Res 68(6):1223–1235

Schipanskii AA, Babarina II, Krylov KA et al (2001) The oldest ophiolites in Earth: late Archean suprasubduction complex of the Irinogorskaya structure in North Karelian greenstoun belt. Dokl Earth Sci 377(3):376–380 (in Russian)

Schoenberg R, Nägler TF, Kramers JD (2000) Precise Os isotope ratio and Re–Os isotope dilution measurements down to the picogram level using multicollector inductively coupled plasma mass spectrometry. Int J Mass Spectrom 197:85–94

Schön R, Wincler G, Kutschera W (2004) A critical review of experimental data for the half-lives of the uranium isotopes ^{238}U and ^{235}U. App Rad Isot 60:263–273

Schwartzman DW (1973) Ar-degassing and the origin of the sialic crust. Geochim Cosmochim Acta 37(11):2479–2497

Seager WR, Shafiqullah M, Hawley JW et al (1984) New K–Ar dates from basalts of the southern Rio Grande rift. Geol Soc Am Bull 95:87–99

Segev A (2000) Synchronous magmatic cycles during the fragmentation of Gondwana: radiometric ages from the Levant and other provinces. Tectonophysics 325:257–277

Sei-Li Kuo and Folinsby P (1977) Geology of lead isotopes in ore deposits spatially related to the Tintina trough, Yukon area. In: Skinner BJ (ed) Economic geology. Stable isotopes as applied to problems of ore deposits 69(6):130–148 (in Russian, translated from English)

Selby D, Creaser RA, Hart CJR (2002) Absolute timing of sulfide and gold mineralization: a comparison of Re–Os molybdenite and Ar–Ar mica methods from the Tintina gold belt, Alaska. Geology 30(9):791–794

Semikhatov MA, Ovchinnikova GV, Gorokhov IM et al (2000) Isotopic age of the boundary between Middle and Late Riphean: Pb–Pb geochronology of carbonaceous rocks of the Lakhandinskii series, East Siberia. Dokl Earth Sci. 372(2):216–221 (in Russian)

Semikhatov MA, Ovchinnikova GV, Gorokhov IM et al (2003) Pb-Pb isochron age and Sr-isotopic signature of the Upper Yudoma carbonate sediments (Vendian of the Yudoma-Maya Trough, Eastern Siberia). Dokl Acad Nauk 393(8): 1093–1097

Sengör AM, Natal'in BA (1996) Paleotectonics of Asia: fragments of a synthesis. In: Yin A, Harrison M (eds) The tectonic evolution of Asia. Cambridge University Press, Cambridge, 21:486–640

Sepkoski JJ Jr (1987) Sepkoski replies. Nature 330:251–252

Shantser AE, Gladenkov YB (1997) History of geological development and Change of paleogeographic environments in the Early Paleogene of West Kamchatka. In: Gladenkov YB (ed) Lower Paleogene of West Kamchatka. GEOS, Moscow, pp 117–128 (in Russian)

Sharkov AV, Bogatikov OA, Kovalenko VI et al (1996) Early Precambrian lower crustal basic granulites and eclogites (an example of formations in the Kola peninsula and Southern Prisayanye). Geol Geophys 37(1):94–112 (in Russian)

Sharma KH, Bal KD, Parshad R et al (1980) Paleo-uplift and cooling rates from various orogenic belts , as revealed from be radiogenic ages. Tectonophysics 70:135–158

Sharpton VL, Dalrymple GB, Marin LE et al (1992) New links between the Chicxulub impact structure and the Cretaceous/Tertiary boundary. Nature 359:819–821

Shatskii VS, Ragozin AL, Sobolev NV (2006) Some aspects of metamorphic evolution ultra-high pressure calc-silicate rocks of the Kokchetav massif. Russ Geol Geophys 47(1):105–118

Shen JJ, Papanastassiou DA, Wasserburg GJ (1996) Precise Re–Os determinations and systematics of iron meteorites. Geochim Cosmochim Acta 60:2887–2900

Sheth HC, Pande K, Bhutani R (2001a) ^{40}Ar–^{39}Ar ages of Bombay trachytes: evidence for a Palaeocene phase of Deccan volcanism. Geophys Res Lett 28:3513–3516

Sheth HC, Pande K, Bhutani R (2001b)^{40}Ar–^{39}Ar age of a national geological monument: the Gilbert Hill basalt, Deccan Traps, Bombay. Curr Sci 80:1437–1440

Shih C-Y, Nyquist LE, Bogard DD et al (1985) Chronology and petrogenesis of a 1.8 g lunar granitic clast: 14321, 1062. Geochim Cosmochim Acta 49:411–426

Shimizu S, Itaya T (1993) Plio-Pleistocene arc magmatism controlled by two overlapping subducted plates, central Japan. Tectonophysics 225(3):139–154

Shirey SB, Walker RJ (1995) Carius tube digestion for low-blank rhenium–osmium analysis. Anal Chem 67:2136–2141

Shukolyukov A, Begemann F (1996) Pu–Xe dating of eucrites. Geochim Cosmochim Acta 60:2453–2471

Shukolyukov A, Lugmair GW (1998) Isotopic evidence for the Cretaceous–Tertiary impactor and its type. Science 282:927

Shukolyukov YA (1982) Products of decay of heavy elements in the Earth. Energoizdat, Moscow (in Russian)

Shukolyukov YA (2003) Isotopic-cosmochemical evidence on natural transport of substance between planets of terrestrial group (review of the literature on a problem). Geochem Int (11):1139–1171 (in Russian)

Shukolyukov YuA, Meshik AP (1995) Chemically fractionated xenon of uranium decay in mantle rocks, gases and terrestrial atmosphere? Geochemistry (8):1072–1093 (in Russian)

Shuto K, Ohki J, Kagami H et al (1993) The relationships between drastic changes in Sr isotope ratios of magma sources beneath the NE Japan arc and the spreading of the Japan Sea back-arc basin. Mineral Petrol 49:71–90

Shuvalov VF, Nikolayeva TV (1985) On age and spatial distribution of Cenozoic basalts in the south of Mongolia. Vestnik of the Leningrad State University. Ser Geol Geogr (14):52–59 (in Russian)

Simanenko VP (1986) Late Mesozoic volcanic arcs of East Sikhote-Alin and Sakhalin. Geol Pac Ocean 1:7–13 (in Russian)

Simkin T, Siebert L (1994) Volcanoes of the World, 2nd edn. Geoscience Press, Inc. Tucson, Arizona

Simonov VA, Mikolaichuk AV, Rasskazov SV et al (2008) Cretaceous–Paleogene within-plate magmatism in Central Asia: data from the Tien Shan basalts. Russ Geol Geophys 49(7):689–705

Size WB (ed) (1987) Use and abuse of statistical methods in the Earth Sciences. Oxford University Press, NY, p 200

Sklyarov EV, Mazukabzov AM, Mel'nikov AI (1997) Metamorphic core complexes of the Cordilleran type, SB RAS Publisher, Novosibirsk (in Russian)

Smedes HW, Prostka HJ (1972) Stratigraphic framework of the Absaroka Volcanic Supergroup in Yellowstone National Park region. US Geological Survey, Professional paper, 729-C

Smelov AP, Timofeev VF, Zaitsev AI (2003) Structure, stages of the North-Asian craton basement formation and Phanerozoic kimberlite magmatism. In: Zinchuk NN (ed) Geologic aspects of mineral resource base of the joint-stock company ALROSA. Mirnyi, pp 186–191 (in Russian)

Smithies RH (2000) The Archean tonalite–trondhjemite–granodiorite (TTG) series is not an analogue of Cenozoic adakites. Earth Planet Sci Lett 182:115–125

Smoliar MI, Walker RJ, Morgan JW (1996) Re–Os ages of group IIA, IIIA, IVA, and IVB iron meteorites. Science 271:1099–1102

Sobel ER, Arnaud N (2000) Cretaceous–Paleogene basaltic rocks of the Tuyon basin, NW China and the Kyrgyz Tian Shan: the trace of a small plume. Lithos 50:191–215

Sobolev NV (1974) Deep inclusions in kimberlite and problem of composition of the upper mantle. Nauka, Novosibirsk (in Russian)

Sobotovich AV, Semenenko VP (1984) Substance of meteorites. Nauk dumka, Kiev (in Russian)

Sokolov BS, Fedonkin MA (eds) (1985) Vendian system. Historical-geological and paleontological substantiation vol 2. Stratigraphy and geological processes. Nauka, Moscow (in Russian)

Solonenko AV (1988) Dependence of maximal amplitudes ratios in S- and P-waves from epicentral distances in the Baikal seismic zone. Geol Geophys (4):135–143 (in Russian)

Song B, Nutman AP, Liu D et al (1996) 3800–2500 Ma crustal evolution in the Anshan area of Liaoning Province, northeastern China. Precambrian Res 78: 79–94

Sorokin AA, Kudryashov NM, Ponomarchuk VA et al (2003) A history of formation of Mongolia–Okhotsk belt on basis of the new geochronological and geochemical data. Isotope geochronology in the decision of problems of geodynamics and ore genesis. Materials II Russian conference on isotope geochronology. The centre of information culture, St Petersburg, pp 480–483 (in Russian)

Spencer EU (1981) Introduction to structural geology. Nedra, St-Petersburg (translated to Russian)

Spencer JE, Reynolds SJ (1989) Middle tertiary tectonics of Arizona and adjacent areas. In: Jenney JP, Reynolds SJ (eds) Geologic evolution of Arizona: Arizona Geol Soc Digest 17:539–574

Spencer JE, Richard SM, Reynolds SJ et al (1995) Spatial and temporal relationships between mid-Tertiary magmatism and extension in southwestern Arizona. J Geophys Res 100(B7):10321–10351

Stacey JS, Kramers TD (1975) Approximation of terrestrial lead isotope evolution by two-stage model. Earth and Planet Sci Lett 26:207–221

Stanley SM, Yang X (1994) A double mass extinction at the end of the Paleozoic era. Science 266:1340–1344

Starik IE (1961) Nuclear geochronology. Publisher of Academii Nauk USSR, Moscow (in Russian)

Steiger RH, Jäger E (1977) Sucommission on geochronology: convention on the use of decay constants in geo- and cosmology. Earth Planet Sci Lett 36: 359–362

Stein HJ, Markey RJ, Morgan JW et al (2001) The remarkable Re–Os chronometer in molybdenite: how and why it works. Terra Nova 13(6):479–486

Stein HJ, Morgan JW, Markey RJ et al (1998) An introduction to Re–Os: what's in it for the mineral industry? SEG Newsletter (32):8–15

Stern MJ, Vogel PC (1971) Relative ^{14}C–^{13}C kinetic isotope effects. J Chem Phys 55(5):2007–2013

Stewart JH, Carlson JE (1976) Cenozoic rocks of Nevada. Nevada Bureau of Mines and geology, Map 52

Stiefenhofer J, Farrow DJ (2004) Geology of the Mwadui kimberlite, Shinyanga district, Tanzania. Lithos 76:139–160

Sukhov VI (1975) Volcanogenic formations of the south of Far East. Nedra, Moscow (in Russian)

Sun S-S, McDonough WF (1989) Chemical and isotopic systematics of oceanic basalts: implications for mantle composition and processes. In: Sounders AD, Norry MJ (eds). Magmatism in the ocean basins. Geological Society Special Publication 42:313–345

Suzuki K, Adachi M (1994) Middle Precambrian detrital monazite and zircon from Hida gneiss on Oki-Dogo Island, Japan: their origin and implications for the correlation of basement gneiss of Southwest Japan and Korea. Tectonophysics 235:277–292

Suzuki K, Adachi M, Kajizuka I (1994) Electron microprobe observations of Pb diffusion in metamorphosed detrital monazites. Earth Planet Sci Lett 128:391–405

Suzuki K, Adachi M, Nureki T (1996) CHIME age dating of monazites from metamorphic rocks and granitic rocks of the Ryoke belt in the Iwakuni area, Southwest Japan. Isl Arc 5:43–55

Suzuki K, Adachi M, Tanaka T (1991) Middle Precambrian provenance of Jurassic sandstone in the Mino Terrane, central Japan: Th–U–total Pb evidence from an electron microprobe monazite study. Sediment Geol 75:141–147

Svetov SA (2003) New data on geochemistry of the oldest (3.05–2.95 By) andesitic associations in the East Fennoscandia. Dokl Acad Nauk 388(5):664–668 (in Russian)

Swanson DA, Cameron KA, Evarts RC et al (1989) Cenozoic volcanism in the Cascade Range and Columbia Plateau, Southern Washington and northwestern Oregon. Chapin CE, Zidek J (eds) Field excursions to volcanic terranes in the western United States, vol 2: Cascades and Intermountain West. New Mexico Bureau of Mines & Mineral Resources, Socorro, pp 1–50

Swisher CC, Grajales-Nishimura JM, Montanari A et al (1992) Coeval $^{40}Ar/^{39}Ar$ ages of 65.0 Million years ago from Chicxulub Crater melt rock and Cretaceous–Tertiary boundary tektites. Science 257:954–958

Sysoev AA, Artaev VB, Kasheev VV (1993) Isotopic mass spectrometry. Energoatomizdat, Moscow (in Russian)

Szurlies M, Bachmann GH, Menning M et al (2003) Magnetostratigraphy and high-resolution lithostratigraphy of the Permian–Triassic boundary interval in Central Germany. Earth Planet Sci Lett 212:263–278

Takahashi M, Ikuya O, Kazuo S (1991) K–Ar dating of welded tuffs in the Kanayama area, eastern Gunma Prefecture, central Japan. Bull Geol Surv Japan 42:167–173

Takeuchi T (1997) K–Ar ages of the Tertiary volcanic rocks in South Sakhalin and their tectonic significance. J Geol Soc Japan 103(1):67–79

Tamaki K, Suyehiro K, Allan J et al (1992) Tectonic synthesis and implications of Japan Sea ODP drilling. Proc Ocean Drill Program Sci Results 127–128:1333–1350

Tapponnier P, Peltzer G, Le Dain AY et al (1982) Propagating extrusion tectonics in Asia: new insights from simple experiments with plasticine. Geology 10: 611–616

Tatsumi Y, Sato K, Sano T et al (2000) Transition from arc to intraplate magmatism associated with backarc rifting: evolution of the Sikhote-Alin volcanism. Geophys Res Lett 27(11):1587–1590

Tatsumoto M, Knight RJ, Allègre CJ (1973) Time differences in the formation of meteorites as determined from the ratio of lead–207 to lead–206. Science 180:1278–1283

Taylor LA, Anand M (2004) Diamonds: time capsules from the Siberian mantle. Chemie der Erde 64:1–74

Tera F, Carlson RW, Boctor NZ (1997) Radiometric ages of basaltic achondrites and their relation to the early history of the Solar System. Geochim Cosmochim Acta 61:1713–1731

Tetley N, McDougall I, Heydegger HR (1980) Thermal neutron interferences in the $^{40}Ar/^{39}Ar$ dating technique. J Geophys Res 85:7201–7205

Thirwall MF (1991) Long-term reproducibility of multicollector Sr and Nd isotope ratio analyses. Chem Geol 94:85–104

Thompson RN, Leat PT, Dickin AP et al (1989) Strongly potassic mafic magmas from lithospheric mantle sources during continental extension and heating: evidence from Miocene minetts of northwest Colorado, U.S.A. Earth Planet Sci Lett 98:139–153

Tikhonov AN, Goncharskii AV (ed) (1987) Incorrect tasks of natural sciences. Moscow State University, Moscow (in Russian)

Tikhonov AN, Samarskii AA (1953) The equations of mathematical physics, 2nd edn. Gostechizdat, Moscow (in Russian)

Titaeva NA (2005) Geochemistry of the natural radioactive decay series. GEOS, Moscow

Tomlinson KY, Hughes DJ, Thurston PC et al (1999) Plume magmatism and crustal growth at 2.9 to 3.0 Ga in the Steep Rock and Lumby Lake area, Western Superior Province. Lithos 46:103–136

Toon OB, Zahnle K, Morrison D et al (1997) Environmental perturbations caused by the impacts of asteroids and comets. Rev Geophys 35(1):41–78

Tozer ET (1988) Towards a definition of the Permian–Triassic boundary. Episodes 11(3):251–255

Travin AV, Aschepkov IV, Yudin D et al (2002) Ar–Ar age of Sayan kimberlite-like rocks. J Conf Abstr Goldschmidt 305

Travin AV, Boven A, Plotnikov AV et al (2001) $^{40}Ar/^{39}Ar$ dating of ductile deformations in the Irtysh share zone (East Kazakhstan). Geochem Int (12):1–5 (in Russian)

Treloar PJ, Rex DC, Guise FG et al (1989) K–Ar geochronology of the Himalayan collision in NW Pakistan: constraints on the timing of suturing, deformation and uplift. Tectonics 8:881–909

Tsekhovskii YG, Akhmet'ev MA (2003) Landscapes, sedimentation with elements of geodynamics in contrast epochs of tectonogenesis (on an example of Eurasia), vol 1. The Laramide epoch. Bulletin of MOIP 78(1):3–15 (in Russian)

Tsuruta K, Takahashi E (1998) Melting study of an alkali basalt JB–1 up to 12.5 GPa: behavior of potassium in the deep mantle. Phys Earth Planet Inter 107:119–130

Turner G (1971) ^{40}Ar–^{39}Ar ages from the lunar marina. Earth Planet Sci Lett 11:169–191

Turner S, Arnaud N, Liu J et al (1996) Post-collision, shoshonitic volcanism on Tibetan Plateau: implications for convective thinning of the lithosphere and the source of ocean island basalts. J Petrol 37(1):45–71

Tveritinov YI, Tveritinova TY, Brandt SB et al (2006) Prospecting of gold mineralization in the south of East Asia and Far East of Russia: aspects of geology and isotopic geochemistry. Institute of the Earth's crust SB RAS, Irkutsk

Vance D, Thirwall M (2002) An assessment of mass discrimination in MC–ICPMS using Nd isotopes. Chem Geol 185:227–240

Van der Beek P, Delvaux D, Andriessen PAM et al (1996) Early Cretaceous denudation related to convergent tectonics in the Baikal region, SE Siberia. J Geol Soc London 153:515–523

Van Kooten GK (1980) Mineralogy, petrology, and geochemistry of an ultrapotassic basaltic suite, central Sierra Nevada, California, U.S.A. J Petrol 21(4):652–684

Van der Voo R, Spakman W, Bijwaard H (1999a) Tethyan subducted slabs under India. Earth Planet Sci Lett 171:7–20

Van der Voo R, Spakman W, Bijwaard H (1999b) Mesozoic subducted slabs under Siberia. Nature 397:246–249

Valley JW, Peck WH, King EM et al (2002) A cool early Earth. Geology 30: 351–354

Vasil'ev EP, Reznitskii LZ, Vishnyakov VN et al (1981) Slyudyanskii crystalline complex. Nauka, Novosibirsk

Vasil'ev YR, Zolotukhin VV, Feoktistov GD et al (2000) An estimation of volumes and a problem of genesis of Permo–Triassic trap magmatism in the Siberian platform. Geol Geophys 41(12):1696–1705 (in Russian)

Venkatesan TR, Kumar A, Gopalan K et al (1997) ^{40}Ar–^{39}Ar age of Siberian basaltic volcanism. Chem Geol 138:303–310

Vernikovsky VA, Pease VL, Vernikovskaya AE et al (2003) First report of early Triassic A-type granite and syenite intrusions from Taimyr: product of the northern Eurasian superplume? Lithos 66:23–36

Villa IM (1997) Direct determination of ^{39}Ar recoil distance. Geochim Cosmochim Acta 61(3):689–691

Vishnevskaya VS, Bogdanov NA, Kurilov DV (2003) The first data on Barremian–Albian radiolarian associations of East Sakhalin. Dokl Acad Nauk 392(6):787–791 (in Russian)

Vladimirov BM, Logatchev NA, Vainer-Krotova GA et al (2003) The Vendian–Cambrian boundary: Rb–Sr isochron age of the final event of alkaline ultrabasic magmatism in the northeastern Sayan region. Dokl Earth Sci 389A(3):346–349

Vöckenhuber C, Oberli F, Bichler M et al (2004) New half-life measurement of ^{182}Hf: improved chronometer for the early solar system. Phys Rev Lett 93:172501-1-172501-4

Völkening J, Walczyk T, Heumann KG (1991) Osmium isotope ratio determinations by negative thermal ionization mass spectrometry. Int J Mass Spectrom Ion Process 105:147–159

Vollmer R, Ogden P, Schilling J-G et al (1984) Nd and Sr isotopes in ultrapotassic volcanic rocks from Leucite Hills, Wyoming. Contrib Mineral Petrol 87:359–368

Vrublevskii VV, Gertner IF, Polyakov GV et al (2004) Ar–Ar isotopic age of lamproite dikes of the Chuya complex. Dokl Acad Nauk 399(4):516–519 (in Russian)

Walderhaug HJ, Eide EA, Scott RA et al (2005) Paleomagnetism and 40Ar/39Ar geochronology from the South Taimyr igneous complex, Arctic Russia: a middle–late Triassic magmatic pulse after Siberian flood-basalt volcanism. Geophys J Int 163:501–517

Wang Y, Chen H (2005) Tectonic controls on the Pleistocene–Holocene Wudalianchi volcanic field (northeast China). J Asian Earth Sci 24:419–431

Wang Y, Mu L, Liu W (1996) Regularity and characteristic of volcanic activity at Wudalianchi, Heilongjiang. 30th international geological congress. Beijing, China

Wasserburg GJ, Jacobsen SB, DePaolo DJ et al (1981) Precise determination of Sm/Nd ratios, Sm and Nd isotopic abundances in standard solutions. Geochim Cosmochim Acta 45:2311–2323

Wasson JT, Choi B-G, Jerde EA et al (1998) Chemical classification of iron meteorites: XII. New members of the magmatic group. Geochim Cosmochim Acta 62(4):715–724

Wasson JT, Ouyang X, Wang J et al (1989) Chemical classification of iron meteorites: XI. Multi-element studies of 38 new irons and the high abundance of ungrouped irons from Antarctica. Geochim Cosmochim Acta 53(3):735–744

Watanabe Y (2002) Late Cenozoic metallogeny of Southwest Hokkaido, Japan. Resource Geol 52:191–210

Webb LE, Hacker BR, Ratschbacher L et al (1999) Thermochronologic constraints on deformation and cooling history of high- and ultrahigh-pressure rocks in the Qinling–Dabie orogen, Eastern China. Tectonics 18(4):621–638

Webster A, Szegoe G (1934) Differential equations in partial derivatives of mathematical physics. Parts 1–2. ONTI Publisher, Moscow–Leningrad (in Russian)

Wegener A (2002) The origin of continents. Int J Earth Sci 91:S4–S17

Weinberg RF, Dunlap WJ (2000) Growth and deformation of the Ladakh batholith, Northwest Himalayas: implication for timing of continental collision and origin of calc-alkaline batholiths. J Geology 108:303–320

White FA, Wood GM (1986) Mass Spectrometry: applications in science and engineering. Wiley, New York

Whyman DA, Kerrich R (2002) Formation of Archean continental lithospheric roots: the role of mantle plumes. Geology 30(6):543–546

Widdowson M, Pringle MS, Fernandez OA (2000) A post K–T boundary (Early Paleocene) age for Deccan–type feeder dykes, Goa, India. J Petrol 41(7):1177–1194

Wilde SA, Valley JW, Peck WH et al (2001) Evidence from detrital zircons for the existence of continental crust and oceans on the Earth 4.4 gyr ago. Nature 409:175–178

Williams HM, Turner SP, Pearce JA et al (2004) Nature of the source regions for post-collisional, potassic magmatism in Southern and Northern Tibet from geochemical variations and inverse trace element modeling. J Petrol 45(3):555–607

Windley BF (1997) The evolving continents, 3rd edn. Department of geology. University of Leicester, UK

Windley BF, Allen MB (1993) Mongolian plateau: evidence for a Late Cenozoic mantle plume under Central Asia. Geology 21:295–298

Wittlinger G, Masson F, Poupinet G et al (1996) Seismic tomography of northern Tibet and Kunlun: evidence for crustal blocks and mantle velocity contrasts. Earth Planet Sci Lett 139:263–279

Wood JA (2004) Formation of chondritic refractory inclusions: the astrophysical setting. Geochim Cosmochim Acta 68(19):4007–4021

Wright N, Layer P, York D (1991) New insights into thermal history from single grain $^{40}Ar/^{39}Ar$ analysis in biotite. Earth Planet Sci Lett 104:70–79

Xu X, Ma X, Deng Q et al (1996) Field trip guide 314 of the 30th international geological congress. Beijing, China

Yang J-D, Sun W-G, Wang Z-Z et al (1996) Sm–Nd isotopic age of Precambrian–Cambrian boundary in China. Geol Mag 133(1):53–61

Yanovskaya TB, Kozhevnikov VM (2003) 3D S-wave velocity pattern in the upper mantle beneath the continent of Asia from Raleigh wave data. Phys Earth Planet Inter 138:263–278

Yarmolyuk VV, Ivanov VG, Kovalenko VI et al (1994) Dynamics of formation and magmatism of Late Mesozoic–Cenozoic South Khangayskii mantle hot spot (Mongolia). Geotectonics 5:28–45 (in Russian)

Yarmolyuk VV, Kudryashova EA, Kozlovskii AM et al (2007) Late Cretaceous–Early Cenozoic volcanism in Southern Mongolia: a trace of the Southern Hangay mantle hot spot. Volcanol seismol 1:3–31 (in Russian)

Yemelyanova TA, Lelikov YP, Syedin VT et al (2003) The geology and the features of volcanism on the Sea of Okhotsk floor. Geol Pac Ocean 22:3–18 (in Russian)

Yin H, Zhang K, Tong J-N et al (2001) The Global Stratotype Section and Point (GSSP) of the Permian–Triassic Boundary. Episodes 24:102–114

Yin Q, Jacobsen SB, Yamashita K et al (2002) A short timescale for terrestrial planet formation from Hf–W chronometry of meteorites. Nature 418:949–951

Yin QZ, Jagoutz E, Verkhovskiy AB et al (1993) ^{187}Os–^{186}Os and ^{187}Os–^{188}Os method of dating: an introduction. Geochim Cosmochim Acta 57:4119–4128

Yogodzhinski GM, Kelemen PB (1998) Slab melting in the Aleutians: implications of an ion probe study of clinopyroxene in primitive adakite and basalt. Earth Planet Sci Lett 158:53–65

York D (1969) Least squares fitting of a straight line with correlated errors. Earth Planet Sci Lett 5:320–324

York D, Macintyre RM, Gittins J (1969) Excess radiogenic ^{40}Ar in cancrinite and sodalite. Earth Planet Sci Lett 7:25–28

Zakharov YD, Rybalka SV (1987) The standards of Permian and Triassic in the Tethyan region. In: Zakharov YD, Onoprienko YI (eds) Problems of biostratigraphy in the Permian and Triassic of East USSR. The Institute of Biology and Soil in Far East Scientific Centre, Academy of Science USSR, Vladivostok (in Russian)

Zakharov YD, Ignat'ev AV, Kotlyar GV et al (1996) Stable isotopes of carbon and relations of Ca–Mg in Permo–Triassic carbonates and mass extinction of organisms. Geol Pac Ocean 15(1):3–15 (in Russian)

Zhai M, Zhao G, Zhang Q (2002) Is the Dongwanzi complex an Archean ophiolite? Science 295:923a

Zhamoyda AI (ed) (2000) Additions to Stratigraphic codex of Russia. VSEGEI, St Petersburg (in Russian)

Zhang M, Suddaby P, Thompson RN et al (1995) Potassic rocks in NE China: geochemical constraints on mantle source and magma genesis. J Petrol 36(5):1275–1303

Zhang L-S, Schärer U (1999) Age and origin of magmatism along the Cenozoic Red River shear belt, China. Contrib Miner Petrol 134:67–85

Zhang Y (2002) The age and accretion of the Earth. Earth Sci Rev 59:235–263

Zhao Y, Liu X, Wang S (1997) Syn- and post-tectonic cooling and exhumation in the Larseman Hills, East Antarctica. Episodes 20:122–127

Zharov AE (2004) Geological structure and Cretaceous–Paleogene geodynamics of Southeast Sakhalin. Sakhalin Publisher, Yuzhno-Sakhalinsk

Zheng Z, Tanaka H, Tatsumi Y et al (2002) Basalt platforms in Inner Mongolia and Hebei Province, northeast China: new K–Ar ages, geochemistry, and revision of paleomagnetic results. Geophys J Int 151:654–662

Zhou M-F, Malpas J, Song X-Y et al (2002) A temporal link between the Emeishan large igneous province (SW China) and the end-Guadalupian mass extinction. Earth Planet Sci Lett 196:113–122

Zhou X-H, Zhu B-Q, Liu R-X et al (1988) Cenozoic basaltic rocks in Eastern China. In: MacDougall JD (ed) Continental flood basalts. Kluwer Academic Publishers, Dordrecht, pp 311–330

Zoback ML (1992) First- and second-order patterns of stress in the lithosphere: the world stress map project. J Geophys Res 97(B8):11703–11728

Zolotukhin VV, Vilenskii AM, Dyuzhikov OA (1986) Basalts of the Siberian platform. Regularities of geology, composition, and genesis of Permo–Triassic effusives. Nauka, Novosibirsk (in Russian)

Zhuravlev AZ, Zhuravlev DZ, Kostitsyn YA et al (1987a) Definition of the Sm–Nd relation for the purposes of geochronology. Geochem Int 8:1115–1128 (in Russian)

Zhuravlev DZ, Pukhtel IS, Samsonov AV et al (1987b) Sm–Nd age of the basement fragments of granite–greenstone belt in Middle Dnepr. Dokl Acad Nauk 294:1203–1208 (in Russian)

Index